Searching for a Mechanism

Searching for a Mechanism

A History of Cell Bioenergetics

John N. Prebble

OXFORD
UNIVERSITY PRESS

OXFORD
UNIVERSITY PRESS

Oxford University Press is a department of the University of Oxford. It furthers
the University's objective of excellence in research, scholarship, and education
by publishing worldwide. Oxford is a registered trade mark of Oxford University
Press in the UK and certain other countries.

Published in the United States of America by Oxford University Press
198 Madison Avenue, New York, NY 10016, United States of America.

Library of Congress Cataloging-in-Publication Data
Names: Prebble, J. N. (John N.), author.
Title: Searching for a mechanism : a history of cell bioenergetics / by John N. Prebble.
Description: New York, NY : Oxford University Press, 2019. |
Includes bibliographical references and index.
Identifiers: LCCN 2018023975 | ISBN 9780190866143
Subjects: LCSH: Bioenergetics—History.
Classification: LCC QH510.P74 2018 | DDC 572/.43—dc23
LC record available at https://lccn.loc.gov/2018023975

1 3 5 7 9 8 6 4 2

Printed by Sheridan Books, Inc., United States of America

In memory of Peter Mitchell (1920–1992), whose genius laid the foundation for the revolution in our understanding of cell bioenergetics.

To Pat

Contents

Preface

The twentieth century saw the elucidation of many of the fundamental problems of the biological sciences. A major achievement was the development of metabolic biochemistry. However, within this general field, one of the most significant challenges was the endeavor to understand the mechanisms of bioenergetics. The complexity of this field became apparent when the links between the mechanisms of various processes in oxidative phosphorylation, photosynthesis, and cellular transport across membranes began to be appreciated. These areas had previously been studied independently, but in the middle of the century their relationships were increasingly understood and the term bioenergetics was often used to cover the emerging field.

Biologists have been concerned with mechanisms for several centuries, but with the development of modern methods, the significance of the term mechanism has been recognized. Philosophers, in a rather different way, have long had an interest in mechanism but this has been revitalized in recent times with the application of this concept to the philosophy of the biological sciences. So the idea of searching for a mechanism seems to be particularly appropriate to the consideration of the history of scientific studies about respiration. These developed through basic issues of metabolism to the crunch question of how cells acquire their energy from the oxidation of foodstuffs such as carbohydrates. Here I trace the history of the search for mechanisms in a group of key biological fields that progressively merged toward the end of the twentieth century. This study is not intended to add to the philosophical ideas on mechanism, but rather to see the way in which mechanisms in one specific area were elucidated over time, from the early seventeenth to the end of the twentieth century.

Few problems have taxed biologists as much as the fundamental mechanisms of bioenergetics. This book is designed partly to explore the often-heated debates particularly in the 1960s and 1970s, when theories and experimental interpretations in bioenergetics were fought over so passionately that the period was known by some as the Ox phos wars. It is also intended to place this important period in biological research in its historical setting. Thus the story of what is now known as bioenergetics is traced from the first experimental studies of respiration and photosynthesis in the enlightenment of the seventeenth and eighteenth centuries to the modern period. The climax of the account comes in the brilliant resolution of the mechanism of adenosine triphosphate (ATP) synthesis in animals, plants, and microorganisms by the proton-translocating ATP synthase at the end of the twentieth century. Such a history reflects the history of biology as a whole as it moved from a primarily observational study to the highly technical investigations of the second half of the twentieth century.

Like most history, writing a science history raises questions about defining the margins of the discussion both in time and in subject matter. A major issue has been where to start. In his history of respiration, David Keilin went back to the second-century Greek physician and philosopher Galen. I have chosen to start with a brief discussion of seventeenth-century ideas, when a truly experimental approach to the study of respiration was initiated. The photosynthesis story is best begun at the beginning of the eighteenth century with the work of Stephen Hales, whose ideas also contributed to the respiration story. I have concluded the story at the end of the twentieth century with a discussion of the enzyme-synthesizing ATP because of its innate interest and because it demonstrated an important result of the search for mechanism. The scope of the book has been slightly narrow, being confined to the discussion of respiration, oxidative, and related aspects of phosphorylation, photosynthesis, and cognate areas, such as facets of membrane transport and specific issues in microbial biochemistry. Fields such as muscular contraction, which might arguably have been included in a discussion of bioenergetics history, have been mostly omitted and similarly some areas of microbial energetics.

Of course, restrictions of space have entailed selecting issues to which particular attention should be given in order to maintain a comprehensible narrative, and this has necessitated some rather difficult choices. Thus judgments have had to be made about which particular scientific contributions should be included, which contributions should be seen as central to the history of bioenergetics, and which might be set aside in order to focus on the major lines of development. This is particularly so in the later part of the twentieth century in which there is a vast amount of material and a wide range of approaches to the subject. Inevitably many rather personal choices have been made. I apologize to those who feel that important aspects of the subject have been omitted or that the work of particular scientists who made valuable contributions has been ignored or should have been given more weight. Another issue that has emerged arises from the nature of the field of bioenergetics that has been noted for its lack of consensus. Although I have covered most of the most obvious disagreements in my discussion I have been advised that in some more recent research, I may not have been sufficiently sensitive to the varying viewpoints. I apologize to those who feel I have misrepresented particular aspects of their interests.

The history has been divided into several phases, each initiated by a major advance often outside the field of bioenergetics and affecting biochemistry more generally. These phases do not have broad applicability as the phases I have recognized for respiration and phosphorylation are not identical with those identified for photosynthesis. After having much in common at the beginning, the two areas of study have tended to proceed separately, partly because the events associated with the light reaction had no counterpart in animal cell respiration. However, in the second half of the twentieth century, the two fields have been more or less merged in the field of bioenergetics. Indeed it is not possible to consider the one without the other as each has made major contributions to the other.

In telling this story I have felt it right to make points through appropriate quotations of the leading workers in the field. This allows the historical figures to be heard. However, in the later period this becomes more difficult; frequently a wide variety of workers are jointly developing the field, and a somewhat random choice has been made. I have also attempted to minimize the technical aspects as much as possible, although it is not possible to describe the development of ideas in a subject such as biochemistry without recourse to a certain amount of chemistry. Further, in the later stages, the use of highly complex techniques has been discussed without much explanation of what is actually involved.

What is the justification for a book on the history of cell bioenergetics? The history of science is a discipline in which it is important that both historians and scientists participate. This story is told by a scientist in the hope that his approach will provide some illumination on an area generally enlightened by contributions of historians and also philosophers. What I have sought to do here is to give a coherent though brief account of the history of those events associated with the energetics of cell respiration and photosynthesis over more than three centuries. There is already a widely acclaimed book by Joseph Fruton on the history of biochemistry as well as the extensive account by Marcel Florkin; both of these present aspects of bioenergetics within a broad discussion of biochemistry but do not give it the detailed attention it deserves. There have been a number of essays on specific events and short periods of biochemical history but there is clearly a need to put all of these within a broad historical perspective. That perspective also illustrates some of the effects of the major developments in the chemical aspects of biological science, the preparation of cell-free systems at the turn of the twentieth century, the ability to describe metabolic pathways, the development of techniques for handling membrane proteins, and so on. However, the justification for the book is in my view the need for a history of bioenergetics comparable to that of Michel Morange's *A History of Molecular Biology*. I hope I have come some way in achieving that goal.

But why single out bioenergetics? Metabolic biochemistry developed only slowly in the earlier part of the twentieth century, but its achievements were substantial—the glycolytic pathway, the citric acid cycle, and so on. The same period began to identify aerobic phosphorylation as a major source of ATP but the mechanism was not apparent. By the 1950s it was clear that there was a challenging problem to solve. The difficulties in finding a solution, the choice between different hypotheses, and the problem of successfully working with membranes provide interesting historical developments not as apparent in other areas of metabolic biochemistry. Such problems were solved by bringing together quite diverse researches, many drawn from cell biology. Thus the roots of this endeavor are to be found in several almost independent lines of research that come together to create the field in the 1950s. Such a story surely needs to be told.

I came into the study of bioenergetics in the 1960s, a most exciting period in the history of this field. I am grateful to those who encouraged me in this early period, particularly Dudley Cheesman, who taught me the value of the historical approach to biochemistry, to my colleagues Peter Zagalsky and particularly John Lagnado, who encouraged me to take up the teaching of bioenergetics in the single Honours biochemistry course at Bedford College, University of London. An abiding inspiration from that period was the occasional lectures of Peter Mitchell that sparked my enthusiasm for the subject. I should also record my appreciation of brief but sound advice on writing science history from my friend of student days, Bob Olby. I am grateful to colleagues who kindly read part or most of the manuscript and provided valuable comment, Bruce Weber and Peter Rich and also Ann Marshall. I also wish to record my appreciation of the detailed and constructive comments of two anonymous reviewers and the staff of the Oxford University Press. Finally, I wish to record my immense debt to my wife, Pat, who has encouraged me over the years and who has had the patience to read and reread my manuscript.

John N. Prebble
School of Biological Sciences
Royal Holloway, University of London
November 2017

Acknowledgments

I wish to express my appreciation to Professor Mikuláš Teich for permission to reproduce several textual extracts from his *A Documentary History of Biochemistry 1770–1940*.

I wish to record my thanks to those who gave me photographs for my earlier book on the subject and that I have reused here, Professor Stanley Bullivant (Fig. 4.1), Professor Humberto Fernández-Morán (Fig. 5.4), Professor Lester Packer (Fig. 5.6), and Dr. Jean Whatley (Fig. 7.1).

I acknowledge the permission to reproduce figures by the American Chemical Society (Fig. 8.5), Elsevier (Figs. 5.6, 6.6, 7.2, and 8.3), Glynn Research Ltd. (Fig. 8.1), Nature Publishing Ltd. (Figs. 7.4, 8.4, and 8.6), Oxford University Press (Figs. 3.9 and 3.10), Pearson Educational Ltd. (Figs. 4.1b and 7.1), Rockefeller University Press (Figs. 5.4 and 5.5), Springer (Figs. 6.1, 6.4, and 8.7), and Wiley (Fig. 3.15).

Abbreviations

$\Delta\psi$	membrane potential
Δp	proton motive force
ADP	adenosine diphosphate
ATP	adenosine triphosphate
ATPase	adenosine triphosphatase, ATP synthase
BChl	bacteriochlorophyll
Bph	bacteriopheophytin
Chl	chlorophyll
CoA	coenzyme A
Cu	copper
Cyt	cytochrome
DCPIP	dichlorophenol indophenol
DCMU	dichlorophenyldimethylurea
DNA	deoxyribonucleic acid
DNP	dinitrophenol
DPN, DPNH	diphosphopyridine nucleotide, oxidized and reduced, respectively (also known as NAD, NADH)
E_m	midpoint potential
EPR	electron paramagnetic resonance
ETP	electron-transport particle
F_o, F_1	the two major components of the ATP synthase (ATPase)
FAD	flavin adenine dinucleotide
Fd	ferredoxin
Fe-S	iron-sulfur center or protein
FMN	flavin adenine mononucleotide
Fp	flavoprotein
GDP	guanosine diphosphate
GTP	guanosine triphosphate
[H]	reducing equivalent
Mn	manganese
NAD, NADH	nicotinamide adenine dinucleotide, oxidized and reduced, respectively (also known as DPN, DPNH, or coenzyme I)
NADP, NADPH	nicotinamide adenine dinucleotide phosphate, oxidized and reduced, respectively (also known as TPN, TPNH, or coenzyme II)
P680, P700, P870	reaction center pigments
PC	plastocyanin

PETP	phosphorylating ETP
Pi	inorganic phosphate
PMF	proton motive force
PQ	plastoquinone
PSI, PSII	photosystems I and II, respectively
Q	quinone (as in Q-cycle), used for ubiquinone also to denote electron acceptor for PSII
SDS	sodium dodecyl sulfate
TPN, TPNH	triphosphopyridine nucleotide, oxidized and reduced, respectively (also known as NADP and NADPH)
UCP	uncoupling protein
X	electron acceptor for PSI
Z	electron donor for PSII

Searching for a Mechanism

1

Introduction

Respiration, Phosphorylation, and Mechanism

Toward the end of the Second World War, just before the development of much modern biology began, the Austrian physicist Erwin Schrödinger (1887–1961) considered, in a small influential book, what were the essential features of life?[1] His prime concern was what might be described as the genetic aspect, but his other major concern was with energy. The former aspect, which has been well documented, became the molecular biological revolution and has been a major subject of interest for historians of biology. Because it developed simultaneously, the bioenergetics revolution was overshadowed by the development of molecular biology. So, it has been largely overlooked although recently the Belgian cell biologist and Nobel Laureate[2] Christian De Duve (1917–2013) drew attention to what he referred to as "the other revolution in the life sciences," that concerning cell bioenergetics.[3] Indeed De Duve felt that the energetic aspect is arguably as fundamental as the information element and in fact preconditions it.[4] This history is an attempt to rebalance the view of biology so that those cellular processes that provide the energy for life are brought into sharper focus.

The story of cell bioenergetics is primarily concerned with the synthesis of adenosine triphosphate (ATP), sometimes referred to as the energy currency of the cell.

Its central theme is the processes of oxidative phosphorylation and photo-phosphorylation, whose mechanisms remained obscure for many years. Indeed, understanding oxidative phosphorylation and photophosphorylation proved to require an appreciation of the energetics of other aspects of cellular processes, particularly those

---

[1] Schrödinger 1944. Schrödinger shared the Nobel Prize in physics in 1933 "for the discovery of new productive forms of atomic theory"

[2] The Nobel Prize in Physiology or Medicine for 1974 was awarded jointly to Albert Claude, Christian De Duve, and George E. Palade "for their discoveries concerning the structural and functional organisation of the cell"

[3] De Duve 2013.

[4] An unusual approach to this relationship between molecular biology and photosynthesis has been provided by Doris Zallen (1993a) who, when considering the bioenergetics of photosynthesis, felt that this field of study should logically be included within molecular biology. This view was based on a rather general set of criteria for defining "molecular biology" that were then seen to cover the molecular side of photosynthesis research since about 1920. This does not sit comfortably with the history of photosynthesis and is not pursued further here.

associated with membranes and membrane transport so that the field can readily be referred to as cell *bioenergetics*. In fact, the term bioenergetics was probably not introduced into the field until Albert Szent-Györgyi published a small book under that title in 1957; it was also similarly used by Albert Lehninger in 1963 and came into more general use at about that time, although some thought it was "too flashy"![5] Nevertheless, after the field had become well established in the 1950s, specialist journals began to appear. A dedicated set of volumes within the journal *Biochimica et Biophysica Acta* was published from 1968 onward, known simply as *BBA Bioenergetics*. In 1973 BBA *Reviews in Bioenergetics* began publication. A totally independent journal, *The Journal of Bioenergetics*[6] was produced from 1970 onward. These journals dealt with problems of photosynthesis as well as oxidative phosphorylation, although specialist photosynthetic journals also emerged around the same time.[7]

1.1 RESPIRATION, PHOTOSYNTHESIS, AND BIOENERGETICS

This history commences with the work of those in the seventeenth century who sought to understand the process of breathing and passes through metabolic biochemistry, concluding with the elucidation of the molecular mechanisms of key enzymes in bioenergetics. Although the story of metabolic biochemistry (which is often taken to include bioenergetics) essentially belongs to the twentieth century, progress in this area cannot be understood without recourse to previous centuries. Thus from the seventeenth century onward it is possible to trace a path of early thinking that eventually laid the groundwork for the dramatic success of twentieth-century studies.

In the seventeenth century, the development of a fruitful experimental approach to science opened new possibilities. A new generation of those perhaps best described as experimental natural philosophers began to pursue physiological issues including the mechanism of respiration. Such activities also expressed themselves in the founding of the Royal Society in 1660 in England and in France the Académie Royale des Sciences in 1666. These early physiologists initially investigated questions concerned with breathing, but this quickly led to others about the role of air and its influence on the properties of blood. It was particularly the chemical revolution at the end of the eighteenth century that combined with these early studies on respiration to provide a serious field of respiration for the nineteenth and early twentieth centuries. However, prior to the twentieth century principally physiologists and chemists led the field. Thus I start the discussion in the seventeenth century, although others such as David Keilin and Marcel Florkin have traced the influence of the classical world, especially Aristotle and Galen, on this period.

The story of photosynthesis can similarly be traced back to the beginning of the eighteenth century when it shares some of the same origins as studies on respiration. However,

---

[5] See Edsall 1973; Lehninger 1965.
[6] This later became the *Journal of Bioenergetics and Biomembranes*.
[7] *Photosynthetica* began publication in 1967 and *Photosynthesis Research* in 1980. Earlier work on photosynthesis had tended to be published in journals of plant physiology.

the development of the field in the nineteenth century is much dominated by questions about the role of light and chlorophyll, thus following a path independent of respiration. The second half of the twentieth century brought the realization that many of the problems faced by those working on photosynthesis are similar to those in oxidative phosphorylation, and this resulted in the two streams coming together, so justifying the treatment of both histories as one.

The history of respiration has been explored by others, including the Cambridge biochemist David Keilin (1887–1963), who first formulated the respiratory chain, and the Belgian biochemist Marcel Florkin (1900–1979); they have probed earlier issues in more detail than I have.[8] However, in general they have not pursued questions relating to bioenergetics, although there is a good introduction in Florkin.

1.2 VITALISM

This history is concerned with those who sought to establish the mechanisms that underlie the process of respiration and later the conservation of metabolic energy as ATP together with those operative in photosynthesis. In all this, there is an underlying assumption that the processes being discussed can be understood in terms of the chemical and physical sciences. The quest for mechanism has been primarily an attempt to explain the relevant cellular processes more or less exclusively in terms of normal laboratory investigation. Thus there is no need to resort to the concept of a vital principle or a vital force in order to explain the operation of living things. Such ideas hold that some processes that are part of living things but are not evident in the nonliving lie beyond the realm of laboratory science. Particularly in the nineteenth century, many scientists invoked the idea of vital forces in order to explain the mysterious properties of living things; such vital forces did not appear to be open to normal scientific investigation. The idea seemed necessary when nineteenth-century advances in the physics and chemistry of the inanimate world did not seem to be capable of application to living things.

The idea that living things possess a quality that cannot be explained in material terms can be traced back to Aristotle. It was suggested that there was something special about living things that distinguished them from the inorganic world, a view present through most of scientific history. In essence it was the question as to whether living organisms could be explained in terms of chemistry and physics without recourse to vitalistic ideas.

During the eighteenth century, the idea of a vital force developed in order to explain the inability to replicate living processes in the laboratory. As physicist and physiologist Hermann von Helmholtz (1821–1894) put it in 1861,

> The majority of the physiologists in the last century [eighteenth century], and in the beginning of this century, were of opinion that the processes in living

---

[8] The discussion here is only brief. Readers are referred to the coverage of nineteenth-century respiration and related issues in Florkin (1972, 1975a), Fruton (1999), Keilin (1966), Needham (1971), and Teich (1992).

bodies were determined by one principal agent, which they chose to call the vital principle.[9]

The precise meaning attached to the terms vital force or vital principle varied from author to author, but it became an integral part of biological and chemical thinking. One of the leading German chemists of the mid-century, Justus von Liebig (1803–1873), professor of chemistry at Giessen, Germany, wrote in his *Animal Chemistry*,

> Viewed as an object of scientific research, animal life exhibits itself in a series of phenomena, the connection and recurrence of which are determined by the changes which the food and the oxygen absorbed from the atmosphere undergo in the organism under the influence of the vital force.[10]

Here vitalistic ideas are used to explain those aspects of the living that cannot be explained by the use of chemistry and physics.

But toward the end of the nineteenth century, the belief that materials were produced in the organs under the influence of the vital force was being challenged. By 1878, Claude Bernard (1813–1878) was strongly opposing vitalism and questioned Liebig's previous statement and asked what is this vital force?

> The chemistry of the laboratory and the chemistry of the living body are subject to the same laws; there are not two chemistries; this Lavoisier said. Only the chemistry of the laboratory is carried out by means of agents and apparatus that the chemist has created while the chemistry of the living being is carried out by agents and apparatus that the organism has created.[11]

Many developments at the end of the nineteenth century and at the beginning of the twentieth century challenged the vitalist's approach, such as the demonstration that the process of yeast fermentation could be shown outside the living cell, the isolation of enzymes, and so on.

Thus, by the beginning of the twentieth century, there was a prevailing view that the functioning of biological systems could be explained in terms of chemistry and physics and that biological systems were fully open to scientific investigation. True, vitalism was not quite dead, but it was certainly in terminal decline.

So even in 1912, Sir Frederick Gowland Hopkins (1861–1947), one of the founders of modern biochemistry, could talk about the "spectre of vitalism" in his classic paper to the British Association.[12] Indeed it was at the end of this paper that Hopkins expressed his

---

[9] Helmholtz 1861, p.120.

[10] Liebig 1842, p.9.

[11] Bernard 1878, p.161. Although Bernard's anti-vitalist view is strongly expressed here in this late lecture, earlier notes suggest that his view on vitalism was much more ambivalent in his earlier years. (See Holmes 1974, p.407.)

[12] Hopkins 1913, p.159.

view on the application of physics and chemistry to the study of life, which he regarded as the mission of the biochemist:

> All of us who are engaged in applying chemistry and physics to the study of living organisms are apt to be posed with questions as to our goal, although we have but just set out on our journey. It seems to me that we should be content to believe that we shall ultimately be able at least to describe the living animal in the sense that the morphologist has described the dead. If such descriptions do not amount to final explanations, it is not our fault. If in "life" there be some final residuum fated always to elude our methods, there is always the comforting truth to which Robert Louis Stevenson gave perhaps the finest expression, when he wrote:
>
> "To travel hopefully is better than to arrive, And the true success is labour."[13]

The mission of the biochemist was clear—to account for life in physical and chemical terms, and if that left something unaccounted for, that did not invalidate the mission.

In the event, a century after Hopkins wrote those words, the success of scientific investigation of livings things has been such that it does not seem reasonable to assume that there is something we cannot investigate in this way. Although the specter of vitalism arises from time to time even now, it is always rejected. Such a cultural shift has been essential in opening the way for the explosion of biochemical and biophysical knowledge that became one of the characteristics of twentieth-century history and of biology in particular.

1.3 HISTORICAL QUESTIONS

There are other critical features that need to be noted. One of these is the way in which the field of bioenergetics achieved coherence in the late 1940s by the merging of several different and to some extent independent lines of investigation. One of these lines, the study of subcellular particles, was investigated by William Bechtel, who saw this process in rather different terms, those of the interfield excursions between biochemistry and cell biology.[14] Once established, the new field continued to draw significantly on other areas of biology, particularly the study of transport across membranes and ultimately the techniques of protein chemistry developed by the molecular biologists. Such a relationship between studies in cell biology and the development of biochemistry is not confined to the twentieth century, and, as will be seen, it occurred in the nineteenth century. But up until the late 1970s, the question hanging over most research in the field concerned the nature of the underlying mechanism of oxidative phosphorylation and photophosphorylation: How was the energy from oxidation of foodstuffs or from light used for the synthesis of ATP? Beyond that time serious questions presented themselves about more detailed aspects of the mechanisms such as those operating in and around the central enzyme of oxidative phosphorylation and photophosphorylation, the ATP

---

[13] Ibid.
[14] Bechtel and Abrahamsen 2007.

synthase, and the manner in which the respiratory chain or the chloroplast electron-transport chain provided energy for the enzyme. The coming together of photosynthetic studies and those of oxidative phosphorylation can be seen in Table 1.1.

Up to the end of the twentieth century, the history of bioenergetics can be divided into five phases or periods. Each is initiated by a significant advance in the field and in changes of the context in which bioenergetic questions were seen. The initial period concerns the origins of the field and is prompted by early experiments seeking to explain the nature of respiration or of photosynthesis. The chemical revolution of the late eighteenth century had a major impact on thinking about respiration and also strongly influenced photosynthetic studies. However, because the course of photosynthetic research diverged early from that of respiration and oxidative phosphorylation, these periods (or phases) need to be considered separately

The first period of respiration studies emerged in the seventeenth century and developed in the eighteenth century, reaching a significant point in the ideas of Antoine Lavoisier (1753–1794). His understanding of respiration, expressed in a number of memoirs given to the Paris Académie Royale des Sciences, gave a new view of the subject. As Culotta, who explored Lavoisier's approach to respiration, noted, "the origin of our present beliefs about respiration is universally found in the work of Antoine Laurent Lavoisier."[15] Although the period is essentially pre-bioenergetic, questions about the nature of respiration and its location were pursued then as they were in later periods. Other issues pursued in the nineteenth century concerned the meaning of Lavoisier's original concept of respiration as slow combustion and the nature of biological oxidation and of energy in biological systems.

The second period, the early twentieth century, began with the discovery of cell-free fermentation and with the birth of biochemistry as a discipline with its own journals. An early series of experimental studies drew attention to the importance of phosphates in intermediary metabolism. The consequences of these researches were complex and multiple but they led to the view that phosphorylations associated with oxidations were a key element in cell bioenergetics. Concurrently, sophisticated spectral studies of Otto Warburg and David Keilin and others brought a fresh understanding of the meaning of respiration and biological oxidation.

The third period commenced with the convergence of several lines of research into the one field of cell bioenergetics, which then began to attract a number of able biochemists. This field emerged in the late 1940s when phosphorylation and oxidation were understood as closely related issues. Simultaneously, the demonstration that the bioenergetic systems under investigation were located in a subcellular particle of eukaryotic cells, the mitochondrion, which could be isolated from other cell materials, strongly stimulated the field. Oxidative phosphorylation was explained by a relatively simple chemical mechanism that predicted the existence of an intermediate substance. Seeking the identity of this substance stimulated much research.

A fourth period of this story arose when the initial searches for the intermediate in oxidative phosphorylation proved unsuccessful. It therefore opened the field to a range of

---

[15] Culotta 1972, p.3.

Table 1.1 History of Cell Bioenergetics

CHRONOLOGY

Date	Oxidative Phosphorylation	Photosynthesis
	Phase 1 **The Nature of Respiration**	**Phase 1** **The Nature of Photosynthesis**
1640–1670	Establishing the physiology of respiration (Hooke, Boyle, Lower, Mayow, the Oxford Physiologists)	
1727, 1772	Early experiments on gases of respiration and photosynthesis (Hales & Priestley)	
1779		Green parts of plants evolve O_2 in the light. (Ingen-Housz)
1780–1790	Respiration as slow combustion (Lavoisier)	
1782		Confirmation of Ingen-Housz view that plant carbon is derived from CO_2 (Senebier)
1804		Assimilation of water and CO_2 (de Saussure)
1837		Starch and chlorophyll in cell granules (Mohl)
1842–	Respiration as oxidation of carbon & hydrogen (Liebig)	
		Phase 2 **Chlorophyll and Its Role**
1845–	Conservation of energy (Mayer, Helmholtz)	
1847	Respiration, a property of cells (Schwann)	
ca. 1850– ca. 1900	The nature of oxidation (Schönbein, Traube, Hoppe-Seyler, Bach)	
1862		Starch synthesized in the light in chloroplasts (Sachs)
1863		Chlorophyll fluorescence described (Stokes)
1870		Formaldehyde proposed as an intermediate in the photosynthetic conversion of CO_2 to sugar (Baeyer)
1878	Slow combustion rejected (Bernard)	

(continued)

Table 1.1 (Continued)

Date	Oxidative Phosphorylation	Photosynthesis
1882		Action spectrum demonstrates role of chlorophyll in evolving O_2 in the light. (Engelmann)
1886	The application of spectroscopy to biological systems. Discovery of myohaematin & histohaematin (MacMunn)	
1897	Discovery of cell-free fermentation (Buchner)	
	Phase 2 *Phosphates, Cytochromes, Oxidation*	
1904		O_2 evolution by wetted leaf powders in the light (Molisch)
1905		The discovery of light & dark reactions (Blackman)
1906	Phosphate, a requirement for fermentation (Harden)	
1909–1913	Biological oxidation as the removal of hydrogen (Thunberg, Wieland)	
1913–1914	Respiration located in cell granules (mitochondria) & associated with iron (Warburg)	
1918–1922		Photosynthesis as the photochemical reduction of CO_2 (Willstatter, Warburg)
1925	Rediscovery of cytochromes (Keilin)	
1929 1929	Discovery of ATP (Meyerhof, Fiske, & Subbarow) Identification of the "Atmungsferment" as a haem-containing enzyme (Warburg)	
1930, 1937	Demonstration of aerobic phosphorylation (Engelhardt, Kalckar)	
1932		Demonstration that most chlorophyll molecules are not involved in photochemistry. Confirmation of light & dark reactions (Emerson)

Table 1.1 (Continued)

Date	Oxidative Phosphorylation	Photosynthesis
1934	Preparation of mitochondria (Bensley)	
1935		Study of bacterial photosynthesis implies photolysis of water (van Niel)
		Phase 3 **Discovery of the Major Aspects of Photosynthesis**
1936		Concept of the photosynthetic unit proposed (Gaffron)
1937–1939		Isolated chloroplasts undertake oxidation– reduction. Photochemical splitting of water; the Hill reaction (Hill)
1939, 1943	Demonstration of phosphorylation linked to respiration (Belitzer, Ochoa)	
1939–1940	Discovery of the respiratory chain (Keilin)	
1941	Central role of ATP in metabolism. (Lipmann)	Using O_2 isotope, evolved O_2 shown to be derived from water in photosynthesis (Ruben)
1943		Demonstration of light energy transfer from carotenoid to chlorophyll (Dutton)
	Phase 3 **Mitochondrial Phosphorylation**	
1946	First theory of oxidative phosphorylation. (Lipmann)	
1948–1949	Mitochondria shown to be locus of oxidative phosphorylation, fatty acid oxidation, & citric acid cycle (Claude, Lehninger)	
1951		Discovery of cytochromes in chloroplasts (Hill)
1951–1954	Identification of regions of electron-transport chain linked to phosphorylation (Lehninger, Slater, Chance etc.)	

(continued)

Table 1.1 (Continued)

Date	Oxidative Phosphorylation	Photosynthesis
1953	First high-resolution pictures of the mitochondrion showing inner & outer membranes (Palade, Sjostrand) Chemical theory for oxidative phosphorylation (Slater)	
1954		Calvin cycle for CO_2 fixation proposed (Calvin) Isolated chloroplasts shown to carry out photosynthesis (Arnon) Demonstration of photophosphorylation in bacterial chromatophores (Frenkel)
1954–		Demonstration of photophosphorylation linked to oxidation reduction in chloroplasts. (Arnon)
1956		Discovery of reaction center P700 (Kok)
1957–1959	Quinone (ubiquinone) shown to function in respiratory chain (Hatefi & Crane)	
1959		Chemical theory adopted for photophosphorylation (Avron, Jagendorf)
	Phase 4 ***Defining the Mechanism***	*Phase 4* ***The Photosynthetic Light Reaction***
1960	Isolation of mitochondrial ATPase (ATP synthase). (Racker)	Z-scheme for photosynthesis proposed (Hill)
1961	Chemiosmotic theory proposed for oxidative phosphorylation & photophosphorylation (Mitchell)	
1961		Evidence for two photochemical reactions (Duysens)
ca. 1962	Isolation of respiratory complexes (Hatefi) Visualization of ATPase (Fernandez Moran) Calcium uptake by mitochondria demonstrated. (Vasington)	

Table 1.1 (Continued)

Date	Oxidative Phosphorylation	Photosynthesis
ca. 1964–1970		Advanced models for chloroplast structure. (Wehrmeyer, Paolillo)
ca. 1965	Respiratory chain shown to include iron sulphur centers (Beinert)	Chloroplast ATPase (ATP synthase) isolated (Racker)
1965	Proposal of conformational theory (Boyer)	
1966	Revised chemiosmotic theory (Mitchell)	
ca. 1966		Proton-based ATP synthesis in chloroplasts in dark (Jagendorf)
1968		Isolation of bacterial reaction center complex (Clayton)
1973	Conformational mechanism for the ATPase proposed (Boyer)	
1974	Demonstration of phosphorylation in artificial vesicles with bacteriorhodopsin & ATPase (Racker & Stoeckenius)	
1975	Proposal of the Q-cycle (Mitchell)	
1978	Nobel Prize for chemiosmotic theory	
	Phase 5 *The Impact of Protein Technology*	*Phase 5* *The Impact of Protein Technology*
1982	Isolation & sequencing of mitochondrial adenine nucleotide translocator (Klingenberg)	Subunits of chloroplast $b_{6}f$ complex separated (Hauska)
1985		Three-dimensional structure of bacterial reaction center determined (Deisenhofer et al.)
1988	Subunit composition of complex III of respiratory chain determined (Capaldi)	
1994	Three-dimensional structure of the ATP synthase (ATPase) determined (Walker)	
1995	High-resolution structure of cytochrome oxidase in *Paracoccus* determined (Iwata et al.)	

(continued)

Table 1.1 (Continued)

Date	Oxidative Phosphorylation	Photosynthesis
1996	High-resolution structure of mammalian cytochrome oxidase (Tsukihara et al.)	
1997	Observation of rotation of γ-subunit of the ATP synthase (Noji et al.)	

Note: In a number of cases the dates are approximate as particular discoveries are often made over a period of time rather than at a particular point in time or covered in a single paper.

more imaginative proposals for the mechanism of the process, drawing on other areas of science. It also broadened the concept of bioenergetics to include membrane-dependent aspects. A plurality of approaches to mechanism emerged and the field became divided, with different workers supporting particular mechanisms, often with some vehemence, a period described by some as the "Ox phos wars" ("Ox phos" is short for "oxidative phosphorylation"). At all events the field was plunged into a state best described as in crisis. A more coherent approach emerged after considerable struggle during the 1970s, culminating in majority support for the chemiosmotic hypothesis of Peter Mitchell, who was awarded the Nobel Prize in 1978.

The fifth period in this history is marked primarily by the impact of techniques drawn from the field of molecular biology. The respiratory chain was now studied in terms of its three main complexes, each of which presented very different problems.

The key development here was the growing ability to handle membrane proteins effectively by use of a range of new methods. In addition, the application of genetic techniques made possible powerful new approaches to understanding the role of these proteins. The solution of the basic problem of bioenergetics by the late 1970s had shown that the link between respiration and phosphorylation was an electrochemical proton gradient. This success clarified the need to solve a set of second-order mechanisms, some of which remained in question at the end of the twentieth century. Techniques for establishing protein structure and also genetic methods brought about new developments in the 1980s and 1990s. Substantiating the mechanisms of various parts of the bioenergetic systems, particularly that of the ATP synthase, the phosphorylation enzyme, produced spectacular conclusions. Thus the structure of the synthase, confirming its predicted mechanism, provided a crowning achievement to the twentieth century's struggle with oxidative phosphorylation. Such success marks an appropriate end point to this story.

The history of photosynthesis does not quite match that of respiration and oxidative phosphorylation. Its story begins concurrently with studies on respiration. Indeed, some of the same investigators and even the same experiments contribute to both histories such as the classic work of Joseph Priestley (1733–1804). However, once the basic outline of the photosynthetic process was established, particularly by Jan Ingen-Housz, the stories diverge. In the second period, the questions that occupied workers were concerned primarily with the role of light and chlorophyll, leading eventually to the division of the process into "light" and "dark" reactions. This long second phase lasted for over a century and made rather slow progress. The third period was initiated by Robin Hill at the end of the 1930s with the introduction of chloroplast preparations and the demonstration that the light reaction could be seen as an oxidation–reduction reaction. The fourth period, from around 1960 until the late 1970s, produced an outline mechanism for photosynthesis correlating the concepts of the pigment complex, a reaction center, a photosynthetic electron-transport chain and photophosphorylation with carbon assimilation. Beyond that point, the fifth period more or less coincided with the fifth phase of the work on mitochondrial bioenergetics, being driven by the same methods and building similar models.

The development of the subject matter has been summarized in Table 1.1, Chronology, which shows the major developments in the oxidative field and in the photosynthetic field.[16] These developments are set out in parallel columns so the progress in each field can be readily seen and compared; it illustrates some of the points just discussed.

The last period in the development of bioenergetics has been highly fruitful, but the elucidation of structures on which so much of the understanding of mechanisms depended proved a relatively slow process. Although there were major achievements before the end of the twentieth century, finding solutions to some of the complex technical problems was delayed until the current century, matters that are beyond the scope of this history. As a result, there is an incomplete aspect to this fifth period whether looked at in terms of chloroplast or mitochondrial bioenergetics.

1.4 PHOSPHORYLATION

The central subjects of cell bioenergetics are oxidative phosphorylation and photophosphorylation. The problem of phosphorylation, which emerged as being similar in the mitochondrion and the chloroplast, has proved to be one of the most challenging faced by biochemists in the twentieth century. Although a first mechanistic proposal was put forward by a Nobel Laureate, Fritz Lipmann in 1947,[17] it was not until 30 years later that a consensus on the basic mechanism was achieved. In the intervening period a more detailed hypothesis, in which a chemical intermediate was proposed and sought, was developed and finally abandoned. Ultimately it was the genius of Peter Mitchell that directed thinking in a novel direction, the chemiosmotic theory, and attracted a Nobel Prize. Even then there were leading workers in the field who felt uncertain about the conclusions reached, that Mitchell's proposal did represent the basic process. The most significant historical issue in this story is probably the attempt to determine the basic mechanism of oxidative phosphorylation and photophosphorylation, an issue that put the field of oxidative phosphorylation, more than in photophosphorylation, into crisis.

The mechanism of the overall process required further refinement in many ways over the ensuing years because detailed aspects had still to be elucidated. These

---

[16] The use of the term field has been discussed and definitions proposed in relation to the biological sciences by Darden and Maull 1977.

[17] Lipmann was awarded a share of the 1953 Nobel Prize in Medicine or Physiology "For the discovery of co-enzyme A and its importance for intermediary metabolism."

included the mechanism of the enzyme catalyzing the synthesis of ATP. Although Mitchell's broad principles had laid an acceptable foundation, his approach proved unsatisfactory, and Paul Boyer put forward proposals for the way in which the enzyme operated, a mechanism confirmed by the structural studies of John Walker. Boyer and Walker enjoyed a share of the 1997 Nobel Prize for Chemistry. There were other aspects of the overall mechanism that at the end of the twentieth century still remained uncertain, but this book will conclude with the achievements of Boyer and Walker.

Oxidative phosphorylation as discussed here operates in bacterial membranes as well as in animal and plant mitochondria. Although studies on bacteria, yeast, and plants have played an important role in understanding the process, the primary development of the subject has proceeded with the mammalian mitochondrial systems, which will be a central theme in the ensuing discussion.

The process of photosynthesis is a significant element of cell bioenergetics, although it has often depended on oxidative phosphorylation for conceptual advances. This can be considered in terms of three questions: first, how energy is harvested by pigments and converted to chemical energy; second, how that energy is used to synthesize ATP and generate reducing power; and third, how these products are used in the conversion of CO_2 into organic carbon compounds. Such questions can also be applied to the small group of bacteria that carry out photosynthesis without O_2 evolution. Here, as in the situation with oxidative phosphorylation in heterotrophic bacteria, the simpler systems of photosynthetic bacteria, particularly the purple bacteria, have proved valuable tools for understanding the process. The second question concerned with ATP synthesis shares many of the underlying principles associated with oxidative phosphorylation, and studies in photophosphorylation have contributed to the understanding of oxidative phosphorylation. Whereas such questions apply to thinking in the middle of the twentieth century, for the nineteenth century, the principal question seems to have been how light energy absorbed by chlorophyll converts CO_2 to carbohydrate and O_2. Workers tended to focus on the photochemical event in which CO_2 or its immediate product was converted to a sugar precursor. Consequently, the fields of photosynthesis and respiration diverged significantly in the nineteenth century.

1.5 MECHANISMS

The way in which philosophers and historians understand science has shifted significantly over the last 50 years. The classic work of Thomas Kuhn, *The Structure of Scientific Revolutions* (first edition 1962), with its central theme of paradigms and paradigm change, influenced a generation of scholars, although it has perhaps been most attractive to scientists. In the history of bioenergetics there appeared a period that is best described as a crisis, when the field lacked an agreed theoretical basis for understanding the nature of the central problem, the mechanism of phosphorylation linked to oxidation–reduction. To understand this situation, the ideas of Kuhn have been invoked to provide illumination of the situation. Kuhn's view of science progression is that normal science, based on some past achievement shared by the scientific community, the paradigm, provides the foundation for research. Research is essentially problem-solving. However, if this foundation can no longer be sustained, then a crisis state is reached, requiring a change of paradigm, a new approach, a novel theory. In the attempts to elucidate the mechanism of phosphorylation in bioenergetics, a crisis was reached but the nature of that crisis was rather more complex than Kuhn's simple examples.

The possibility that the biological sciences may well need to be considered in a manner different to that of the physical sciences has led to a philosophy of biology. Both Glenman and Darden and coworkers have proposed that much biology might best be understood in terms of seeking a mechanism. Although there is a long history to this approach, such a view was recently developed by Glenman, in which mechanism was defined in causal terms:

> A mechanism underlying a behavior is a complex system which produces that behavior by the interactions of a number of parts according to direct causal laws.[18]

A major contribution to mechanistic discussion came from Darden and coworkers, in which mechanisms were described in terms of entities and activities in which these entities were engaged. Their definition of mechanism is slightly different, taking a more explanatory approach:

> Mechanisms are entities and activities organized is such a way that they are productive of regular changes from start or set up to finish or termination conditions.[19]

Thus *mechanism* acquires a technical meaning in relation to biological science, and this has been explored by Darden in various contributions. For her, discovery is not an irrational act followed by logically characterized justification (or falsification); it is in itself a reasonable process. In the field of biochemistry, seeking a mechanistic explanation is an attractive approach to understanding the progress of research, and this has been adopted as the theme for considering the story of cell bioenergetics.

1.6 THE RELEVANCE OF CELL BIOENERGETICS TO THE QUESTION OF MECHANISM

This study is not concerned with contributing to the philosophical debate on mechanism directly but rather to look at the approach taken by biochemists to research on a particular biochemical problem or set of problems. The central question of cell bioenergetics can be posed as follows. How do cells convert the energy from breakdown of foodstuffs into chemical energy that becomes available to drive the cell's energy-dependent processes? A similar question arises in plants. How do plants convert light energy into chemical energy? Although seemingly rather different, the answers to these two questions turn out to have essentially the same process of phosphorylation at their core. There were several mechanisms proposed for phosphorylation in order to provide a basis for such answers.

---

[18] Glenman 1996, see p.52.
[19] Machamer, Darden, and Craver 2000, p.3.

They were remarkable particularly because of their diversity. They drew their inspiration from basic chemical processes through to morphological and structural biology. In this respect they reflected the disciplinary parentage of biochemistry as discussed by Robert Kohler, who examined the way in which biochemistry emerged as a discipline in the United States and Europe from medical, chemical, or physiological origins.[20] As a result I am concerned here to seek a historical description and an analysis of a search for a key set of mechanisms in biochemistry that is of central relevance to understanding the biology of cells.

The examination of the mechanisms involved in cell bioenergetics shows that there is a significant spatial aspect involved. This is apparent at the beginning of the nineteenth century and takes on different guises as the field progresses through to the late twentieth century. Similarly, the mechanisms proposed moved through a series of layers as they move to more and more detailed chemical levels, reflecting the growth of biochemistry as it emerged toward the end of the nineteenth century and established itself at the beginning of the twentieth.[21] Toward the end of the twentieth century the new molecular biology makes the major contribution.

There is another issue associated with this search for mechanism. This is the way in which two fields of research have related to each other, the studies of respiration and mitochondrial ATP synthesis and the studies on photosynthesis. Although there is a loose link between the two in the early and middle of the eighteenth century, essentially the two fields developed independently for nearly two centuries. It was only around the middle of the twentieth century that both groups of workers began to realize that these two fields would probably share the same basic mechanistic principles. This realization led to cross-fertilization between the two fields, although it seems that photosynthesis drew more on oxidative phosphorylation than the reverse. The way in which these two streams of research proceeded can be seen in Table 1.1, Chronology. As one example, theories developed for the basic mechanism of oxidative phosphorylation were also applied to photosynthesis, although there was a delay in applying Slater's chemical theory to photophosphorylation. Consequently, Chapter 8 discusses advances in both oxidative phosphorylation and photosynthesis together since most of the problems being addressed require similar solutions. However, the history of the two fields will need to be followed independently for much of the period covered here.

[20] Kohler 1982.
[21] See Kohler 1982.

2

From Physiology to Biochemistry
Respiration and Oxidation From 1600 to 1900

This chapter, covering the seventeenth through the nineteenth centuries, is concerned with the exploration of respiration in the whole organism and developing the understanding gained through initial physiological concepts to the beginning of ideas about enzymes. Thus it moves from physiology to biochemistry of the respiratory process. The period was concerned primarily with a discussion of the nature of respiration, although serious questions about the process of oxidation were being addressed by the chemists but with limited success. These studies provided the basis for understanding the development of cell bioenergetics that followed. Indeed it is not possible to appreciate the way in which the twentieth-century understanding of bioenergetics developed without a knowledge of the earlier period. It was in the seventeenth century that experimental philosophers began to discuss and to obtain evidence for the physiological mechanisms such as respiration that operated in the mammalian body. Their approach was a relatively simple one, and it is easy in retrospect to credit them with greater achievements than are justified. Nevertheless they provided an understanding which could be supported, rejected, or amended in the period that followed.

In this chapter, I discuss only four main strands of chemical–physiological reasoning that emerged from enlightenment science and that are foundational for twentieth-century bioenergetics. These themes are first and principally the understanding of respiration itself, mainly a physiological discussion but including both its location in the organism and its relationship to the problem of body heat; such considerations also cover concepts of energy. The other three themes all became major issues for nineteenth-century biologists. Of these three, the nature of biological oxidation, following on the discovery of O_2, was the most important, not for providing answers but for generating the questions for the twentieth century. The other two are technical advances of great significance for the twentieth century: the use of spectroscopy to investigate biological pigments and the discovery of cell-free metabolic systems.

An underlying issue is the nature of respiration and linked to this, body heat, seen originally as innate heat. The understanding of respiration was frequently associated with the understanding of combustion, and the analogy between respiration and combustion was often employed as a model. Older views compared the vital flame, thought to be located in the heart, to the burning of a lamp. However, a late medieval view regarded respiration as cooling the blood. The source of heat was often thought to lie with the blood, and one view saw blood as inherently warm. These views were partially resolved in the seventeenth century with the beginnings of the application of an experimental approach to biological problems. This led some significant physiologists to conclude that atmospheric air was involved in the physiology of the body in the same way as in combustion. Breathing was seen as the process responsible for providing animal heat, thus rejecting the idea of innate heat and undermining ideas of "fermentation" in the blood. After the chemical revolution in the eighteenth century linked in major part with the nature of gases, the idea of slow combustion and also the location of the process prompted much thought during the nineteenth century.

2.1 INITIATION OF THE EXPERIMENTAL STUDY OF RESPIRATION

Perhaps one of the greatest of the early physiologists was William Harvey (1578–1657) who became physician to the English king, Charles I. Harvey had studied initially in Cambridge and then in Padua under Fabricius (ca. 1533–1619). It was in Padua that his interest in circulation was stimulated. His principal book on animal circulation, published in 1628, *Exercitatio Anatomica de Motu Cordis et Sanguinis in Animalibus* [*Anatomical Treatise on the Movements of the Heart and the Blood in Animals*], gave an experimental analysis of the operation of the heart and the blood and promoted interest in such physiological studies. The historian Robert Frank commented, "Yet his book began the overthrow of a medical and physiological system . . . that had endured largely unchanged for almost a millennium and a half."[1]

Harvey had a significant influence on those who considered physiological questions later in the century. His later views on respiration were dependent on his belief that blood was naturally hot (if it cooled it was not blood). Therefore inspiration served to temper the blood sent from the lungs to the heart and expiration to expel waste products.

A group of English physicians and natural philosophers with interests in respiration and related matters set the course in the middle of the century for those who followed. Indeed, Holmes identifies the years from 1660 to 1674 as the crucial period when "a brilliant series of investigations concerning the function of the lungs in animals" provided the basis for later studies in the eighteenth century. However, he also notes that for an extensive period after about 1674 there was little further progress.[2] Among those who provided this stimulation to the physiology of respiration were Robert Hooke (1635–1703), Robert Boyle (1627–1691), Richard Lower (1631–1691), and particularly John Mayow (ca.1641–1679), all of them associated with the University of Oxford. Frank saw that

Such cooperative scientific activity was possible because Commonwealth and Restoration Oxford provided institutional support, intellectual resources and a

[1] Frank 1980, p.1.
[2] Holmes 1985.

congenial environment for a core of scientific leaders, such as William Petty, John Wilkins, Thomas Willis and Robert Boyle who in turn recruited large numbers of Oxonians to an interest in the new experimental philosophy.[3]

This group of experimental philosophers, who had wide-ranging interests, was associated with the foundation of the Royal Society in 1660.

Thomas Willis (1621–1675), a physician, was professor of natural philosophy at Oxford from 1660, and also involved in the early development of the Royal Society. His view of animal heat was that "fermentation"[4] in the blood produced heat. Later he espoused the view of his contemporaries, viewing heat production in the blood as a re-action between aerial niter and sulfurous particles (see subsequent discussion). It was Willis who engaged Robert Hooke as an assistant when the latter went to Oxford in the mid-1650s. Subsequently, Hooke worked for Boyle, also at Oxford, on the latter's vacuum or air pump.

Of particular note was Hooke's experimental demonstration that it was the supply of air to the lungs, rather than their movement, that was essential for life. The experiment was designed to refute the argument that the movement of the lungs was necessary for life because they promoted the circulation of the blood. By means of bellows he supplied air to the lungs of a dog that were kept full and immobile. He concluded from his experiment that "the bare Motion of the lungs without fresh air contributes nothing to the life of the animal" and that his animal died because of "the want of a sufficient supply of fresh air."[5] This experiment endorsed the role of air in respiration. Indeed, for Hooke, as for Boyle, Lower, and Mayow, breathing was the process responsible for maintaining body heat by a process analogous to combustion.

Robert Boyle, born in County Cork, Ireland, in 1627, settled in Oxford in 1654. He is especially remembered for his work on the relationship between pressure and volume of a gas. The basis for much of his work on the physical properties of air was the vacuum pump he constructed with Hooke. He viewed air as composed of corpuscles that were elastic because air would expand into a greater volume. If air was compressed, it was ca-pable of expanding again. Boyle's view of respiration was modeled on his understanding of combustion. It was aerial niter combined with sulfurous particles that constituted combustion. He also used his vacuum pump to show that in an evacuated vessel, a candle would not burn and a sparrow or a mouse would die.

Richard Lower went up to Christ Church College, Oxford, from Cornwall in 1649 and later also worked as an assistant to Thomas Willis. Later his studies of circulation, using vivisection with dogs, led him to conclude that the heart was a muscle that propelled the blood through the circulation system. Of particular interest was his observation on blood color that led him to conclude that the blood had taken up some of the air in the lungs. Lower was elected a fellow of the Royal Society (FRS) in 1667.

---

[3] Frank 1980, p.43.
[4] An intense motion of Particles or the Principles of every body.
[5] Hooke 1666.

2.2 JOHN MAYOW'S *TRACTATUS QUINQUE*

In some ways the most interesting of this group of experimental philosophers was John Mayow, born in Cornwall and baptized toward the end of 1641. The youngest of this group, he went up to Wadham College, Oxford, in 1658 and was elected to a Jurist's fellowship at All Souls in 1660, resigning the fellowship in 1678. He obtained a degree in civil law at Oxford, but probably not in medicine, although he practiced medicine in Bath. From Oxford he visited London on several occasions and shared his interests in respiration and combustion with Hooke. He was elected a FRS in 1678 but died the following year at the early age of 38.

He published two major works (both in Latin): the first, in 1668, the *Tractatus Duo,* was composed of an essay on respiration and a medical essay on rickets. In 1674 he published his major work, *Tractatus Quinque Medico-Physici.* This was a revised version of the essays on respiration and rickets together with a substantial treatise on *Sal Nitrum and Nitro-Aerial Spirit* and treatises *On Muscular Motion* and *On the Respiration of the Foetus in the Uterus and in the Egg.* The work was reprinted in the Hague in1681.

The significance of Mayow's work has been much debated. Assessments have varied widely, from the view that he discovered O_2 (discredited) to the idea that much of the work for which he has been given credit was due to the work of Boyle, Hooke, and Lower. Frank felt that "Mayow's major contribution was the logic that tied the elements of his system together" and that (referring to the *Tractatus Duo*), he

> reported several interesting and original experiments, but its central conception of an aerial nitre posited nothing that would have been unknown to someone who was acquainted with the Oxford group.[6]

Similarly, the influence of his writing on seventeenth- and eighteenth-century thinking has been debated but his work was certainly of significant interest to and was quoted by Stephen Hales.

Mayow's views were based on a system in which combustion involved the combination of saline sulfurous particles with igneo-aerial particles (aerial niter). This approach to combustion was also applied to respiration, in which Mayow carried out numerous experiments as well as noting the results of those carried out by Boyle, Hooke, and Lower. Mayow was clear from his experiments and those of others that a portion of air was used to support life by respiration and that this material passed into the blood. He wrote

> It is quite certain that animals in breathing draw from the air certain vital particles which are also elastic. So that there should be no doubt at all now that an aerial something absolutely necessary to life enters the blood of animals by means of respiration.[7]

And elsewhere,

> With respect, then, to the use of respiration, it may be affirmed that an aërial something essential to life, whatever it may be, passes into the mass of the blood.[8]

[6] Frank 1980, p.230.
[7] Mayow 1674, pp.73–74.
[8] Mayow 1674, p.204.

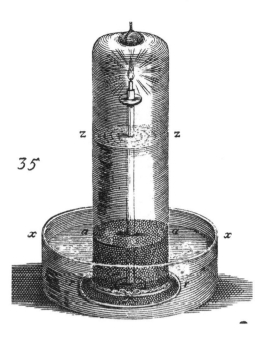

Figure 2.1 The Use of an Inverted Glass Over a Water Trough Designed to Study Changes in Air
The system uses a platform inside the jar on which to mount a candle (as here) or a rat. This experimental arrangement was used by Hales, based on Mayow, and developed further by Priestley. This version is taken from Hales's *Vegetable Staticks* 1727.

One of the experiments Mayow described involved placing a small animal or a candle in an inverted glass over water. (See Fig. 2.1 for a later version of this type of experiment.) Here he observed a loss of air from the system that he endeavored to measure. He also considered that respiration and combustion were similar processes and wrote

> Air is deprived of its elastic force by the breathing of animals very much in the same way as by the burning of flame. And indeed we must believe that animals and fire draw particles of the same kind from the air.[9]

These particles were, for Mayow, aerial niter.

Mayow was also impressed by Lower's observations on the color of blood flowing to and away from the lungs:

> Blood which entered the lungs with a dark colour, returns from them more florid and ruddy as arterial blood is, as was observed by the illustrious Lower in vivisections.[10]

[9] Ibid., p.75.
[10] Ibid., p.102.

It is important to realize that many of Mayow's views, while being shared by the Oxford Group of physiologists, were at odds with the general view at the time. For example, Mayow wished to distance himself from the Vital Flame of René Descartes, as pointed out by Everitt Mendelsohn:

> In the development of his theory of respiration and combustion, Mayow was quite consciously constructing a rational explanation in order to avoid the necessity of having "recourse to an imaginary Vital Flame that by its continual burning warms the mass of the blood."[11]

2.3 STEPHEN HALES'S *VEGETABLE STATICKS*

Following the impressive work in the mid-seventeenth century, little progress was made after 1674 until well into the eighteenth century. However, Stephen Hales (1677–1761) known primarily for his work on plants and initiation of thinking on photosynthesis,[12] did play a significant role in the respiration story. Hales entered Corpus Christi College, Cambridge, in 1696 and read theology and became perpetual curate in Teddington.[13] Here he remained, carrying out his experiments in the physical sciences and biology. He was elected FRS in 1718. He summarized much of his work in *Vegetable Staticks,* published in 1727.[14] Hales was aware of the work carried out three-quarters of a century earlier and indeed repeated Mayow's experiment, using his version of the inverted glass-over-water system (see Fig. 2.1):

> . . . I repeated Dr. Mayow's Experiment, to find out how much air is absorbed by the breath of Animals inclosed in glasses, which he found with a mouse to be 1/14 part of the whole air in the glass vessel.
> I placed on the pedestal under the inverted glass a full grown *Rat.* At first the water subsided a little . . . caused by the heat of the Animal's body. But after a few minutes the water began to rise and continued rising as long as the rat lived which was about 14 hours . . . the quantity of elastick air which was absorbed was 73 cubick inches, above 1/27 part of the whole, nearly what was absorbed by a Candle in the same vessel.

Mayow had used a rat and found 1/27 part of the air was consumed and in another experiment with a young cat, 1/30 part. In a following experiment Hales describes breathing into a bladder in which 1/13 part was absorbed. Unlike Mayow, who saw the process as

---

[11] Mendelsohn 1964, p.62. The words quoted by Mendelsohn are from Mayow 1674, p.108.
[12] See Chapter 6 for Hales's contribution to the early studies of photosynthesis.
[13] A small town on the Thames near Richmond, west of London.
[14] Hales published *Haemastaticks* in 1733.

involving the removal of aerial niter, Hales interpreted the process as demonstrating the effects of respiration on the elasticity of the air.[15]

The quality of Hales's experimental work is impressive. Conceptually he may not have advanced the field greatly, but his work was translated into Dutch, German, French and Italian.

2.4 RESPIRATION AND COMBUSTION

Although the chemical revolution brought about by Antoine Lavoisier marks the initiation of the modern study of respiration, earlier work in the eighteenth century provided a basis for the new understanding. In addition to the work of Stephen Hales, whose interests were essentially physical rather than chemical, Joseph Black (1728–1799) and particularly Joseph Priestley (1733–1804) made significant contributions.

Black's contribution concerned the chemical discovery of fixed air (CO_2) and owed something to Hales's experiments, including the name fixed air.[16] Black appears to have concluded that respiration converted ordinary air to fixed air and that the process generated heat. In experiments published in 1756, he noted the properties of fixed air as follows:

> I had discovered that this particular kind of air, attracted by alkaline substances, is deadly to all animals that breath it by the mouth and nostrils together . . . . And I convinced myself that the change produced on wholesome air by breathing it, consisted chiefly, if not solely, in the conversion of part of it into fixed air. For I found that by blowing through a pipe into lime water, or a solution of caustic alkali, the lime was precipitated, and the alkali was rendered mild. I was partly led to these experiments by some observations of Dr. Hales.[17]

Thus Black had identified CO_2 (fixed air) as a product of animal respiration.

It was Joseph Priestley's prodigious work on gases that may have been an inspiration for some of Lavoisier's work. As Holmes, in his in-depth study of Lavoisier, noted,

> By 1772, Joseph Priestley had become the most indefatigable investigator of the new "species" of airs. Priestley, like Black, brought this growing new dimension of chemistry quickly to bear on the phenomenon of respiration.[18]

His best-known experiment (repeated numerous times and in various ways) demonstrated the ability of plant material to regenerate air fit for breathing again after a candle had

---

[15] Hales 1727, pp.232–233.
[16] See Guerlac 1957.
[17] Black 1803, vol. II, p.87.
[18] Holmes 1985, p.5.

burned out in it or a mouse had been respiring and died.[19] Nevertheless, he interpreted his experiments in terms of the then-current phlogiston theory.[20] It is noteworthy that Priestley, like others before him, was aware of the similarity between combustion and respiration, as the following comment illustrates:

> That candles will burn only a certain time, in a given quantity of air is a fact not better known, than it is that animals can live only a certain time in it; but the cause of the death of the animal is not better known than that of the extinction of flame in the same circumstances; and when once any quantity of air has been rendered noxious by animals breathing in it as long as they could, I do not know that any methods have been discovered of rendering it fit for breathing again. It is evident, however, that there must be some provision in nature for this purpose, as well as that of rendering the air fit for sustaining flame; for without it the whole mass of the atmosphere would, in time, become unfit for the purpose of animal life; and yet there is no reason to think that it is, at present, at all less fit for respiration than it has ever been.[21]

Such a discussion summarizes much of Priestley's thinking about respiration and is also a preface to his experiment with the "sprig of mint" (see Chapter 6). Toward the end of the eighteenth century, the nineteenth-century approach to respiration was initiated by Antoine Lavoisier (1743–1794), arguably the father of modern chemistry and a man of wide interests. In addition to being a chemical scientist, Lavoisier was involved in preparing a geological map of France and had qualified as a barrister. However, he soon decided to devote his time to his chemical interests although he was also involved in collection of government taxes, an activity that provided an income. He was elected to the Académie Royale des Sciences in 1768 and was one of the most distinguished victims of the French Revolution, being sent to the guillotine in 1794.

The earlier work discussed in the preceding sections influenced the thinking of Lavoisier. Although already familiar with the work of Hales, it seems that toward the end of 1772, he started to read other published work, including that of Priestley, which he found of great interest. In 1777 he read his first paper on respiration to the Académie (published in 1780) where he discussed, among other things, the conversion of eminently respirable air into chalky acid air in the lungs. In 1783, together with Pierre Simon Laplace (1748–1827), he published a memoir on heat. Here he regarded respiration as

---

[19] Priestley 1772, see pp.168–169. See also Chapter 6 for a comment on this experiment, which used the experimental approach shown in Fig. 2.1.

[20] The phlogiston theory was formulated by the Bavarian chemist Ernst Stahl (1660–1734) and dominated chemical thinking until the time of Lavoisier, although the ideas are older. Phlogiston is the inflammable principle in combustible bodies and is released on combustion. "Dephlogisticated air" was the name given by Priestley to O_2. Air in which combustion had occurred was "phlogisticated."

[21] Priestley 1775, p.70.

slow combustion but localized the process in the lungs, reflecting the contemporary view. For Lavoisier, respiration provided the heat used to maintain body temperature:

Respiration is therefore a combustion, a very slow one to be sure, but moreover perfectly similar to that of carbon. It takes place inside the lungs, without giving off any perceptible light, because the matter of fire, once liberated, is forthwith absorbed by the moisture of these organs. The heat developed in this combustion is imparted to the blood that flows through the lungs, whence it spreads throughout the animal's body. Thus the air we breathe serves two purposes equally necessary to preserve life: it removes from the blood the base of fixed air whose excess would be very harmful; and the heat which this reaction gives to the lungs makes up for our continual loss of heat to the atmosphere and surrounding bodies.[22]

Such a view was endorsed and clarified in a later memoir on animal respiration by Lavoisier and Armand Séguin (1767–1835). Now the slow combustion concept was clearly set in a general framework of understanding of animal respiration. The fuel for the combustion is considered and respiration is seen as the source of animal heat:

Thus, to confine myself to simple ideas that everyone can grasp, I would say that animal respiration is nothing else but a slow combustion of carbon and hydrogen, which takes place in the lungs, and from this point of view animals that respire are veritable lamps that burn and consume themselves.

In respiration, as in combustion, it is the air of the atmosphere which furnishes the oxygen and the caloric. In respiration it is the blood which furnishes the combustible; and if animals do not regularly replenish through nourishment what they lose by respiration, the lamp will soon lack its oil; and the animals will perish, as a lamp is extinguished when it lacks its combustible.[23]

Lavoisier understood respiration as slow combustion entirely comparable with that of a burning lamp but in which the fuel was drawn from nutrition, not oil. Carbon and hydrogen were burnt to form carbonic acid and water. But his views on respiration were not built on the back of his theory of combustion; rather, his understanding of combustion and of respiration were developed in tandem. Holmes points out

That Lavoisier's investigation of respiration was not an application of a chemistry of combustion previously worked out, that he developed his theory of respiration and his theory of combustion in intimate association with one another from the time he first became interested in processes which fix air.[24]

---

[22] Lavoisier and de La Place 1783; see Guerlac 1982, p.37.
[23] Lavoisier and Séguin 1789, quoted in Holmes 1985, p.449.
[24] Holmes 1985, pp.467–468.

Indeed, Lavoisier felt that his mature understanding of respiration depended on a wide range of his work.

The idea that the mechanism of respiration could be understood in terms of combustion was explored and debated during much of the nineteenth century. In the mid-century, thinking about respiration began to move firmly from consideration of the whole organism in the direction of more chemical concepts. Physiological chemists of the period endeavored to define what in reality the combustion model meant. There was also a second issue raised in the preceding quotations from Lavoisier, that of the location of the respiration in the lungs. The question of the site of respiration in the animal became a serious subject for experimentation and discussion.

By the mid-nineteenth century, chemists were beginning to grasp more clearly what respiration might involve. Justus Liebig, one of the founders of organic chemistry, endorsed the view of slow combustion. Liebig engaged in extensive analysis of the elemental composition of compounds of biological origin and devised a method for measuring carbon and hydrogen in organic compounds. He viewed respiration as the oxidation of carbon and hydrogen in food to carbonic acid (CO_2) and water, which gave rise to heat:

> . . . it is especially carbon and hydrogen which by combination with oxygen serve to produce animal heat. In fact observation proves that the hydrogen of the food plays a not less important part than the carbon.

He went further, commenting that

> the production of carbonic acid is dependent on a formation of water, and that these two operations cannot be imagined as separate from each other.[25]

He regarded the oxidation of fat as an important part of the food oxidized and considered the fuel of the animal body providing heat as carbohydrate and fat. Despite his application of chemistry to living organisms, he nevertheless had recourse to vitalistic ideas to account for the chemical behavior of living organisms.

2.5 THE LOCATION OF RESPIRATION

The biochemical historian Joseph S. Fruton (1912–2007) has suggested that one of the main questions concerning nineteenth-century physiologists was "In what parts of the organism does the conversion of oxygen into carbon dioxide and water occur?"[26] It soon became clear that there were problems about Lavoisier's view that respiration took place in the lungs. If that were true then the lungs should be at a considerably higher temperature than other tissues, whereas observations indicated that this was not the case. An alternative approach was required.

[25] Liebig 1852. See pp.16 and18.
[26] Fruton 1999, p.235.

For a while workers such as Carl Ludwig (1816–1895) at Leipzig, who made a major contribution to the knowledge of the circulation system, and even Claude Bernard in the 1850s believed that respiration was located in the blood.

The opposing idea that respiration took place in the tissues was probably initially due to Lazaro Spallanzani (1729–1799),[27] professor of natural history at the University of Pavia. He noted that animals without lungs undertook respiration, taking up O_2 and releasing carbonic acid. Additionally, it was tissues like brain, flesh, liver, and skin that were all able to carry out respiration. Despite the significance of this work it seems to have had little influence and was probably a little in advance of its time.

Similar but more considered views came from a pioneer of modern biology, the professor of anatomy at Louvain, Belgium, Theodor Schwann (1810–1882). When he was developing his cell theory, which he stressed was based on argument rather than observation, he considered O_2–CO_2 exchange to be an innate characteristic of cells. Thus, Schwann regarded respiration as a fundamental property of cells:

> Oxygen or carbonic acid in a gaseous form or lightly confined is essentially necessary to the metabolic phenomena of the cells. The oxygen disappears and carbonic acid is formed, or *vice versa*, carbonic acid disappears and oxygen is formed. The universality of respiration is based entirely upon this fundamental condition to the metabolic phenomena of the cells.[28]

These views, initially published in the late 1830s, also failed to receive appropriate serious consideration.

It was not until after the middle of the century that respiration began to be seen more widely as a property of tissues. Justus von Liebig's son Georg Liebig (1827–1903) found that respiration continued in muscle tissues where the blood had been replaced with water, leading to the conclusion that the tissues, not the blood, were the site of respiration. Both Moritz Traube (1826–1894) and particularly Eduard Pflüger (1829–1910) of the University of Bonn strongly criticized the earlier proposal that the blood was the site of respiration. The latter argued, as Traube had done, that respiration was a property of cells and that the O_2 associated with respiration diffused from the blood to the tissues. In his classic 1875 paper he commented on respiration in a range of biological systems, particularly invertebrates, where "oxygen is led directly to the cell without intervention of the blood." From a discussion of a range of biological systems he concluded thus:

> What comparative physiology makes known to us clearly, ascending from the simplest forms of the animal kingdom to the higher creatures, ... that with the first moment of the development of the embryo the oxygen absorption and formation

---

[27] Spallanzani 1807. Spallanzani's work was published posthumously by Jean Senebier (1742–1809). The first volume, *Mémoires sur la respiration*, was published in 1803 but does not deal with tissue respiration. Further material from Spallanzani was published in three volumes in 1807.

[28] Schwann 1847, p.200.

of carbonic acid begin, thus at a time when *neither blood nor blood vessels* exist, when thus only cells can consume the oxygen and form carbonic acid.[29]

The conclusion that respiration was a property of the cells themselves echoed the views of a number of biologists of the late nineteenth century, but not all.

In the 1880s, the significance of cellular respiration was underlined by experiments of Paul Ehrlich (1854–1915). He used dyes, particularly indophenol, to stain tissues and noted their oxidation and reduction. He also injected a mixture of α-naphthol and *p*-phenylenediamine (later named the nadi reagent for short) that in the tissues was oxidized to indophenol blue and in some cases reduced to the colorless form. Ehrlich proposed a new theory of physiological combustion but did not seem to have been aware of the involvement of enzymes in these systems. Subsequently, following further studies of this system by others, the term indophenol oxidase was applied to the enzyme responsible for the oxidation. It became a subject for more detailed examination during the earlier part of the twentieth century, but this work underlines the view in the late nineteenth century that respiration is a property of cells and tissues. Thus respiration was now becoming an intracellular phenomenon, and this raised questions of the enzymes that might be associated with it.

2.6 THERMODYNAMIC QUESTIONS

Lavoisier had seen respiration as being the source of body heat, but in the 1840s a deeper understanding of the thermodynamic aspects of the process led to new models. Studies of muscular contraction directed attention to the use of the energy of respiration for mechanical work and also served to underline the location of respiration in the tissues. The law of the conservation of energy was developed within biological models, and physiologists became aware of the importance of the concept of energy (often referred to as force at the time).

Robert Mayer (1814–1878), a physician, seems to have developed many of his ideas when traveling as a ship's doctor on a voyage to the East Indies in 1840 on the *Java*. Mayer's major contribution was his understanding of the conservation of energy, but initially, in regard to the combustion theory, he argued that there was a fixed relationship between heat and work. He made this point in *Remarks on the Mechanical Equivalent of Heat* (1851), where his conception of energetic issues is well expressed:

> The physiological combustion theory takes as its starting point the fundamental principle that the amount of heat that arises from the combustion of a given substance is an invariable quantity . . . the living organism, with all its mysteries and marvels, is not capable of generating heat out of nothing.
>
> . . . then it cannot be assumed that the sum total of the heat produced by it [the organism] . . . could ever turn out to be *greater* than the chemical effect taking place. The combustion theory unless it wishes to renounce itself from the

---

[29] Pflüger 1875, p.274; translation Teich 1992, p.130.

beginning, has therefore no alternative but to assume that the *total* heat developed by the organism, partly directly, partly mechanically, agrees quantitatively with (or is equal to) the combustion effect.

He concluded that "an invariable quantitative relationship between heat and work is consequently a postulate of the physiological combustion theory."[30] Mayer's ideas on heat and work as just stated may not quite have reached the law of the conservation of energy, but this development followed. They certainly extended the understanding of Lavoisier's view of respiration as slow combustion, but this work did not immediately make a significant impact on the scientific community. There were, however, other concurrent influences that endorsed these advances, particularly the work of James Joule (1818– 1889) in Manchester, England, on the equivalence of mechanical work and heat.

The physicist Hermann von Helmholtz (1821–1894), who also started his career in medicine, almost concurrently developed similar ideas to those of Mayer. He succeeded in firmly establishing the notion of the conservation of energy. In his memoir, *The Conservation of Force*, delivered in Berlin in the summer of 1847, he begins with the assumption that it is impossible to create force (energy) continuously out of nothing. He concludes with a biological application:

> . . . the question of the conservation of force reduces to the question whether the combustion and transformation of the substances which serve as food generate a quantity of heat equal to that given out by animals. According to the experiments of Dulong and Despretz, this question can, at least approximately, be answered in the affirmative.[31]

Later, in 1861 in a lecture on *The Application of the Law of the Conservation of Force* delivered in London, he commented "at first sight it seems very remarkable and curious that even physiologists should come to such a law." He went on to consider the human body as an engine using food for energy:

> . . . If we take any thermodynamic engine, we find that the greatest amount of mechanical work which can be gained by chemical decomposition or chemical combination is only an eighth part of the equivalent of the chemical force and seven eighths of the whole are lost in the form of heat. And this amount of mechanical work can only be gained if we have the greatest difference of temperature which can be produced in such a machine. In the living body we have no great difference in temperature and in the living body the amount of mechanical work which could be gained if the living body were a thermodynamic engine, like the steam engine or the hot-air engine, would be much smaller than one eighth. Really we

---

[30] Mayer 1851. Quoted in Caneva 1993, p.240.
[31] Helmholtz 1847, p. 48. In the 1820s, César Despretz (1792–1863) and later Pierre Dulong (1785–1838) measured gas exchange and heat production of animals. The former concluded that respiration was the principal source of animal heat although the latter was less certain because the results were not entirely satisfactory as indicated by the use of the word "approximately."

find from the great amount of work done that the human body is in this way a better machine than the steam engine, only its fuel is more expensive than the fuel of steam engines.

He added that "we find that the laws of animal life agree with the laws of the conservation of force [energy]."[32] Such ideas were not greatly developed in the biology of the nineteenth century, but they nevertheless provided a background to better appreciation of the process of respiration. They also provided a basis for beginning to think about the relation between biological oxidation and the energetics of the cell. The formulation of such questions in terms of mechanism needed to await molecular developments in biochemistry.

There were, however, more immediate concerns in relating the mechanical action of muscles to respiration. Liebig, in his book *Animal Chemistry* in 1842, had already said that

> The contraction of muscles produces heat; but the force necessary for the contraction has manifested itself through the organs of motion, in which it has been excited by chemical changes. The ultimate cause of the heat produced is, therefore, to be found in the chemical changes.[33]

It should be noted that it was widely believed that it was the protein in muscle that was oxidized. In contracting muscle tissue, Helmholtz found that without blood, heat was still produced. Further, O_2 was not essential for the contracting process, leading to the view that O_2 must be bound to a protein in the tissue. As the understanding of energy was developed in the middle of the nineteenth century, the appreciation of muscle heat was expanded.

For some time, inspired by contemporary engineering successes, muscular contraction was understood in terms of the heat engine as a model. Such a view lasted at least to the end of the century and included the suggestion that temperatures as high as 140° might be reached locally! However, for respiration itself the model of slow combustion continued to be a guiding theme.

Although the impact of the new studies in thermodynamics was limited in the second half of the nineteenth century, a number of physiologists sought to understand the nature of body heat. A German physiologist, Max Rubner (1854–1932), worked initially in Munich and then in Marburg and finally in Berlin. He showed a relationship between the heat produced by an animal and the heats of combustion of carbohydrates, fats, and protein fed to an animal, taking into account losses in urine and feces. After his move to Marburg, Rubner built a calorimeter that measured the heat produced by a dog over a 24-hour period while also measuring the animal's metabolism. His experiments and calculations confirmed the view that metabolism was the only source of heat in the

---

[32] Helmholtz 1861, p.119.
[33] Liebig 1842, p.31.

animal body.[34] He also came close to showing that the law of conservation of energy applied to living systems.

The mid-nineteenth-century interest in energy served to broaden the understanding of respiration and, toward the end of the century, to direct attention to the chemistry of oxidation.

2.7 BERNARD'S CRITICISM
OF SLOW COMBUSTION

Claude Bernard, professor of physiology at the Collège de France and often seen as the founder of modern experimental medicine, initially regarded respiration as a property of the blood, but later he concluded the process occurred in the tissues. Although he held Lavoisier in high regard he felt that the situation in the 1870s required a revision of the slow-combustion model because it ignored the growing knowledge of and interest in enzymes or ferments. In a series of lectures delivered in 1878, he pointed out that

> They [the chemists] equated the chemical processes that take place in the organism to a direct oxidation, to a fixation of oxygen upon the carbon within the tissues. In a word they believed that organic combustion had as its prototype the combustion that takes place outside living beings, in our fireplaces and in our laboratories. Quite to the contrary there is within the organism perhaps not a single one of these phenomena of this alleged combustion that takes place by direct fixation of oxygen. All of them make use of the services of special agents, of the ferments for example.[35]

He accepted the current view that O_2 was added to the materials to be oxidized but felt nothing was known about the mechanism of O_2 utilization itself. It was, however, clear to him that ferments (enzymes) would be involved.

While regarding the precise role of O_2 as unknown, Bernard felt that Lavoisier's model of respiration as slow combustion was misleading, although valuable at the time it was developed. Comparison with the process of fermentation, which he discussed in the same lecture, led him to take a different view:

> Combustion is not direct in the organisms, and the production of carbonic acid which is such a general phenomenon in vital manifestations is the result of a true organic destruction, of a cleavage similar to those produced by the fermentations. These fermentations are moreover the dynamic equivalent of combustion: they fulfil the same purposes in the sense that they engender heat and are consequently a source of the energy that is necessary for life.[36]

[34] Rubner 1894. Thermochemical studies were developed further from around 1911 onward by A. V. Hill and Otto Meyerhof (see Chapter 3 and Florkin 1975b).
[35] Bernard 1878, p.120.
[36] Bernard 1878, p.124.

Bernard sees respiration not only as a source of heat but also as a source for the "energy of life." In this he was ahead of many of his contemporaries, who seemed to have assumed, but not questioned, a role for heat.

Indeed, knowledge of thermodynamics was developing and was yet to have a significant impact on biology. Hence many workers in the late nineteenth century appear to have made the tacit assumption that heat was associated with the formation of cellular constituents. Bernard's contemporary and colleague Louis Pasteur (1822–1895) summarized his view of metabolism and heat formation as follows:

> The aerobic organism makes the heat that it needs by combustions resulting from the absorption of oxygen gas; the anaerobic organism makes the heat that it needs by decomposing so-called *fermentable* matter belonging to the group of explosive substances that are able to release heat by their decomposition.[37]

Such a comment seems to carry the traces of Lavoisier's views on slow combustion that, while unacceptable to Bernard, remained influential. However, these issues began to clarify at the turn of the twentieth century in the writings of Franz Hofmeister (1850–1922).

2.8 HOFMEISTER'S INTEGRATION OF CELL BIOLOGY

Until 1896 Hofmeister had been professor of pharmacology in Prague, where he had contributed significantly to the study of proteins. Then in 1896 he accepted the chair of physiological chemistry at Strasbourg, France, in succession to Felix Hoppe-Seyler (1825–1895). Fruton notes that Hofmeister regarded solutions to the most fundamental physiological questions to be dependent on progress in three fields: protein chemistry, enzyme action, and intermediate metabolism. He has been described as seeing biochemical problems with the eyes of a biologist. This is demonstrated in an important essay on the chemical organization of the cell that deals with the integration of enzyme action in metabolic processes.[38] It also succeeded in bringing together the chemical views of life with those of the cell morphologists and is an indication of the thinking that was developing at the end of the nineteenth century.

> Here [in the organism] the nutrient material introduced as source of energy, before it is transformed into definite end products, undergoes a whole series of alterations which proceeding in parallel and in sequence, can be of very diverse chemical nature, and of very unequal energetic significance. Further while with the steam engine only the heat formed from the chemical energy is activated, so that it is quite immaterial from what fuel it comes, for the animal machine the substances that go to make nutrients are of the greatest significance.[39]

[37] Pasteur 1879, p.141, quoted in Fruton 1972, pp.359–360.
[38] Fruton 1985 provides a biographical view of Hofmeister and his work.
[39] Hofmeister 1901, p.581; translation Teich 1992, p.504.

Here the heat engine is no longer seen as a useful model for respiratory energetics. The organism is seen as a more complex machine, and a more refined model must be sought, particularly because metabolism proceeds by multiple steps. For Hofmeister, understanding the protoplasm requires the contribution of both the biochemist and the morphologist:

> If on the one side the morphologist strives to elucidate the structure of the protoplasm up to its finest details, on the other hand the biochemist endeavours, with his apparently grosser yet more penetrating aids, to ascertain the chemical performance of the same protoplasm, on the whole this still concerns only two different sides of the same coin. The one has in mind as final aim a horizontal and vertical projection, as detailed as possible, of the protoplasm structure. The other the description of the total in protoplasm operating processes by means of a connected chain of chemical and physical formulae. To make so far as the facts permit, these two widely divergent sets of ideas compatible with each other will be the laborious but not unrewarding task of the future.[40]

Hofmeister's perceptive comments on the relationship of structure and function set the course for the twentieth century's investigation of metabolism with its multiple steps and the need to link it with morphology. In essence it is a charter for the investigation into cell bioenergetics. Metabolism is to be seen as a process linked to the structure of the cell. This is perhaps the first clear example of the influence of cell biology on the development of the mechanisms associated with respiration; the twentieth century saw several more.

2.9 O$_2$ AND OXIDATION

The consideration of respiration in the nineteenth century brought forward more detailed questions about the role of O$_2$ and the process of slow combustion. For Lavoisier, oxidation implied the addition of O$_2$ to substances, a principle that prevailed through the nineteenth century. The puzzling fact was the ready oxidation of carbohydrates, fats, and proteins within the living organism but not outside. This led to a view that "activated" O$_2$ was involved. A variety of proposals were made in an attempt to determine how molecular O$_2$ reacted with organic substances in living systems. A few examples will serve to demonstrate the range of possibilities considered.

A role for ozone was suggested by Christian Schönbein (1799–1868), professor of chemistry at Basle, Switzerland, in the mid-century who contributed to the discovery of this form of O$_2$. He regarded certain substances as able to react with atmospheric O$_2$ to produce an ozonized form capable of carrying out oxidation at room temperature. He concluded that O$_2$ reacted with ozone, forming ferments to carry out cell oxidations.

---

[40] Hofmeister 1901, p.614; translation Teich 1992, p.506.

Traube held rather similar views on this subject. He thought that ferments that take up O_2 transfer it to other substances that thereby undergo oxidation. He commented

> . . . the analysis given here of the chemical processes of muscle respiration is essentially an application of the theory of slow combustion
> . . . It arises from the assumption that the muscle fibre, or rather the fibrous body contained in it is a *vital decay ferment* which transfers the oxygen taken up out of the blood on to the substance dissolved in the muscle fluid, itself suffering no damage in this process.[41]

Thus O_2 supplied by the blood is transferred to organic substances in the soluble part of the system to undertake oxidation.

Hoppe-Seyler regarded oxidation as the release from tissues of an activated hydrogen, "nascent" hydrogen, that then split molecular O_2 into two atoms, one reduced to water, the other liberated in active form that would then react with substances, leading to their oxidation.

This approach was developed by Alexei Bach (1857–1946) and by Carl Engler, who simultaneously produced similar proposals. Bach, initially at the University of Kiev, worked with Paul Schutzenberger in Paris at the Collège de France and then with Robert Chodat in Geneva before becoming director of various institutes in Moscow.[42] He developed his peroxide theory in which one of the two bonds in O_2 is cleaved and a peroxide formed by attaching to the oxidized substance. This peroxide would then be able to oxidize other substances. Such principles that emphasized the addition of O_2 to organic substances in the cell remained the basis for understanding oxidation into the twentieth century. Eventually it was overthrown by the work of Heinrich Wieland and Thorsten Thunberg, who brought about a revolution in the approach to this problem. Oxidation was to be seen as the removal of hydrogen (or electrons), not the addition of O_2. The view had to be revised when in 1955 oxygenases, enzymes that added one or two atoms of O_2 to the substrate, were discovered.

The nineteenth century saw a long and complex debate on the nature of oxidation, and a variety of different proposals were made. Palladium and platinum had been shown to catalyze oxidations of organic compounds and this, among other things, had led the Swedish chemist Jöns Jacob Berzelius (1779–1848) to develop his view of catalysis. This was rapidly extended to the view that biological ferments catalyzed biological reactions, providing a basis for the enzymology of the twentieth century. The possible importance of iron in blood was noted by Liebig, but Wilhelm Spitzer (1865–1901), who studied the indophenol oxidizing system in cells discovered by Ehrlich, in 1897 wrote

> The known oxidation processes which are brought about by animal cells isolated from the organism are due to the presence in these cells of specifically active nucleoproteins or, more precisely, of organically bound iron which mediates the oxygen transport.[43]

[41] Traube 1861; translation Teich 1992, p.126
[42] Kretovich 1983.
[43] Spitzer 1897, p.653; quoted in Keilin 1966, p.130.

He was the first to demonstrate the involvement of iron in oxidation, a theme developed at the beginning of the twentieth century.

The interest in oxidation by molecular O_2 can be seen as an inevitable consequence of the role ascribed to O_2 in respiration and its location in cells. It also prompted the need to consider a role for enzymes.

2.10 SPECTROSCOPY, HEMOGLOBIN, AND ANIMAL PIGMENTS

Visible–ultraviolet spectroscopy was developed in the nineteenth century and became a major tool in the development of metabolic studies, particularly of the respiratory chain. Although Isaac Newton (1641–1726) described the visible spectrum that could be observed with the use of a prism in the seventeenth century, it was not until the nineteenth century that this observation began to be used successfully in scientific research. Following the pioneering work, first of Joseph von Fraunhofer (1787–1826), who observed absorption lines in the spectrum, and second of Robert Bunsen (1811–1899) and Gustav Kirchhoff (1824–1887) who introduced a spectroscope, spectroscopic techniques became available to chemists and biologists. The English physicist George Stokes (1819–1903) and particularly Felix Hoppe-Seyler (1825–1895), professor of physiological chemistry at Strasbourg, were among the first to introduce spectroscopic methods into physiological chemistry.

This new experimental approach provided a means of examining a number of biological issues. One of these was the color of blood that had interested physiologists at least since Lower's observations in the seventeenth century. Lavoisier had noted that the red color of blood resulted from its combination with "eminently respirable air." Mayer had contemplated the phenomenon during his voyage on the *Java*, and Justus Liebig noted in his *Animal Chemistry* that

> The globules of arterial blood contain a compound of iron saturated with oxygen, which, in the living blood, loses its oxygen during its passage through the capillaries.[44]

The conversion of red arterial blood to dark-red blood was considered by Bernard in the 1850s as resulting from an increase in the proportion of CO_2 to O_2. However, it was Hoppe-Seyler who confirmed an earlier suggestion that the red pigment contained O_2 (oxyhemoglobin) in a loosely bound form, the dark red lacking the O_2. Both Hoppe-Seyler and Bernard showed that carbon monoxide would combine with hemoglobin, replacing the O_2 and giving a vivid red color.

In the 1860s, Hoppe-Seyler and also Stokes applied spectroscopic techniques to the study of hemoglobin in which they found two absorption bands in the visible spectrum whereas removal of O_2 gave only a single diffuse band (Fig. 2.2). This subsequently enabled Hoppe-Seyler to show that cyanide specifically inhibited respiration while not

---

[44] Liebig 1842, p.257.

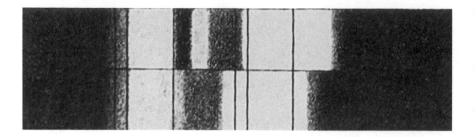

Figure 2.2 Spectrum of Hemoglobin as Seen by George Stokes in 1864

The red end of the spectrum is to the left and blue to the right. Oxyhemoglobin, shown in the upper spectrum, has two main absorption bands between the D and E Fraunhofer lines (with centers at 579 and 542 nm in the green region of the spectrum). Hemoglobin (reduced hemoglobin, lower spectrum) has one diffuse band with its center around 559 nm. Reproduced from Stokes 1864, figs. 1 and 2.

inhibiting O_2 binding to hemoglobin. These studies underpinned the now-prevailing view that respiration occurred in the tissues and that "for the body of rapidly respiring vertebrates, hemoglobin is therefore only a convenient carrier with large capacity."[45] Such work resulted in a growing interest in the use of spectroscopy in the study of biological systems.

However, it was a physician, Charles A. MacMunn (1852–1911) of Wolverhampton, England, who championed the use of spectroscopy in medical science. His first book, *The Spectroscope in Medicine* (1880), described forms of the spectroscope and their use in medical analysis. His other work, published posthumously, *Spectrum Analysis Applied to Biology and Medicine* (1914), covers similar ground updated with more modern equipment. Spectroscopic techniques developed in the late nineteenth century were one of the key advances stimulating progress in biochemistry in the first half of the twentieth century (see Fig. 2.3).

In the tissues he looked at, MacMunn observed lines in the visible spectrum that he believed had previously been seen by Henry Clifton Sorby (1826–1908), who had been working on mollusks. MacMunn was able to detect these absorption lines in a variety of preparations and noted that his four-banded spectrum belonged to the reduced state (Fig. 2.4). He attributed them to pigments he called histohaematin and myohaematin. His results were outlined in a paper to the Royal Society in 1886 in which he concluded that his pigments "are capable of oxidation and reduction and are hence respiratory" and added

> These observations appear to me to point to the fact that the formation of CO_2 and absorption of oxygen takes place *in the tissues* themselves and not in the blood.[46]

[45] Pflüger 1875, p.275; see Fruton 1972, p.283.
[46] MacMunn 1886, p.294.

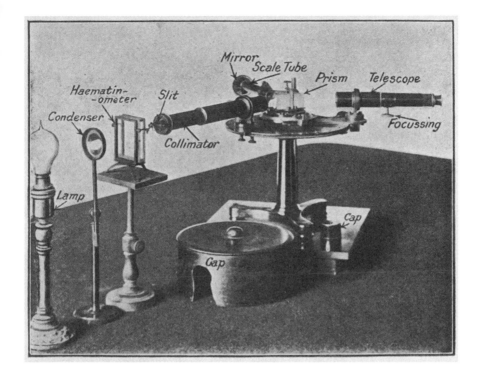

Figure 2.3 MacMunn's Spectroscope

This was the instrument used to identify the lines in the tissue spectra that MacMunn attributed to histohaematin. Reproduced from MacMunn 1914, fig. 15, p.77; photo courtesy of the Wellcome Library, London.

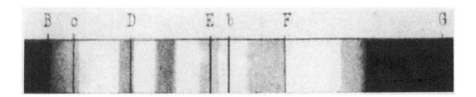

Figure 2.4 The Spectrum of MacMunn's Histohaematin as Seen in Rabbit Liver

The absorption bands lie first to the left of the D line, at approximately 605 nm, the second and third between D and E at around 560 and 550 nm, and a rather faint band to the left of the E line at about 520 nm. Reproduced from MacMunn 1886, plate 12.

Although he argued that his pigeon-breast muscle was free of muscle hemoglobin (myoglobin) his spectrum was in fact a mixture.[47] In Hoppe-Seyler's laboratory, myoglobin was found in extracts of the muscle, and MacMunn's work was discredited on the grounds that myohaematin was a hemoglobin degradation product and for other

---

[47] For an analysis of the shortcomings of MacMunn's work on histohaematin and myohaematin see Florkin 1975a, pp.219–222 and particularly Keilin 1966, pp.86–116.

reasons. Hoppe-Seyler even dismissed MacMunn's argument that myohaematin could be found in insect muscle where there was no hemoglobin; sadly, the weight of Hoppe-Seyler's opposition was too great and MacMunn's work was largely discredited and forgotten. As MacMunn commented in his last work,

> The name of Hoppe-Seyler has prevented the acceptance of the writer's views. The chemical position is undoubtedly weak, but doubtless in time this pigment will find its way into the text-books. Once and for all, it is not a derivative of haemoglobin.[48]

MacMunn's work was not seriously considered again until 1925 when Keilin rediscovered these pigments, which he named cytochromes. Keilin's own view of the matter was set out in a letter in 1923 to Sir Ray Lankaster:

> Hoppe-Seyler was wrong in generalizing the few errors of MacMunn and applying his criticism to the whole of his work, and he was obstinate in neglecting MacMunn's observations on other animals devoid of haemoglobin. In this history, we see a case, unfortunately not uncommon in science of how the weight of an authority may paralyse the development of a problem for almost 40 years.[49]

Later Warburg made a rather similar assessment.[50]

2.11 CELL-FREE SYSTEMS

There is no more significant observation for the development of the field of metabolic biochemistry than that carried out at the end of the nineteenth century by Eduard Buchner (1860–1917). Buchner was working on fermentation in Munich from around 1893. His older brother Hans (1850–1902) was pursuing studies in immunology and also working in Munich. He wished to find a method for the preparation of bacterial proteins in their native state for immunological purposes. Eduard was developing a method for obtaining yeast juices involving grinding brewer's yeast with sand and kieselguhr and then pressing out the juice with a hydraulic press. It is interesting to note that the authorities discouraged Eduard's work on grinding yeast cells on the grounds that "nothing will be achieved by this"!

Hans decided to try this method for preparation of microbial proteins by using yeast but found that the resulting juice was unstable. He then decided to try a number of preservatives, including glucose. When Eduard arrived in the laboratory in the summer of 1896 he noticed that an extract in 40% glucose was generating

---

[48] MacMunn 1914, p.73.

[49] Letter, Keilin to Sir Ray Lankester, September 7, 1923 (D. Keilin Papers, file 223, Cambridge University Library archives).

[50] "It shows how dangerous it is when people allow themselves to be influenced by false objections. MacMunn remained silent, and the result was that nothing more was heard of histohaematins during the next 33 years." Warburg 1949, p.63.

a stream of bubbles. He recognized this as fermentation but only slowly came to the conclusion that this was cell-free fermentation. The yeast juice, free of cells, was opalescent yellow and capable of fermenting a similar range of sugars to those fermented by whole yeast cells giving bubbles of CO_2. Eduard Buchner concluded in his classic paper,

> It is established that an apparatus as complicated as the yeast cell is not required to institute the fermenting process. Rather, the carrier of the fermenting activity of the press juice must be regarded to be a dissolved substance, undoubtedly a protein. This will be called *zymase*.[51]

Buchner took the view that he was dealing with a single enzyme, zymase.

The work also served to settle a major dispute between Liebig, who argued that fermentation was a chemical process in the cell, and Pasteur, who regarded the process as a vital process of a living yeast cell. However, it would be wrong to assume that Buchner's work slotted neatly into a progression of nineteenth-century physiological chemistry. As Kohler has pointed out,

> Thus for nearly twenty years before 1897, the existence of an alcoholase was not an active scientific problem, and twenty years is a long time in a science as active as cell biology. How then was the thread picked up again by Buchner twenty years later? Cell-free fermentation was first observed in yeast extracts prepared by Eduard's brother Hans Buchner, the immunologist, to test a new method of getting proteins from bacteria for immunization. Unlike the fermentation tradition, the germ theory of disease and immunity had enjoyed a period of tremendous expansion from 1880 on, and beginning about 1890 was undergoing a radical development from a cellular to a chemical-physiological science. The discovery of zymase was a result of this new approach to medical science, and in turn stimulated similar revolutionary changes in the biochemical study of cell life.[52]

The new results were debated in the years immediately following the demonstration, and although some found difficulty in repeating them, acceptance followed early in the twentieth century.[53] The significance of Buchner's achievement was summarized by Arthur Harden, who subsequently developed the work on yeast juice:[54]

> As in the case of so many discoveries, the new phenomenon was brought to light, apparently by chance, as the result of an investigation directed to quite other ends,

---

[51] Buchner 1897; translated by H. C. Friedmann, p.27.
[52] Kohler 1971, see p. 40. It should be noted that Teich disputes Kohler's assessment that studies on the nature of fermentation were not an active problem in the 1880s and 1890s. In evidence he cites the work of Emil Fischer with Lindner published in 1895 (Teich 1992, pp.44–45).
[53] For an account of the acceptance of Buchner's experiments, see Kohler 1971, 1972.
[54] See the next chapter.

but fortunately fell under the eye of an observer possessed of the genius which enabled him to realize its importance and give to it the true interpretation.[55]

Buchner was awarded the Nobel Prize in Chemistry for 1907, "For his biochemical researches and his discovery of cell-free fermentation."

This chapter has briefly charted the path from studies of respiration in the organism to the dawn of the modern biochemical approach. The background to the latter part of this chapter is the impact of the chemical approach on biology, a pervasive group of concepts that in due course transformed the approach to respiration and bioenergetics and brought about the foundation of biochemistry as a field of study.[56] Around 1900 major changes began to take place in physiological chemistry and chemistry resulting from several areas of experimentation, particularly the development of the understanding of enzymes and the demonstration of fermentation by zymase. As Kohler commented,

> It is no accident that the word "biochemistry" . . . suddenly caught the public fancy around 1902–1905. It was about that time that a profound change was taking place in the way the chemical activities of living cells were explained: for each vital chemical reaction, biochemists were coming to expect a specific intra-cellular *enzyme*. This belief is the hallmark of the new biochemistry and marks a major departure from the nineteenth century belief that the reactions of living cells were carried out by the whole cell protoplasm.[57]

Such a change in outlook opens the way for a new approach to chemical processes occurring in cells. The years leading up to the First World War saw the first successes in the biochemistry of truly intracellular processes coupled with an early understanding of the role of enzymes. These advances created a field of study out of which the twentieth-century understanding of bioenergetics could develop. These major events of the turn of the century were marked both by the first biochemical appointments within universities and the creation of the major journals: the *Journal of Biological Chemistry* (1905), the *Biochemical Journal* (1906), and the *Biochemische Zeitschrift* (1906), which later became the *European Journal of Biochemistry* and then the *FEBS (Federation of European Biochemical Societies) Journal*.

---

[55] Harden 1932, p.18.
[56] The importance of enzyme theory in driving the foundation of biochemistry in the period around 1900 is dealt with by Kohler 1973.
[57] Kohler 1973, p.184.

3

Relating Phosphorylation, Respiration, and Oxidation

1900–1945

This chapter looks at two main fields of research that emerged from nineteenth-century studies and that began to raise questions about the nature of bioenergetic mechanisms, although the major themes did not come together until after 1945. Answers to questions pursued in the nineteenth century now began to be sought much more in molecular terms.

3.1 RESOLVING NINETEENTH-CENTURY QUESTIONS

First, issues concerning the nature of biological oxidation by O_2 were pursued by Thunberg and Wieland, who countered the earlier views of Bach and others, although such views remained in contention for some considerable time. Further resolution of the problem was provided by the work of Szent-Györgyi and Hans Krebs. The role of O_2 in respiration was now partially clarified by the work of Otto Warburg (1883–1970). This work eventually linked with that of David Keilin (1887–1963), who reinitiated the work begun by MacMunn on histohaematin and myohaematin in the 1880s, leading eventually to the formulation of a simple respiratory chain. These studies gave rise to an understanding of the basic principles of cell oxidation.

Second, it was the experiments on the fermentation of Buchner's yeast juice that led to the discovery of the importance of phosphate in fermentation and in due course to the discovery of a number of phosphate compounds found to be important in the breakdown of glucose. Meyerhof's studies on muscle and those of several other workers including Gustav Embden raised the question of energy in metabolism and the role of phosphate compounds. Later Warburg solved the riddle of precisely why glycolysis and fermentation needed phosphate. The relationship of this work to the studies of O_2 consumption began to emerge late in this period, when phosphorylation was found to be associated with O_2 consumption. Hence respiration, whose mechanism had seemed to be akin to a slow combustion in the late eighteenth and early nineteenth centuries, began to acquire what Claude Bernard had foreseen, an enzymatic biochemical mechanism, albeit rather unclear but broadly definable.

3.2 ACHIEVEMENTS OF THE FIRST HALF OF THE TWENTIETH CENTURY

Before the work described in this chapter is outlined, it is desirable to outline the overall outcome of metabolic studies that provide the background to the discussion. The first half of the twentieth century was a time when the principal pathways for the breakdown of carbohydrates to water and CO_2 generating substantial cellular energy were elucidated.[1] It was this work involving glycolysis, the citric acid cycle, and the respiratory chain that provided the bedrock from which an understanding of molecular bioenergetics developed. The significant advances were a growing understanding of phosphate metabolism and of biological oxidation. The broad understanding of the oxidation of carbohydrate to CO_2 and water was the remarkable achievement of this period, and it was summarized by Hans Krebs in the early 1950s as a connected series of pathways that, taken together, achieved the production of energy from carbohydrate oxidation. The overall metabolic framework of Krebs's account[2] can be seen in Figures 3.1–3.4.

In Figure 3.1 the result of the studies of carbohydrate oxidation up to about 1950 is summarized. The pathways involved are the glycolytic pathway converting sugar to pyruvate; the oxidation of pyruvate through the citric acid cycle, producing the reducing equivalents shown as [H]; finally [H] is oxidized to water. The oxidation of glucose sugar in the glycolytic pathway could be balanced by reduction of pyruvate to lactate if not oxidized by O_2 in the respiratory chain. Many of the detailed questions remained to be resolved after 1950, but the elucidation of these pathways provided the biochemical culture for the ideas and experimental support for bioenergetics.

As will be shown in Section 3.3, much of the initial work on yeast metabolism was driven by an investigation of why phosphate was important in fermentation, a question that later led to questions about phosphorylation as a theme in bioenergetics. Although the early work led to the discovery of sugar phosphates, the pathway shown in Figure 3.2 shows that the phosphate in sugar phosphates is derived from adenosine triphosphate (ATP), not from inorganic phosphate. The role of inorganic phosphate is in the phosphorylation of glyceraldehyde 3-phosphate, leading to ATP synthesis, a process described as substrate-level phosphorylation. The term is also used to describe the ATP synthesis in the phosphoenolpyruvate-to-pyruvate step.

The understanding of biological oxidation underwent a revolution early in the twentieth century. It was particularly the pathway that ultimately emerged as the citric acid cycle (Fig. 3.3) that confirmed the validity of the new approach to cellular oxidation. Thus oxidation proceeds by removal of hydrogens that, during the period under consideration, were shown to be transferred to the pyridine nucleotide, nicotinamide adenine dinucleotide (NAD, earlier known as diphosphopyridine nucleotide [DPN]). This was then oxidized by the respiratory chain in which the energy available

---

[1] The breakdown of lipid involved many of these processes but the oxidation of fatty acids by the β-oxidation pathway became clear a little later, although the understanding of this pathway had its historical roots at the beginning of the century. The urea cycle was worked out by Krebs in the period around 1930 and was another contribution to the emerging picture of metabolism.
[2] See Krebs 1953.

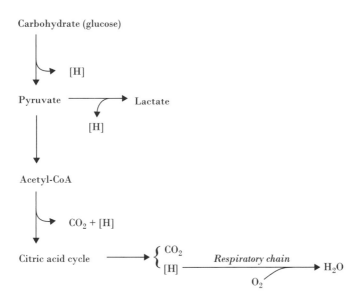

Carbohydrate (glucose)

[H]

Pyruvate ——————→ Lactate

[H]

Acetyl-CoA

$CO_2 + [H]$

Citric acid cycle ——————→ { CO_2 *Respiratory chain* —————→ H_2O
                              [H] O_2

Figure 3.1 Summary of the Metabolic Oxidation of Carbohydrate at Around 1950
(See Krebs 1953.)

was conserved by the synthesis of ATP (Fig. 3.4). An exception to this was the oxidation of succinate to fumarate for which the flavoprotein enzyme concerned was a part of the respiratory system (see Fig. 3.4). It should be noted that the oxidation of α-ketoglutarate was coupled to the synthesis of an ATP analogue, guanosine triphosphate (GTP) in a substrate-level phosphorylation. Many of the details of the citric acid cycle were worked out after 1950, but the overall pathway was clearly understood by that date.

The respiratory chain (Fig. 3.4) began to be understood during the first half of the twentieth century but the details of its operation remained to be elucidated at least up until the mid-1970s and later. That understanding was foundational for the field of bioenergetics in which most of the ATP synthesized in aerobic cells is associated with the respiratory chain.

3.3 YEAST AND ANIMAL JUICES: THE IMPORTANCE OF PHOSPHATE

A key step in the development of bioenergetics in the early part of the twentieth century was the discovery of the importance of phosphates in metabolism and particularly in fermentation. This arose out of work by Arthur Harden (1865–1940), a chemist, together with William J. Young (1878–1942) at the British Institute for Preventive Medicine (later the Lister Institute for Preventive Medicine). Following up Buchner's demonstration of fermentation in cell-free yeast juices, they showed the importance of phosphate in the fermentation process. As part of a series of experiments designed

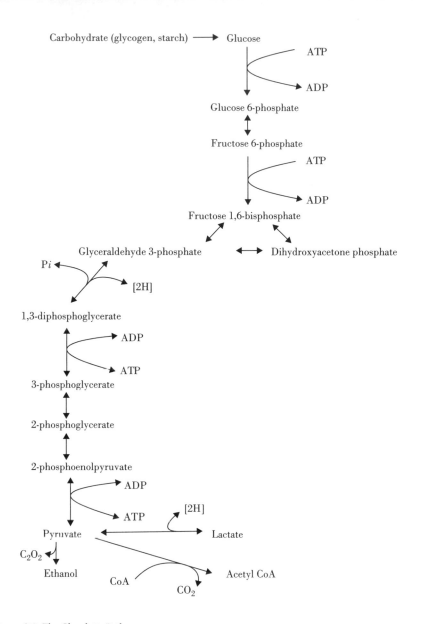

Figure 3.2 The Glycolytic Pathway

The pathway is the major route in most cells for the breakdown of carbohydrate. In animals with oxygen deficiency, lactate is produced whereas in oxidizing conditions acetyl CoA is generated. In yeast fermentation the product is ethanol. Note that the sugar phosphates are cleaved to form two molecules of triose (glyceraldehyde 3-phosphate and 3-hydroxyacetonephosphate).

to test the effects of various proteins on fermentation by yeast juice, they tested the effect of boiled yeast juice. This turned out to stimulate fermentation by ordinary yeast juice substantially although the boiled juice itself was inactive in fermentation.[3]

[3] It appears that Buchner and Rapp had earlier made a similar observation in a single experiment, a point that Harden was aware of.

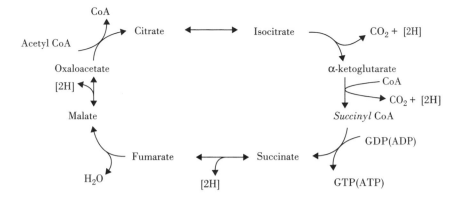

Figure 3.3 The Citric Acid Cycle (Krebs Cycle)

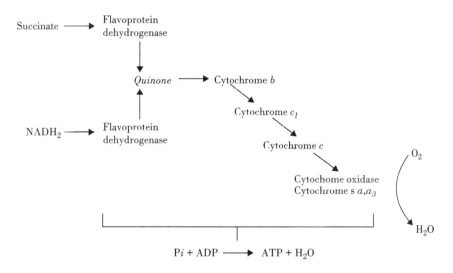

Figure 3.4 The Respiratory Chain as Understood Around 1950

Note that the function of a quinone was not discovered until around 1960 (see Chapter 4) and that the position of the b-cytochrome remained questionable until at least the 1970s. See Slater 1953.

Dialyzing[4] the yeast juice showed that the dialysate (although inactive itself) also stimulated fermentation. Searching for the active component of the boiled or dialyzed yeast juice led them to discover that phosphate strongly stimulated

[4] Dialysis is a process in which small molecules can be removed from a preparation by passing them through a membrane through which large molecules cannot pass. Thus dialyzed yeast juice has lost the small molecules but retained the enzymes and other proteins.

fermentation.[5] In a later paper, studying the effect of phosphate on fermentation, they concluded thus:

> When a soluble phosphate is added to a fermenting mixture of glucose and yeast-juice the following phenomena are to be observed: (1) The rate of fermentation is at once greatly increased. (2) This acceleration lasts for a short time and the rate then falls off, and returns approximately to its original value. (3) During this period the extra amount of carbon dioxide evolved and alcohol produced are equivalent to the phosphate added. (4) The phosphate is . . . probably present as a salt of a hexose phosphoric acid.[6]

This led to a search for phosphates formed in fermenting yeast juice that resulted in the discovery of fructose 1,6-diphosphate (Harden and Young ester, now fructose bisphosphate)[7] and subsequently two monophosphates. Glucose 6-phosphate was found by Robert Robison in 1922 (the Robison ester), and later a second monophosphate was identified, fructose 6-phosphate. Although Harden himself regarded the sugar phosphates as side reactions in fermentation, this work initiated and stimulated the study of organic phosphates in intermediary metabolism.

The work on yeast juice inspired work with press juices of muscle by Gustav Embden (1874–1933), physiological chemist and professor at Frankfurt am Main. From about 1912, Embden's group found that this juice gave rise to lactic acid but the process was not influenced by adding glucose or glycogen:

> We rather believe that by the observation that muscle press-juice under certain conditions forms equimolar amounts of lactic acid and phosphoric acid, taken together with the fact that hexose phosphoric acid as the only one of all the substances investigated increases the extent of this lactic acid and phosphoric acid formation, it becomes extremely probable that also the lactacidogen is to be regarded as a carbohydrate phosphoric acid, or indeed contains a carbohydrate phosphoric acid complex in its molecule.[8]

These results confirmed the fact that muscle juices were capable of producing lactic acid and also phosphate but the source of the lactic acid was unclear, and Embden postulated a substance he called lactacidogen, which he thought could be a sugar phosphate. Progress with muscle juice proved slow until the work of Otto Meyerhof (1884–1951) opened a way forward.

Meyerhof was born in Hanover, Germany, and educated at the University of Heidelberg, where he obtained a doctorate in psychiatry. His early interests were in psychology and philosophy, but he was persuaded by Otto Warburg to pursue cellular physiology. In 1913

---

[5] Harden and Young 1906. Together with Euler, Harden received the Nobel Prize in Chemistry in 1929 for "their investigations on the fermentation of sugar and fermentative enzymes."
[6] Harden and Young 1908, p.299.
[7] Ibid.
[8] See Embden et al. 1914, p.42; translation Teich 1992, p.219.

he joined the physiological laboratory at the University of Kiel in Germany, eventually becoming assistant professor. In 1924 he was invited to the Kaiser Wilhelm Institute for Biology in Berlin-Dahlem and in 1929 became head of the department of physiology at the Kaiser Wilhelm Institute for Medical Research in Heidelberg. Of Jewish extraction, he had to leave Germany in 1938 and eventually moved, through France, arriving in the United States in 1940.[9]

Embden's muscle-juice preparation did not allow for the type of experiments carried out with yeast juice. However, in 1926 Meyerhof developed a preparation with properties similar to those of yeast juice by crushing frog or rabbit muscle in cold isotonic potassium chloride or distilled water.[10] It was this material that enabled Meyerhof and his group to isolate important phosphate compounds that laid the basis for appreciating the significance of these substances in bioenergetics. Work in Meyerhof's laboratory and elsewhere showed that the pathway for carbohydrate breakdown in muscle (glycolysis; see Fig. 3.2) was essentially the same as fermentation in yeast except that the end product in the former was lactic acid and in the latter alcohol (ethanol). Particularly significant was the demonstration of the importance of relatively low-molecular-weight heat-stable dialyzable substances found in both yeast and muscle juices. In the mid-1920s, such substances began to be identified and they confirmed the importance of organic phosphates in intermediary metabolism. These compounds are considered further in Section 3.9.

3.4 THUNBERG, WIELAND, AND THE NATURE OF BIOLOGICAL OXIDATION

It is probably no exaggeration to say that there was a revolution in the concept of biological oxidation that started early in the twentieth century. As previously noted (see Chapter 2), most nineteenth-century thinking assumed that oxidation would involve the addition of O_2 to the organic substrate. Heinrich Wieland (1877–1957), an organic chemist who succeeded Richard Willstatter as professor of chemistry at the University of Munich in 1921, developed a new approach to oxidation theory in the field of pure chemistry. This depended on the activation of hydrogen rather than O_2. Thus oxidation could consist of the removal of hydrogen rather than the addition of O_2. In 1913 he extended his views to cover some biological reactions.

Of particular significance to Wieland's views were Thunberg's observations on the respiration of finely chopped muscle, published in 1909. Thorsten L. Thunberg (1873–1952), professor of physiology in Lund, Sweden, tested a large number of organic acids to see their effect on respiration. Whereas oxalic acid and malonic acid had shown some stimulation of respiration, succinic acid had a significant effect. In the 1920s, Juda Quastel at Cambridge University demonstrated the competitive inhibition of the succinic acid dehydrogenase by malonic acid. These observations initiated a long period of investigation of succinate metabolism lasting more than 50 years.

[9] Peters 1954.
[10] Meyerhof 1926.

$$CH_2^- COOH \qquad\qquad CH_2^- COOH$$
$$| \qquad\qquad\qquad\qquad\quad ||$$
$$CH_2^- COOH \qquad\qquad CH_2^- COOH$$

2H

Succinic acid   Fumaric acid

Figure 3.5 Oxidation of Succinic Acid to Fumaric Acid Involves the Loss of Two Hydrogens

Initially, Federico Battelli (1867–1941) and Lina Stern (1878–1968) showed that the net O_2 uptake corresponded to one atom of O_2 per molecule of succinic acid and Hans Einbeck (b. 1875) identified the product as fumaric acid;[11] further, conversion to malic acid was also observed. Thunberg concluded that succinic acid was oxidized by the activation of its hydrogen (Fig. 3.5).

> We perceive the process in the system succinic acid-enzyme-oxygen or in the system succinic acid-enzyme-methylene blue in the way that the enzyme robs the succinic acid of two hydrogen atoms, which hydrogen atoms it then passes over to oxygen or methylene blue.[12]

An important element of Thunberg's work was his "Thunberg tube" technique (see Fig. 3.6) whereby an evacuated tube at constant temperature is used to mix muscle preparation and methylene blue, which is reduced (becomes colorless) in the presence of succinate. Because methylene blue was believed to be reduced (decolorized) by hydrogens, Thunberg regarded the oxidation as proceeding by the removal of hydrogens. The enzyme in muscle was named succinic dehydrogenase. The technique was used to demonstrate a number of other "dehydrogenases" in muscle tissue, including those for malic, succinic, citric, lactic, α-ketoglutaric, and glutamic acids.[13]

The evidence in favor of Wieland's and Thunberg's view of oxidation grew during the 1920s. It should be noted that this approach to oxidation was not without its critics, but Wieland became emphatic about his interpretation in a book on oxidation:

> A closer examination of those substances which are oxidized in the cell shows that, in general, they are substances which undergo the chemical changes involving loss of hydrogen, that is, by dehydrogenation. Limiting ourselves to the chief energy-supplying foods, we have in this class carbohydrates, amino-acids, the higher fatty acids, and glycerol. There is no known example among them of an unsaturated compound in the case of which it is necessary to assume direct addition of oxygen, that is, additive oxidation.[14]

[11] Einbeck 1914.
[12] Thunberg 1918, p.185. See Teich 1992, p.148, for the English translation.
[13] For an account of Thunberg's work and its relationship to Wieland's views, see Thunberg 1930; for a short biography see Young 1953.
[14] Wieland 1932, p.26.

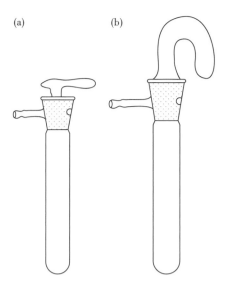

(a)   (b)

Figure 3.6 Thunberg Tube

The left-hand tube (a) is the original form and (b) represents the developed form. The reactants could be placed in the bottom of the tube and in the hollow stopper and mixed after evacuation through the side arm and bringing the tube to the desired temperature in a water bath. Methylene blue (included with the reactants) could be reduced by the reaction but not reoxidized in the reduced pressure. Reproduced from Ernest Baldwin 1947, *Dynamic Aspects of Biochemistry*, Cambridge University Press, fig. 13, p.91.

Such a situation remained true until the mid-1950s. Nevertheless, uncertainty over the nature of oxidation persisted, and in 1932 the fourth edition of Harden's book on fermentation[15] cites Bach's and Wieland's theories as alternatives.

The question of cellular oxidation was given substantial clarification by events that occurred in the 1930s. A number of studies including those of Thunberg had provided a background to the work of Albert Szent-Györgyi (1893–1986) who worked in Hungary at the University of Szeged as professor of medical chemistry. He measured O_2 uptake in minced pigeon-breast muscle. He found that several dicarboxylic acids (succinic, fumaric, malic, and oxaloacetic) all had a catalytic effect on O_2 uptake, that is, they stimulated much more uptake of O_2 than was necessary for their oxidation.[16] This led him to the idea that they were involved in the transfer of hydrogen from a donor to succinate (Fig. 3.7), which could then be oxidized by what he referred to as the Warburg–Keilin system of cytochrome and *Atmungsferment* (see Section 3.5). He explained:

> The theory is this: the C_4 dicarboxylic acids are a link in the respiratory chain between foodstuff and the WK system. Their function is to transfer the hydrogen of the foodstuff to cytochrome and to reduce by this hydrogen its trivalent iron again to the divalent form.[17]

[15] Harden 1932.
[16] Annau et al. 1935.
[17] Szent-Györgyi 1939, p.170.

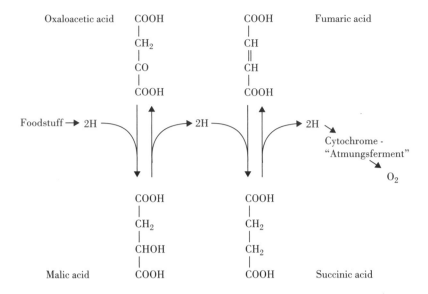

Figure 3.7 Szent-Györgyi's Scheme for the Oxidation of Foodstuffs by a Pathway Involving the Four-Carbon Dicarboxylic Acids

The pathway is based on two couples, oxaloacetate/malate, which receives the hydrogen [H] from the oxidation of "foodstuffs," and fumarate/succinate, which receives a hydrogen from the malate and transfers it to the respiratory system of Keilin and Warburg for reduction of oxygen to water. This can be compared with the citric acid cycle that emerged from the work of Hans Krebs, as shown in Figure 3.8 and also Figure 3.3 (based on Szent-Györgyi 1939).

Following on this work, Hans Krebs (1900–1981) substantially developed the ideas of Szent-Györgyi. Born in Hanover, Germany, Krebs spent some of his early career working with Warburg at the Kaiser Wilhelm Institute for Biology in Berlin-Dahlem.[18] However, in 1933 when he was working in Freiberg, he felt it unsafe to remain in Germany and moved to Cambridge, England, and then in 1935 to Sheffield, England. Krebs carried out an extensive study on the respiration of minced pigeon-breast muscle using a variety of organic compounds, much of the work employing the Warburg apparatus (see Section 3.5) that he had learned to use in Warburg's laboratory. In 1937 he published his paper[19] proposing the citric acid cycle (later also known as the tricarboxylic acid cycle and the Krebs cycle, Fig. 3.8, but see also Fig. 3.3). It was based on four sets of experimental observations on pigeon-breast muscle. First, citrate, isocitrate, and α-ketoglutarate were rapidly oxidized (Carl Martius and Franz Knoop had shown a metabolic pathway from citrate to α-ketoglutaric acid[20]). Second, citrate had a catalytic effect on respiration of the muscle that was comparable to that of succinate and fumarate. Third, Krebs showed that citrate could be synthesized from oxaloacetate. Fourth, succinate could be formed

---

[18] For an autobiography see Krebs 1981. For a detailed analysis of Krebs's work leading to the formulation of the citric acid cycle, see Holmes 1993.

[19] Krebs and Johnson 1937.

[20] See Martius 1937.

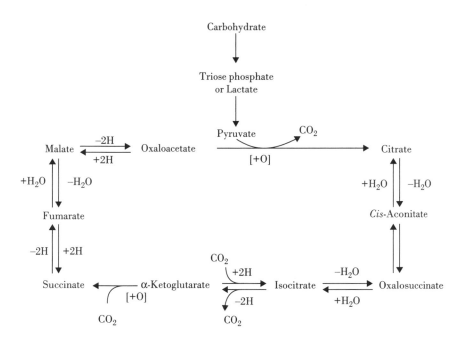

Figure 3.8 The Citric Acid Cycle as Conceived by Krebs in the Early 1940s

oxidatively from fumarate even when the succinate dehydrogenase was inhibited by malonate. Krebs concluded that

> The quantitative data suggest that the "citric acid cycle" is the preferential pathway through which carbohydrate is oxidized in animal tissues.

Although there were doubts about the role of citrate in the cycle during the 1940s, this discovery became the basis for the understanding of metabolic cell oxidations. The relationship of this pathway to respiration and oxidative phosphorylation (summarized in Figs 3.1, 3.3, and 3.4) are considered in the next chapter. Nevertheless, the pathway put the view of biological oxidation as removal of hydrogens in an unassailable position.

3.5 WARBURG'S *ATMUNGSFERMENT*

Otto Warburg has been described as one of the great pioneers of biology but he is also known for some major disputes. He was descended from several generations of bankers, and his father, Emil Warburg, became professor of physics in Berlin. His early interests were in chemistry and he was a student under the distinguished organic chemist and Nobel Laureate, Emil Fischer. He then developed an interest in medicine and particularly cancer. He worked initially in Heidelberg until the outbreak of the First World War in 1914, when he joined an elite cavalry regiment and saw service on the front line; he was decorated with the Iron Cross, First Class. He felt that his war service enabled him to know the realities of life outside the laboratory.[21] After the war he returned to the Kaiser Wilhelm Institut für Biologie in Berlin-Dahlem to resume his work, which included his conviction that iron was involved in respiration and also interests in photosynthesis, especially the issue of quantum efficiency. During the Second World War he remained in Berlin, being classified as quarter Jewish. He was allowed to continue his work almost unimpeded, although why this was so is not clear.

Warburg's early research included work on rapidly developing fertilized sea-urchin eggs, particularly at the Naples Zoological Station. He studied the respiration of the disrupted eggs initially with titrimetric methods. He found that the O_2 uptake of his preparations, which contained some iron, was stimulated by the addition of ferrous and ferric salts. Interestingly, he could separate out granules from his material that appeared to be the location of the respiration, indicating that the process was structurally bound, a conclusion also drawn from experiments with liver.[22] The significance of these latter findings did not become clear until more than 30 years later with the work of Claude and others on small and large granules (mitochondria) from cells in which the latter were shown to be the locus of respiration. He also noted the sensitivity of the system to inhibition by cyanide and urethane.[23] He concluded that

> On the ground of the facts communicated I advance the theory that the oxygen respiration in the egg is an iron catalysis; that in the respiration process the oxygen consumed is primarily taken up by the dissolved or adsorbed ferrous ion.[24]

In 1910, Warburg visited the laboratory of the physiologist Joseph Barcroft in Cambridge, where he was impressed by the apparatus Barcroft and John Scott Haldane had devised to measure blood gases. Warburg developed the principles of Barcroft and Haldane to devise a manometric method of measuring the rate of gas exchange in a small shaken reaction vessel at constant temperature that subsequently became known as the Warburg apparatus. Over the first half of the twentieth century, this system was elaborated by Warburg and others for the study of respiration, CO_2 release, and any reaction that could be coupled to a reaction involving gas uptake or evolution (see Fig. 3.9). It became a major tool supporting the advance of metabolic biochemistry, and it provided a key instrument for the work of Hans Krebs on the citric acid cycle. Krebs, a student of Warburg, learned the methodology working in Warburg's laboratory in the 1920s.[25]

During the first half of the 1920s, Warburg showed that iron-containing charcoals (blood charcoal) could catalyze oxidations by molecular O_2. In a series of elegantly designed experiments, he showed that oxidation was associated with iron bound to nitrogen. As he later recalled,

[21] For a biography of Warburg, see Krebs 1972.
[22] Warburg 1913.
[23] Urethane is a narcotic known to inhibit respiration at relatively high concentrations.
[24] Warburg 1914, p.253; translation Teich 1992, p.145.
[25] For an account of working in Warburg's laboratory as seen by Krebs, see Holmes 1991, pp.131–139.

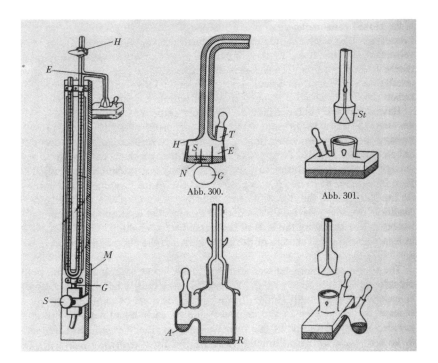

Figure 3.9 The Warburg Apparatus

The left-hand side shows a manometer with a reaction flask attached. The remainder of the figure shows a variety of reaction flasks that could be attached to the manometer side arm. The apparatus was designed to measure pressure changes in the small flask attached to the side arm (E). The flask was maintained at constant temperature in a water bath (not shown) and shaken. The pressure changes in the flask were measured with the manometer. The flask had one (or more) side arm so that reactants could be mixed at the desired time after temperature equilibration. The flask could have a separate central compartment that could be used for a small amount of potassium hydroxide that absorbed all the CO_2 from the gaseous phase. Taps allowed for starting the internal pressure at atmospheric pressure and, if necessary, replacing atmospheric gas with another gas (such as nitrogen). Reproduced with permission from Holmes 1991, *Hans Krebs*, Oxford University Press (fig. 22, p 144)

The result of the study of the charcoal model containing iron bound to nitrogen was that this combination produced a biological oxidation like that of the oxidative deamination of amino-acids, an oxidation which is reversibly inhibited by narcotics and irreversibly and specifically inhibited by cyanide. Who could believe that this was only by chance in agreement with the behaviour of cell respiration?[26]

Warburg interpreted his results as showing that the O_2 reacts with iron and by doing so initiates the oxidation of organic substances:

Molecular oxygen which is used up by the respiration of aerobic cells never reacts directly with the biological substrates, but always and exclusively with divalent

[26] Warburg 1949, pp.32–33.

iron combined in a complex. Iron of a higher valency is thereby formed, and this is reduced back again to the state by organic substances.[27]

Hence Warburg regarded respiration as involving an enzyme responsible for O_2 metabolism, which he referred to as the *Atmungsferment*, the respiratory enzyme.

In the light of Bernard's observations on the effect of carbon monoxide on hemoglobin,[28] in 1926 Warburg tested the effect of carbon monoxide on the respiration of yeast cells. He discovered that carbon monoxide not only inhibited respiration but the inhibition was reduced by increased O_2 pressure. He concluded that carbon monoxide "attaches itself to the enzyme molecule at the point which under normal conditions is the site of the oxygen reaction."[29] He also noted Haldane's conclusion that "carbon monoxide has a poisonous action on rats apart from its combination with hemoglobin."[30] The English physiologist from London University, Archibald Vivian Hill (1886–1977) while visiting Warburg's laboratory drew the latter's attention to the fact that carbon monoxide inhibition of hemoglobin was reversible by light, as shown by Haldane and Smith.[31] Warburg then found that the carbon monoxide inhibition of respiration was also substantially decreased by illumination. By determining the effect of different wavelengths of monochromatic light on the carbon monoxide inhibition, Warburg was able to construct an absorption spectrum of the carbon monoxide–binding pigment involved (see Fig. 3.10). He concluded that it was an iron porphyrin. The immense skill and ingenuity required for achieving this result in the late 1920s should not be underestimated. The work identifying the nature of the *Atmungsferment*, the respiratory enzyme of respiration, led to the award of a Nobel Prize to Warburg.in 1931.[32] Significantly it brought enzymology to bear on the uptake of O_2 in respiration.

3.6 KEILIN'S CYTOCHROME

A different and highly significant approach to the study of respiration emerged from the work of David Keilin. Keilin[33] was born in Moscow, brought up in Warsaw, and then studied in Liège and Paris, where he developed interests in entomology. In 1915 he became a research assistant to the Quick professor of biology at Cambridge, George H. F. Nuttall (1862–1937), a parasitologist. Already with an international reputation in

---

[27] Ibid., p.47.

[28] In the 1850s, Claude Bernard showed that the poisonous effects of carbon monoxide were due to its action on hemoglobin by displacing O_2 bound to the protein.

[29] Warburg 1949, p.74.

[30] Haldane 1927, p.1071.

[31] Haldane and Smith 1896.

[32] The 1931 Nobel Prize for Physiology was awarded to Otto Warburg "for his discovery and mode of action of the respiratory enzyme."

[33] Contrary to almost all biographical information on Keilin, his daughter, Dr. Joan Keilin-Whiteley confirmed to Professor Slater that Keilin was born into a Russian-Jewish family in Moscow who were moved west in one of the periodic pogroms and settled in Warsaw. (Letter, Professor Slater to the author, July 29, 2004).

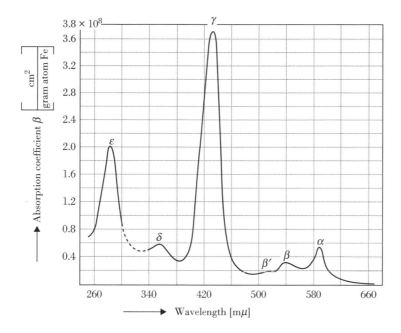

Figure 3.10 Warburg's Absorption Spectrum for the Carbon Monoxide Compound of the Oxygen-Transporting Enzyme, Atmungsferment

It was this spectrum that persuaded Warburg that the *Atmungsferment* was an iron porphyrin because of its close similarity to absorption spectrum of carbonyl hemoglobin. The absorption maxima of the latter were mostly at slightly shorter wavelengths. Reproduced with permission from Warburg 1949, *Heavy Metal Prosthetic Groups*, Oxford University Press (fig. 5.5, p.136).

entomology, Keilin continued his studies on insects and moved with Nuttall to the new Molteno Institute when the latter became its director. In 1931 Keilin succeeded Nuttall as Quick professor and director of the Institute.

Studying the fate of oxyhemoglobin in horse bot-fly larvae with a microspectroscope, Keilin observed four absorption bands in the spectrum of the thoracic muscles.[34] These absorption bands disappeared on oxidation of the preparation but reappeared on standing (see Fig. 3.11). He realized these were almost identical to the absorption bands described in the 1880s by MacMunn but felt that the name "cytochrome," cell pigment, would be more appropriate. He found similar absorption bands in a wide range of animals, plants, and microorganisms. He identified three separate components of cytochrome—*a, b,* and *c*—corresponding to three of the bands; the fourth was identified as linked to all three components. The oxidation was inhibited by cyanide and the reduction by urethane, leading to the conclusion that cytochrome was part of the mechanism of respiration. He

[34] The decisions that led Keilin to undertake this research are described in chapter 8 of Keilin 1966. Much of Keilin's work depended on a modified microscope, the direct-prism spectroscopic ocular (manufactured by Zeiss) and, in the early years, the modified Hartridge reversion spectroscope, both designed to give simple spectroscopic information on turbid preparations.

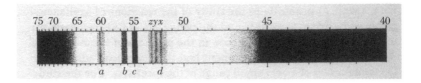

Figure 3.11 Keilin's Absorption Spectrum of Cytochrome for the Honey Bee Thoracic Muscle as Published in 1925

The spectrum was obtained using a microspectroscope (a combination of microscope and spectroscope). Keilin's wavelength calibration is shown on the top edge with the red (here labeled 65–75) at the left-hand end and blue (40–45) at the right-hand end. The four absorption bands of cytochrome, *a, b, c, d,* seen in the reduced but not in the oxidized system, are marked at the bottom. The cytochrome bands were named (*a–d*) according to their position in the spectrum. This observation can be compared with MacMunn's results discussed in the previous chapter. Reproduced from Keilin 1966, *The History of Cell Respiration and Cytochrome,* Cambridge University Press, fig. 5, p.154.

further deduced from his experiments that the components of cytochrome were heme compounds.[35]

Following the publication of his first paper on cytochrome in 1925, Keilin's work over the next 8 years was devoted to trying to elucidate further the nature and role of these pigments, and numerous papers on cytochrome followed. As Keilin subsequently wrote,

> Between 1926 and 1933 my work on cytochrome was proceeding along two main lines: (1) The determination of the nature of its three components, cytochromes *a, b* and *c*; and (2) the study of the mechanism of its oxidation and reduction in yeast cells and in cell-free heart-muscle preparations.[36]

He soon concluded that cytochrome acted between the oxidases and the dehydrogenases. The oxidase was referred to as indophenol oxidase, an enzyme identified by its ability to oxidize indophenol with O_2 as originally described by Ehrlich in the 1880s (see Section 2.5). This enzyme appeared to Keilin to have the same properties as Warburg's respiratory enzyme, "Warburg's respiratory ferment is a polyphenol or indophenol oxidase system."[37] One component only, cytochrome *c*, could be isolated in water and was oxidizable by the indophenol oxidase and O_2; the system was inhibited by cyanide, sulfide, and by carbon monoxide in the dark. Such results led Keilin to conclude that

> The catalytic system reconstructed from the oxidase of heart muscle and cytochrome *c* of yeast cells behaves, therefore, like a true respiratory system of the cell.[38]

It was perhaps inevitable that two such different approaches to the study of the respiratory use of O_2 as pursued by Keilin and Warburg should produce different understandings

---

[35] Keilin 1925.
[36] Keilin 1966, p.166.
[37] Keilin 1927.
[38] Keilin 1930, p.443.

of the terminal step in respiration.[39] Referring to the latter's *Atmungsferment*, Keilin pointedly commented that it was "inconceivable to ascribe the whole of the respiratory system to the activity of only one type of enzyme to the exclusion of the other type." Warburg disagreed. In his 1929 Herter lecture he responded to Keilin's view that the *Atmungsferment* [respiratory enzyme] was identical with the indophenol oxidase:

> If one classifies the oxidases according to their effect, one has in the extracts at different times different oxidases: glucose oxidase, alcohol oxidase, indophenol oxidase and so on . . . . Therefore if different oxidases occur in an extract of one kind of cell they are not enzymes which were preformed in the living cell, but transformation and decomposition products of a substance uniform in life.[40]

Warburg's criticism of Keilin's work is implicit in a comment in his Nobel lecture:

> It is still not possible to answer the question whether the MacMunn hemins form part of the normal respiratory cycle, i.e., whether respiration is not a simple iron catalysis but a four-fold one.[41]

In 1934, Keilin was joined by E. F. Hartree (1910–1993) as a collaborator and the two of them resolved the argument by describing a further cytochrome, a_3 which reacted directly with O_2 and mediated the oxidation of the other cytochromes. They identified cytochrome a_3 with cytochrome oxidase (indophenol oxidase previously shown to oxidize cytochrome[42]) and especially with Warburg's respiratory enzyme. They concluded that they were

> in agreement with the main results obtained by Warburg who has demonstrated that a haematin compound, which with CO gives an absorption spectrum showing bands at about 590 and 432 mµ plays an essential role in cellular respiration. Component a_3 can therefore be identified with Warburg's respiratory or oxygen transporting enzyme. [43]

Thus the dispute between Warburg and Keilin on the relationship of the former's respiratory enzyme and the latter's cytochrome chain was settled. Keilin, who was not awarded a Nobel Prize (to the disappointment of many) had embarked on a program of research which led to the discovery of the respiratory chain but this achievement was not easily gained. As a letter to Britton Chance suggests, many distinguished workers had not been supportive of Keilin's work on cytochrome and intracellular oxidation reactions, especially during the first period (1925–1933):

---

[39] A flavor of this argument, which continued over several years, can be gained by comparing Warburg 1949, pp.166–172, with Keilin 1966, pp.194–203.

[40] Warburg 1930, p.357.

[41] Warburg 1932, p.255.

[42] Keilin and Hartree 1938.

[43] Keilin and Hartree 1939, p.188.

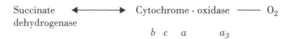

Succinate
dehydrogenase
Cytochrome - oxidase —— O_2

b c a a_3

Figure 3.12 Simple Respiratory Chain for the Oxidation of Succinate
This represents Keilin's view of the system where cytochrome was seen as singular with three main components. (See Keilin and Hartree 1940.)

My first papers in this field (1925, 27, 29, 30), contrary to my expectations, antagonized many workers who believed that they had unraveled the mechanism of cellular respiration by postulating either the existence of an omnipotent iron (based on the charcoal-iron model), or of a direct reaction of hydrogens of activated substrate molecules with O_2, or of the formation by glutathione of a direct link between activated substrate and the molecular oxygen.[44]

The problem of the role of O_2 in respiration had attracted a variety of proposals mostly before Warburg had described his *Atmungsferment*. They included those based on a central role for iron and on the view of oxidation as removal of hydrogens, and inevitably they had led to dispute.

Much of Keilin's work had been carried out on heart muscle but at the end of the 1930s a refined method for preparing the ox-heart-muscle oxidizing system, based on earlier work of Battelli and Stern in 1912, was developed. This became a standard method for biochemical studies of muscle respiration (the Keilin–Hartree muscle preparation).[45] The preparation made possible a study of the oxidation of succinate by the cytochrome system. Keilin and Hartree described a bound succinic acid oxidizing system that was able to reduce the cytochrome system. They commented that it

is of special interest as it represents so far the only known complex intracellular oxidation system which in a cell-free preparation behaves exactly as the natural respiratory system of the cell.[46]

It constituted a respiratory chain composed of succinic dehydrogenase, cytochromes *b*, *c*, *a*, a_3 that reduces O_2 (Fig. 3.12). Such a simple respiratory chain provided a first, very general answer to the question of what the molecular mechanism of respiration was (a more advanced version can be seen in Fig. 3.4 although this still remained to be substantially revised in later years).

However, knowledge of the structure of cytochrome molecules was required for a better understanding of their function. Most cytochromes were firmly bound to the particulate parts of the cell, defying attempts at isolation, and it was decades before their structures could be studied. In contrast, cytochrome *c* could be successfully solubilized, opening up

[44] Letter, Keilin to Chance, March 2, 1961, attached to a letter, Keilin to Slater, March 6, 1961. EC Slater archive Noord-hollands archief, Haarlem, The Netherlands.
[45] Keilin and Hartree 1938.
[46] Keilin and Hartree 1940, p.277.

the possibility of determining some of its structural properties. In 1930 Keilin described the separation of this cytochrome from respiring baker's yeast cells, obtaining "a strong transparent solution of a deep red color," and showed that this material retained its natural properties.[47] Studies on cytochrome *c* were also pursued by the Swedish biochemist Hugo Theorell (1903–1982), who had graduated in medicine in 1924 from the Karolinska Institutet in Stockholm and had worked in Warburg's laboratory in the early 1930s. In 1937 he was appointed professor in the biochemistry department of the Nobel Medical Institute.[48] He obtained a relatively pure preparation of the cytochrome *c* protein,[49] but a simpler preparative method was devised by Keilin shortly afterward.[50] Experiments on the structure of cytochrome *c* by Theorell led to the conclusion that the porphyrin[51] part of the molecule was protoporphyrin and that it was covalently linked to a cysteine group in the protein. The 1930s had been a time when separation of proteins by electrophoresis (separation dependent on the charge on the protein) was being developed.[52] Theorell used electrophoresis to prepare the cytochrome in a higher state of purity in the 1940s.[53] This new preparation enabled him to elucidate aspects of the structure, particularly on the binding of the heme and the role of histidine residues in the protein. Thus it was concluded that the heme was covalently linked to the protein. Theorell studied the cytochrome as one of a group of proteins that contained heme groups, particularly cytochrome *c*, hemoglobin, peroxidase, and catalase.[54] It was for work that included this group that he was awarded the Nobel Prize in 1955.[55] Later in the 1960s cytochrome *c* was crystallized and a full study of the protein structure undertaken.

3.7 DPN (NAD) AND ITS OXIDATION

Harden and Young obtained from yeast juice a dialyzable, heat-stable fraction that stimulated the fermentation of sugars. Hans von Euler (or Euler-Chelpin,[56] 1873–1964) and Karl Myrbäck (1900–1985), who collaborated in Stockholm in the 1920s, studied this fraction, which they termed cozymase.[57] They regarded it as an adenine-containing nucleotide. However, by the mid-1930s, Warburg's group had already identified a coenzyme involved in glucose phosphate oxidation by blood cells as a triphosphopyridine nucleotide

---

[47] Keilin 1930.

[48] See Dalziel 1983 for a short biography of Theorell.

[49] Theorell 1936.

[50] Keilin and Hartree 1937.

[51] The heme is composed of a porphyrin with an iron in the middle of the ring.

[52] Arne Tiselius introduced electrophoresis to successfully separate proteins around 1937.

[53] Theorell and Åkesson 1941.

[54] Theorell 1947.

[55] Theorell was awarded the 1955 Nobel Prize in Physiology or Medicine "for his discoveries concerning the nature and mode of action of oxidative enzymes"

[56] Hans von Euler, born in Augsberg, Germany, became the professor of general and organic chemistry in the Royal University of Stockholm in 1906. In 1929 he shared the Nobel Prize in Chemistry with Arthur Harden for his work on alcoholic fermentation.

[57] Euler and Myrbäck 1923.

(TPN) composed of adenine and nicotinamide with three phosphates. In 1936 they identified Euler's cozymase as a very similar compound designated diphosphopyridine nucleotide (DPN).[58] It was shown to be reversibly reduced by two hydrogens. The compound was known for several years as coenzyme I but after the Second World War, the name given by Warburg was used. Following the *Report of the Commission on Enzymes of the International Union of Biochemistry* in 1961, the compound became known as nicotinamide adenine dinucleotide (NAD and, when reduced, NADH). Consequently, coenzyme I, DPN, and NAD refer to the same compound.

Coenzyme I was known to be necessary for fermentation by yeast and for glycolysis in muscle where it underwent oxidation and reduction. Warburg's work, previously mentioned, with the oxidation of glucose phosphate linked to reduction of triphosphopyridine nucleotide (TPN, coenzyme II, or NADP) from blood cells also resulted in the isolation of the *yellow enzyme*. Work on this enzyme was pursued by the Swedish biochemist Hugo Theorell, who worked in Warburg's laboratory from 1933 to 1935. The enzyme was shown to be a flavoprotein, having a flavin phosphate that could be reversibly detached from the protein. It catalyzed the oxidation of TPN coupled to the reduction of cytochrome *c*. This reaction, in which the activated hydrogen of the substrate was transferred to a cytochrome, could be seen as a short respiratory chain. This drew attention to a possible role for flavoproteins and the pyridine nucleotides (coenzymes I and II) in the reduction of the cytochrome system.

The term diaphorase was originally used to describe an enzyme that oxidized reduced coenzyme I (DPNH) and reduced the dye methylene blue.[59] The dye was a commonly used reagent in oxidation studies in which it became colorless on reduction but was readily reoxidized in air. Such an enzyme could be the sought-after candidate for linking DPN to the respiratory chain, although the initial diaphorases described were not able to reduce cytochrome.

Probably the first enzyme of this type was described by Euler's group, who gave it the name diaphorase. It was obtained from muscle and shown to be a flavoprotein.[60] However, the identity of this enzyme proved a problem throughout the 1940s and 1950s, and the name died out toward the end of the 1950s when the enzyme became better understood. Bruno Straub (1914–1996), working in Keilin's Molteno Institute in the late 1930s, isolated a flavoprotein from muscle that appeared to be the elusive enzyme linking DPN with the respiratory chain.[61] The matter was controversial and confused by the fact that several flavoprotein enzymes that reacted with oxidized or reduced coenzyme I existed. Straub's enzyme was eventually shown to be a component of a different system, the α-ketoglutarate dehydrogenase. The respiratory enzyme was not isolated until much later when Thomas P. Singer (1921–1999) and his colleagues isolated a flavoprotein, a DPNH (NADH) dehydrogenase, with the appropriate properties.[62] Even then

[58] See Warburg and Christian 1936.
[59] The term diaphorase is not well defined, and it has been used to cover several similar enzymes.
[60] See Adler, Euler, and Günther 1939. Similar observations were made by Dewan and Green 1938.
[61] Straub 1939. The story of early flavoprotein research and the role of flavoproteins as oxidizing enzymes is told in Florkin 1975a. See pp.370ff.
[62] Ringler, Minakami, and Singer 1963.

the nature of the enzyme proved to be much more complex than just a simple flavoprotein transferring hydrogens to the respiratory chain; the elucidation of its structure and catalytic activity proved a major challenge. The succinate dehydrogenase, another flavoprotein that oxidizes succinate and reduces the respiratory chain (as demonstrated by Keilin and Hartree in 1940) was isolated in 1954 by Singer's group.[63]

3.8 MUSCLE, LACTIC ACID, AND ENERGY

The study of muscle metabolism stems from the observations of Walter Fletcher (1873–1933) and Frederick Gowland Hopkins (1861–1947) in Cambridge in 1907.[64] These studies were based on earlier work of a German physiological chemist from Strasbourg, Felix Hoppe-Seyler (1825–1895) who had identified deficient oxidation as a factor in lactic acid accumulation in muscle. Among Fletcher and Hopkins's observations was the disappearance of lactic acid in aerobic conditions in isolated frog muscles. This work became the basis for studies of muscle metabolism in the early part of the twentieth century. In 1914, A. V. Hill, working in London, was measuring the heat produced in muscle. From his experiments, he concluded that the total amount of lactic acid could not be accounted for in terms of oxidation; the major part was resynthesized to carbohydrate. Only the remainder was oxidized.[65] This result was subsequently confirmed by Meyerhof.

Otto Meyerhof in Berlin examined the role of lactic acid in muscle with a view to understanding the mechanism of muscular contraction. As Peters described Meyerhof's studies on this point,

> All this work arose from the desire to place upon a firm basis the "lactic acid"
> theory[66] of contraction according to which muscular contraction was due to the
> effect of the liberation of lactic acid upon the colloidal apparatus.[67]

In 1923 he shared the 1922 Nobel Prize for Physiology with Hill, with whom Meyerhof had begun collaboration shortly after the First World War.[68] Meyerhof's work brought together the energy concepts developed in the nineteenth century together with the developing biochemistry of the twentieth century. As he said in his 1922 lectures,

> The problem of the activity of muscle has accompanied physiology, so to speak,
> from its cradle . . . . For it is nothing else but the question, how chemical energy

[63] Singer 1954.

[64] Fletcher and Hopkins 1907.

[65] Hill 1914.

[66] A rather general relationship between lactic acid and muscular contraction had been considered for some time but Meyerhof had developed his lactic acid theory of contraction whereby lactic acid formed from glycogen produced a physicochemical change of the muscle protein directly associated with the contractile process.

[67] Peters 1954, p.181.

[68] The prize was awarded to Meyerhof "for his discovery of the fixed relationship between the consumption of oxygen and the metabolism of lactic acid in the muscle" and to Hill "for his discovery relating to the production of heat in the muscle."

in the animal body is transferred into mechanical work. As a fact the Heilbronn physician, Robert Mayer conceived the law of the conservation of energy first from work on the human body. Helmholtz demonstrated with the thermopile constructed by himself, the relation between heat and mechanical response of the muscle. At the same time, the chemists were searching for substances which might serve as "the source of muscular energy."[69]

In the course of these studies Meyerhof, influenced by Hill's work, concluded that the lactic acid that was produced in muscle anaerobically from carbohydrate and that disappeared aerobically was not all oxidized in a subsequent aerobic phase but that most was resynthesized to glycogen. Only some was burnt (oxidized), and this provided energy for the synthetic activity. In 1927, he was still unclear whether the increased respiration was due to lactic acid oxidation or carbohydrate oxidation, but he regarded the energy from respiratory oxidation as driving the synthesis of glycogen from lactic acid: "lactic acid is reconverted to sugar, that is to glycogen, by means of energy furnished by the accompanying oxidation of carbohydrate."[70] Thus, in the 1920s, Meyerhof was fully aware that respiratory oxidation provided energy for synthesis although there was little suggestion as to the processes whereby respiration would generate usable energy or how it might be used.

3.9 ADENOSINE TRIPHOSPHATE AND MUSCLE PHOSPHATES

The significance of phosphates in the metabolism of carbohydrates by yeast juice and muscle juice was discussed earlier. Low-molecular-weight substances, such as coenzyme I, were found to be important components of these systems. In the 1920s, further compounds were discovered. Philip Eggleton (1903–1954) and Grace Eggleton (1901–1970), working in Hill's department at University College London used improved methods to detect a new compound they named phosphagen. Cyrus Fiske (1890–1978) and Yellapregada Subbarow (1895–1948), working in the Harvard Medical School, independently found phosphagen and identified it as creatine phosphate.[71] The Eggletons found that phosphagen disappeared during muscle activity but in aerobic conditions, isolated muscle would resynthesize phosphagen. They did not try to interpret this result although in retrospect it was a demonstration of aerobic phosphorylation (see subsequent discussion). They noted that "in the presence of oxygen phosphagen rapidly reappears and an exactly equivalent amount of inorganic phosphate is lost."[72]

Einar Lundsgaard (1899–1968) obtained his doctorate in medicine in 1929 at the University of Copenhagen and subsequently became professor of physiology at that

---

[69] Meyerhof 1924, p.61. The work of Mayer and Helmholtz is discussed in Section 2.6.
[70] Meyerhof 1927, p.533.
[71] Fiske and Subbarow 1927.
[72] Eggleton and Eggleton 1927a, p.161. See also 1927b.

university.[73] At that time there was considerable interest in iodinated compounds following the discovery of iodine in the thyroid hormone, thyroxine. Lundsgaard was investigating the effects of iodinated amino acids and other iodinated compounds, and this led to his key discovery that iodoacetate blocked the formation of lactic acid[74] but not muscular contraction. More important, he showed that in muscle treated with iodoacetate, phosphagen could be broken down to provide energy for muscular contraction. This led him to this conclusion (in contradistinction to Meyerhof's view):

> One must reckon that the phosphagen splitting directly yields the energy of contraction. The role of the lactic acid formation which is a more general anaerobic energy-yielding process demonstrable in almost all tissues, must then be sought . . . in this that it brings about a continuous resynthesis of the phosphagen.[75]

Lundsgaard's work on phosphagen caused a major revision in the understanding of muscle biochemistry[76] that was described as a revolution in muscle physiology and biochemistry by Hill[77] and, significantly, forced Meyerhof to totally reinterpret his understanding of the role of lactic acid. Lundsgaard also extended the view of aerobic phosphorylation by showing, as the Eggletons had done, that phosphagen can be synthesized from creatine and phosphate in O_2. As Lundsgaard wrote in 1932,

> Even if the lactic acid formation is blocked [by iodoacetate], an aerobic resynthesis of the phosphagen can take place, thereby making the oxidation energy available to the contraction mechanism. Therefore, this observation makes possible the assumption that a normal muscle, in which oxidation is sufficiently rapid relative to the rate of work, is able to perform work without any intermediary lactic acid formation.[78]

Such a view implies that oxidation by molecular O_2 can supply the energy for the phosphorylation of creatine, that energy then being available for muscular activity. On the basis of this work Lundsgaard was later described as "the first to herald oxidative phosphorylations."[79] However, Lundsgaard did not pursue further the role of O_2 or the

[73] For a biographical note on Lundsgaard, see Kruhøffer and Crone 1972.

[74] Iodoacetate blocks the glycolytic pathway and, as was later shown, inactivates a key enzyme in the pathway, the glyceraldehyde 3-phosphate dehydrogenase converting 3-phosphoglyceraldehyde to 1,3-diphosphoglyceric acid (see the glycolytic pathway in Fig. 3.2). The overall effect is to block both lactic acid formation and glycolytic ATP synthesis.

[75] Lundsgaard 1930; translation Teich 1992, p.231.

[76] Lundsgaard visited Meyerhof at Heidelberg, and the effect of Lundsgaard's work on that laboratory is recorded in Lipmann 1971, see p.349.

[77] Hill 1932.

[78] Lundsgaard 1932; English translation Kalckar 1969, p.349.

[79] Kruhøffer and Crone 1972, p.5.

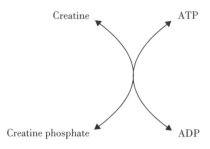

Creatine   ATP

Creatine phosphate   ADP

Figure 3.13 Creatine Phosphate Can Be Formed Reversibly From ATP and Creatine

nature of phosphorylation but his student, Herman Moritz Kalckar, investigated these issues a few years later.

One of the many developments to emerge from Meyerhof's group had been the discovery in 1929 of ATP by Karl Lohmann (1898–1978), who worked in Meyerhof's laboratory, and simultaneously and independently by Fiske and Subbarow.[80] The compound obtained from fresh muscle was seen as a cofactor[81] in the production of lactic acid from glycogen in muscle, and its addition was necessary for activity in dialyzed muscle juice. The relationship of ATP to phosphocreatine was soon seen in terms of a synthesis of the latter at the expense of ATP (Fig. 3.13). The situation was summarized by Meyerhof and Lohmann, who came to believe that

> The endothermic and not spontaneous synthesis of phosphocreatine can take place through a coupling of this process with the exothermic and spontaneous breakdown of adenylpyrophosphate whilst the resynthesis of the adenylpyrophosphate from adenylic acid and inorganic phosphate is brought about through the energy of lactic acid formation.[82]

Thus ATP was seen as being a possible source of energy in muscle and a cofactor in the conversion of carbohydrate to lactic acid. As we will see, these studies that related the role of ATP to the process of carbohydrate breakdown created the basis for Vladimir Engelhardt to link ATP with aerobic phosphorylation.

The realization of the importance of aerobic conditions in aspects of metabolism was extended by the Polish biochemist Jakob Karol Parnas (1884–1949), who studied the release of ammonia from adenine nucleotides under anaerobic conditions. He found that in the presence of O_2, the level of adenine nucleotide remained relatively high as compared with that in anaerobic conditions. He concluded

---

[80] For a full account of the discovery of ATP see Maruyama 1991.

[81] Early in the study of fermentation and glycolysis, substances essential for the processes were separated by dialysis. Although, except for phosphate, their identity remained obscure they were known not to be proteins or to have enzymic activity and were regarded as cofactors.

[82] Meyerhof and Lohmann 1932; translation Teich 1992, p.232.

That in the oxidative recovery processes, deamination of other substances brings about a resynthesis of adenine nucleotide from inosinic acid; and in this way a system is maintained which is over and over renewed and conserved, readily available for anaerobic instantaneous splitting of ammonia.[83]

Here again, as with the Eggletons and with Lundsgaard, there is an indication of aerobic phosphorylation but it remains unrecognized.

By the beginning of the 1930s, there was an appreciation that aerobic metabolism provided the energy for synthesis of carbohydrate from lactic acid and that oxidation could provide the energy for the phosphorylation of creatine (see Fig. 3.13). All this was strongly illuminated by an emerging understanding of the role of ATP, which was seen simply as a cofactor, but an appreciation of its importance in muscle metabolism was dawning.

3.10 AEROBIC ATP SYNTHESIS: ENGELHARDT AND KALCKAR

The growing view that phosphorylation occurred in the presence of O_2 was supported by the work of the Russian biochemist Vladimir Engelhardt (1894–1984), often described as responsible for the "discovery" of oxidative phosphorylation. Appreciation of the role of phosphate in metabolism and the significance of ATP in muscle metabolism prompted Engelhardt, working at Kazan in the Russian Soviet Union, to investigate phosphate metabolism in red blood cells.[84] In the late 1920s, Eleazar S. G. Barron and George A. Harrop Jr. had investigated both glycolysis and respiration in mammalian and avian red cells. Although both showed glycolytic activity, only the avian cells showed respiratory activity. Using methylene blue, which mediated oxidation, they found that even the mammalian cells showed good levels of respiration.[85]

In 1930, Engelhardt measured phosphate levels during glucose breakdown under aerobic and anaerobic conditions in mammalian and avian cells.[86] He found that ATP concentrations were maintained during the natural respiration of nucleated avian cells but in anaerobic conditions ATP was decomposed. Cyanide, a potent inhibitor of respiration also inhibited the maintenance of ATP levels. He obtained similar results on ATP concentrations in his experiments with mammalian cells and methylene blue,[87] but it is not clear that Engelhardt appreciated that he was in fact measuring glycolytic phosphorylation; he appears to have believed he was studying aerobic metabolism.[88] He interpreted these results as showing that ATP was broken down in anaerobic conditions but was being actively synthesized during respiration. Further, he thought that the products of ATP breakdown promoted respiration. He concluded that ATP had the same role in

---

[83] Parnas (1929); translation Teich 1992, p.228.
[84] A detailed consideration of the experiments described in these papers is given in Slater 1981.
[85] Harrop Jr. and Barron 1928; Barron and Harrop Jr. 1928.
[86] Engelhardt and Braunstein 1928.
[87] Engelhardt 1930; Engelhardt and Ljubimova 1930.
[88] See Prebble 2010 for a discussion of this issue.

respiration as Meyerhof and Lohmann had described in muscle glycolysis, in which ATP was synthesized as part of the associated coenzyme system.[89] He wrote

> One appears justified in attributing to adenyl phosphoric acid a coenzyme function in respiration and the adenylpyrophosphate as a component of the respiratory coenzyme complex.[90]

Engelhardt's work was an advance on previous studies, and he did have a claim to have discovered oxidative phosphorylation, as Slater has noted. [91] Engelhardt's contemporaries do not appear to have recognized his work as a major discovery; the author of a review on nucleic acids did record Engelhardt's observations but they were linked to those on ATP as a cofactor.[92] The identification of a cofactor for respiration seemed but a minor advance in a world where major discoveries on glycolysis, on the urea cycle, and on enzymes were being made. As Engelhardt later remarked,

> The role of respiration was reduced to nothing more than the removal of the anaerobically formed products, lactic acid or alcohol . . . a kind of scavenging function.[93]

A rather similar comment was made by Herman Kalckar (1908–1991), a student of Lundsgaard's in Copenhagen, who commented that the early work on aerobic phosphorylation did not attract much attention owing to

> [t]he lack of appreciation of the importance of phosphorylation in cellular physiology in most leading biochemical and physiological circles at that time. It was not uncommon to consider phosphocreatine and ATP as nice "cellular buffers" and phosphorylation as an artefact found in broken cells.[94]

It should also be noted that at this stage ATP was one of two phosphate compounds (the other being creatine phosphate) identified as synthesized in association with respiration.

Phosphorylation linked to O_2 metabolism became much more clearly established by the work of Kalckar in Copenhagen. Around 1934, Lundsgaard transferred his interests from muscle physiology to questions of glucose transport in the intestine and kidney.

---

[89] Lohmann 1931.

[90] Engelhardt 1932, p.368. It should be noted that the concept of a coenzyme was of a small-molecular-weight (nonprotein) molecule, a cofactor, necessary for enzyme action. Thus ATP was seen as a necessary factor for some reactions in glycolysis.

[91] Slater 1981, 1984.

[92] See Cerecedo 1933, p.121.

[93] Engelhardt 1975. See p.62.

[94] Kalckar 1966, p.4. The latter part of this comment referring to buffers was probably based on Sacks 1943–1944. The point is noted by Florkin 1975a, who regards it as an attack on the phosphate cycle originally arising from the work of Lundsgaard but elaborated by Lipmann in his 1941 review.

Theories at that time led him to look at glucose phosphorylation and dephosphorylation, which seem to have encouraged Kalckar to take up studies on phosphorylation and dephosphorylation, particularly in kidney, which possessed a strong phosphatase. Kalckar was influenced in the choice of kidney, rather than pigeon-breast muscle, by the ready availability of kidney tissue within the department, thereby avoiding the need to kill pigeons, an unattractive activity for Kalckar. In addition, Fritz Lipmann, who became Kalckar's mentor, strongly encouraged the interest in phosphorylation.

Initially, Kalckar discovered that there was "a coupled reaction between oxygen consumption and phosphorylation" of glucose.[95] Inhibition of respiration with cyanide also inhibited phosphorylation. In the light of work by Albert Szent-Györgyi (1893–1986),[96] these experiments were extended by use of a range of substances including dicarboxylic acids, which stimulated both respiration and phosphorylation. He also examined various phosphate acceptors, which included adenylic acid, carbohydrates, glycerol, and pyruvic acid. The conclusions from these studies were that some substances, particularly the dicarboxylic acids that promoted phosphorylation might do so by providing phosphate acceptors.[97] In summary,

A coupling between the phosphorylations and the respiration in kidney tissue undoubtedly exists since the phosphorylation and the respiration are inhibited equally by cyanide and stimulated by different dicarboxylic acids.[98]

Although the link between phosphorylation and respiration was thus established, the initial phosphate acceptor remained unclear.

Kalckar's papers were noted by Carl Cori (1896–1984) and Gerty Cori (1896–1957) at the Washington University School of Medicine, St. Louis. The Coris were seeking a system for glucose phosphorylation in glycogen synthesis and reexamined the aerobic phosphorylation of glucose described by Kalckar. After advice on the need for shaking the tissue in order to observe the phosphorylation,[99] they confirmed the earlier work and concluded that

It is evident that phosphorylation of glucose in kidney extracts is not dependent on the dehydrogenation of glyceraldehyde phosphate [phosphorylation driven by the glycolytic pathway] but on the oxidation by molecular oxygen of any one of a number of dicarboxylic acids.[100]

---

[95] Kalckar 1937.
[96] The Hungarian biochemist Albert Szent-Györgyi had proposed that the dicarboxylic acids played a role in respiration as hydrogen carriers; see Annau et al. 1935. The experiments but not their interpretation were important in Krebs's formulation of the citric acid cycle. (See Section 3.4.)
[97] Kalckar 1939a.
[98] Kalckar 1939b.
[99] See Kalckar 1991, p.10.
[100] Colowick, Welch, and Cori 1940.

3.11 PHOSPHORYLATION LINKED TO RESPIRATION: BELITZER AND OCHOA

Although Kalckar's work attracted attention, it was in many respects work of the Russian Vladimir Aleksandrovich Belitzer (1906–1988) at the University of Moscow that put the phenomenon of aerobic phosphorylation beyond reasonable doubt. Probably the key to the impact made by his work with Elena Tsybakova was the measurements of the P/O ratios,[101] which were substantially higher than those of Kalckar so that aerobic phosphorylation could not be confused with glycolytic phosphorylation. Importantly, he tentatively concluded that adenylic acid was the primary substance phosphorylated, even though most of the experiments concerned creatine phosphate formation.

Belitzer interpreted the earlier work, including Kalckar's, as showing a strictly respiratory synthesis of phosphate esters such as phosphagen (creatine phosphate) and ATP. With this viewpoint and together with Elena Tsybakova, he set out to investigate the matter further using muscle tissue from pigeon breast and rabbit heart. They made measurements of phosphorylation against O_2 uptake and concluded that the P/O ratio lay between 2 and 3.5. They estimated that this phosphorylation was some ten times greater than might be obtained from glycolysis, showing that this aerobic phosphorylation could not be confused with glycolytic phosphorylation. They also noted that whereas in glycolysis, metabolism was "unconditionally" (obligatorily) coupled to phosphorylation, in respiration it was "conditionally" linked. That is, respiration could occur without phosphorylation whereas glycolysis could not occur without phosphorylation. Elsewhere Belitzer demonstrated a dependence of respiration on creatine phosphorylation in muscle.[102] Although higher levels of phosphorylation were obtained with creatine, they concluded that there was nothing to "prevent one from assuming that the adenylic system is a primary acceptor of phosphate in the synthesis of phosphagen":

> In general, the results of the investigation supported the assumption that adenylic acid is capable of playing the role of acceptor of phosphate in phosphorylations linked with respiration.[103]

Hence adenylic acid phosphorylation was seen as the primary event in phosphagen synthesis.

By the end of the 1930s, respiration linked to phosphorylation had been firmly established by the work of Kalckar and Belitzer and shown to be separate from glycolytic phosphorylation. They also observed the phenomenon of coupling and uncoupling. Indeed some, such as David Green, have credited Kalckar and Belitzer with the discovery of oxidative phosphorylation whereas others regarded them as simply initiating the study of the process.[104]

---

[101] The P/O ratio is the number of phosphorylations per *atom* of O_2 consumed. This became the standard measure of phosphorylation in the 1940s and thereafter; I have therefore used it here. Belitzer used a P/O_2 ratio that gave values numerically twice those of the P/O ratio.

[102] Belitzer 1939.

[103] Belitzer and Tsybakova 1939. An English translation appears in Kalckar 1969, pp.211–227.

[104] See Ziegler et al. 1956; Slater 1966.

The Spanish biochemist Severo Ochoa (1905–1993) obtained his MD degree from the University of Madrid. During the 1930s he worked with Meyerhof in Heidelberg and also in Madrid, London (Medical Research Council), and Oxford. He left Oxford for New York in 1941. He was awarded a Nobel Prize in 1959.[105] During his time at Oxford he carried out studies on oxidative (aerobic) phosphorylation that complemented the work of Belitzer. He made apparently very accurate measurements of the number of phosphorylations per O_2 atom taken up (the P/O ratio) in systems catalyzing phosphorylation linked to O_2 metabolism. Significantly, he was the first to use the term oxidative phosphorylation in the title of a paper devoted to a study of pyruvate oxidation in brain.[106] Because Ochoa's preparations contained an enzyme that broke down ATP, he reasoned that his initial P/O ratio, around 2.0, was in fact low because of some ATP destruction. He therefore measured ATP breakdown alongside phosphorylation so that he could correct his P/O ratio for losses. The average result from ten experiments was P/O = 3.0 (with a range of 2.3–4.0),[107] a figure that became the basis for attempting to elucidate the mechanism of oxidative phosphorylation in the following two decades. The understanding of the mechanism of oxidative phosphorylation up into the 1960s was based on glycolytic phosphorylation, and this would predict an integer for the P/O ratio. After much debate this figure ultimately had to be abandoned, although it strongly influenced the interpretation of experiments and the understanding of oxidative phosphorylation for many years. In retrospect it can be seen as misleading.[108]

3.12 LIPMANN: THE SIGNIFICANCE OF PHOSPHORYLATION

At the beginning of the 1940s, Fritz Lipmann (1899–1986) emerged as a key figure in the development of cell bioenergetics. Lipmann was educated initially in Koenigsberg (then in Germany, now in Russia) studying medicine and then chemistry. He joined Meyerhof's laboratory in 1927, worked there until 1930, and, after a visit to America, moved to Copenhagen, where he lived from 1932 to 1939. He then moved to Massachusetts General Hospital in Boston. He was awarded the Nobel Prize in 1953.[109]

Lipmann was probably the first to think comprehensively about the importance of phosphate in metabolism and to develop an understanding of ATP and its role in the energetics of the cell. Two lines of experimentation had shown the importance of phosphorylation linked to oxidation. The first we have considered particularly in relation to the work of Kalckar and Belitzer. The second concerned phosphorylation in glycolysis.

---

[105] The Nobel Prize in Physiology or Medicine for 1959 was awarded to Severo Ochoa and Arthur Kornberg "for their discovery of the mechanisms in the biological synthesis of ribonucleic acid and deoxyribonucleic acid."

[106] Ochoa 1940.

[107] Ochoa 1943.

[108] This issue is pursued in Chapters 4 and 8.

[109] The Nobel Prize in Medicine or Physiology for 1953 was shared by Hans Krebs "for his discovery of the citric acid cycle" and Fritz Lipmann "for his discovery of co-enzyme A and its importance for intermediary metabolism."

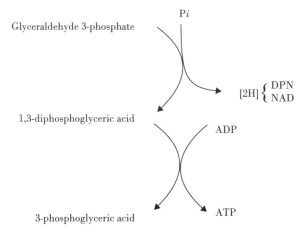

Glyceraldehyde 3-phosphate

Pi

1,3-diphosphoglyceric acid

[2H] $\left\{\begin{array}{l} \text{DPN} \\ \text{NAD} \end{array}\right.$

ADP

3-phosphoglyceric acid

ATP

Figure 3.14 The Oxidation of Glyceraldehyde 3-Phosphate in Glycolysis Linked to ATP Synthesis
The oxidation involves the removal of two hydrogens that are transferred to NAD (DPN or coenzyme I). The inorganic phosphate (Pi) is initially used to form the second phosphate in the diphosphoglyceric acid and then transferred to ADP to form ATP.

Dorothy Needham and Raman Pillai at Cambridge found a link between phosphorylation of adenylic acid by inorganic phosphate (Pi) and oxidation in the glycolytic breakdown of sugar phosphate (see Fig. 3.2).[110] This was clarified in 1939 by Otto Warburg and his colleagues who reported on the mechanism of a key reaction in glycolysis, glyceraldehyde phosphate oxidation. They showed the way in which inorganic phosphate was used in glucose breakdown (glycolysis) and linked to the synthesis of ATP.[111] It was this description of the coupling between triose phosphate (glyceraldehyde phosphate) oxidation and inorganic phosphate uptake that Kalckar believed would be the clue to understanding other energetic couplings such as that in aerobic phosphorylation:

> The clarification of the coupling between the triose phosphate oxidation and the uptake of inorganic phosphate represents one of the greatest advances in modern biology . . . . For the first time since this recent discovery of the Warburg school, a complete description of a biological coupling is possible. By following this new line a clarification of other energetic couplings is to be expected.[112]

The key elements of this reaction were first that the oxidation step was coupled to a phosphorylation of the substrate with inorganic phosphate and second, that the phosphate was transferred to adenosine diphosphate (ADP) to form ATP (Fig. 3.14).

[110] Needham and Pillai 1937a, 1937b. Note that Meyerhof 1937 simultaneously came to similar conclusions with slightly different experiments.
[111] Warburg and Christian 1939; Negelein and Brömel 1939. It should be noted that Warburg and Christian's mechanism for the phosphorylation, although influential, was subsequently shown by Racker to need significant revision; see Racker 1965, pp.17–29.
[112] Kalckar 1941, p.131, but see also pp.122–131.

These experiments provided a strong background to Lipmann's thinking about biological phosphorylation. They reinforced the idea, put forward by Belitzer, that ATP should be the first formed phosphorylated compound in aerobic phosphorylation; respiration linked to phosphorylation was the addition of phosphate to ADP to form ATP. Lipmann himself had studied the metabolism of pyruvate by a bacterium. He found that when pyruvate was oxidized, ADP was phosphorylated to ATP. Subsequently he showed the intermediate in this process to be acetyl phosphate.[113] These observations that provided an example of phosphorylation similar to that involving glyceraldehyde phosphate oxidation studied by Warburg's group strengthened the view (developed during the 1930s) that phosphorylation is associated with oxidation. Lipmann's thinking about the role of phosphate, particularly in terms of energy transformations in the cell, was revealed in his classic 1941 review on phosphate bond energy.[114] He considered those bonds, which the cell uses to convey energy from breakdown systems (like glycolysis) to energy-consuming systems such as muscular contraction, as energy-rich phosphate bonds, which he denoted as ~ph.[115] In contrast, other phosphate bonds that he regarded as not energy rich were denoted as –ph. The use of Lipmann's squiggle (~) became customary in biochemistry for at least two decades as a means of conveying some thermodynamic aspects of metabolism.[116] It was extensively used in discussions of oxidative phosphorylation to convey the conservation of energy from oxidation in the respiratory chain. It was, however, seen as an unsatisfactory terminology by some chemists.[117]

In considering the role of ATP in metabolism, Lipmann described a phosphate cycle in which inorganic phosphate was used for formation of an organic phosphate that in turn was converted to a high-energy phosphate by an associated oxidation–reduction (see Fig. 3.15).

This high-energy phosphate was used for energy-requiring reactions and the inorganic phosphate released. The key high-energy phosphate was ATP, which subsequently was dubbed the energy currency of the cell. Quantitatively, the most important generator of ATP was oxidative phosphorylation and, to a much lesser extent, glycolysis. Thus ATP was seen as being central to metabolism.

The energy provided by the respiratory chain in oxidative phosphorylation was demonstrated by the work of Eric Glendenning Ball and others. They found that the oxidation–reduction potentials of the components of the respiratory chain showed a progressive increase from DPN and TPN through the flavoproteins and cytochromes to O_2 (see Fig. 7.3 in Chapter 7). From such increases in potential it was possible to calculate the amount of energy available during respiration.[118] Prior to Ochoa's estimate of the

---

[113] Lipmann 1939, 1940.

[114] Lipmann 1941.

[115] Energy-rich bonds were those now understood as having a high negative standard free energy of hydrolysis ($-\Delta G^{o'}$). The difference between ~ph and –ph was not clearly defined. The sugar phosphates had "low-energy" phosphate bonds whereas ATP and similar compounds, creatine phosphate and phosphoenolpyruvate, had "high-energy" bonds.

[116] Lipmann designated ATP as ad–ph~ph~ph, distinguishing between the energy available from hydrolysis of the first phosphate from that of the other two.

[117] See Banks and Vernon 1970.

[118] Ball 1939, 1975.

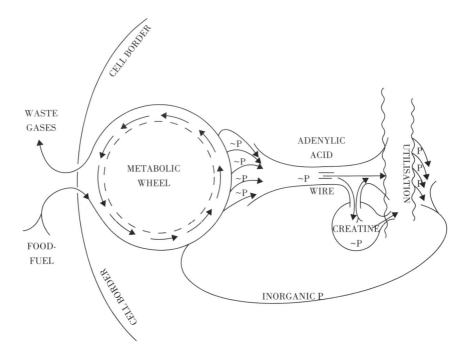

Figure 3.15 The Phosphate Cycle of Fritz Lipmann

Lipmann thought in terms of his high-energy phosphate bonds as ~P and cells as generating ~P to drive metabolism. The model is an analogy with an electric dynamo generating current (phosphate energy) for use in cellular processes. Thus the metabolic wheel (dynamo) generates high-energy phosphate (~P) that is carried as ATP to the point of ATP utilization releasing inorganic phosphate. This returns to be comverted to ~P (ATP) by metabolism. Creatine phosphate (creatine~P) serves to store a charge of ~P (see Fig. 3.13). Reproduced with permission from Lipmann 1941, *Metabolic Generation and Utilization of Phosphate Bond Energy* (Advances in Enzymology 1, Interscience), Wiley, fig. 1, p.122.

P/O ratio, Lipmann estimated that the value of the ratio would be between 2 and 4,[119] so endorsing the process as the principal source of ATP in the cell.

From the growing knowledge of metabolism, phosphorylation in general, and particularly of phosphorylation in glycolysis, Lipmann recognized that in fact ATP was the energy currency of the cell. Such a step was essential to understanding both the nature and the importance of oxidative (aerobic) phosphorylation. His views strongly influenced the bioenergetics of the post–Second World War era, and in 1946 he proposed a simple theory of oxidative phosphorylation that reflected some of the arguments that emerged in his 1941 review.[120] The review might be seen as the culmination of the search for the significance of phosphate in metabolism with which metabolic biochemistry started the century.

By the mid-1940s, oxidative phosphorylation had become a recognized part of cell metabolism with a key role in providing energy through the synthesis of ATP. However, the molecular basis for this view remained less than substantial.

---

[119] This was an extension of calculations carried out by Meyerhof.
[120] Lipmann 1946.

4

Emergence of the Field of Cell Bioenergetics
1945–1960

After the Second World War, studies in the area of bioenergetics changed radically. In part this was due to the much-increased importance attached to scientific research and to the return of workers previously engaged in the war effort. But it was not until after 1945 that bioenergetics came together as a coherent field of study with an identifiable group of practitioners. The emergence of the new field was stimulated by the merging of several disparate lines of investigation that progressively became relevant to each other and integrated. This convergence opened up stimulating research possibilities and new opportunities.

4.1 EMERGENCE OF A NEW FIELD

Foremost among these opportunities was Belitzer's somewhat tentative recognition that aerobic phosphorylation was linked to the respiratory system, which raised the whole question of how these two processes were related. Of similar significance was the realization that oxidative phosphorylation was a property of cell organelles, the mitochondria. As the nature and detail of the complex web of intermediary metabolism began to be unraveled, the detailed relationship of metabolic processes such as the Krebs citric acid cycle to the respiratory phosphorylation emerged.[1] Although these were the issues, together with further elucidation of the respiratory chain, that occupied bioenergeticists in the late 1940s and the 1950s, a major underlying question emerged and rose to dominate the field. What was the mechanism of oxidative phosphorylation? In the 1950s a very large number of publications aspired to clarify this question, but they did not lead to its resolution.

Toward the end of the period, workers began to discover that another area of biology would need to be taken into account if oxidative phosphorylation was to be fully understood—the energetics associated with the transport of substances across biological membranes. In essence, this period was stimulated by and dependent on the merging of strands of biological research: respiration, phosphorylation, subcellular structure (mitochondria), general intermediary metabolism, and membrane transport.

---

[1] It was noted in the previous chapter that the work of Kalckar had demonstrated a relationship between the metabolism of dicarboxylic acids and aerobic phosphorylation.

This coalescence created both an identifiable field of cell bioenergetics and a group of scientists who recognized themselves as a coherent interest group.

4.2 THE MITOCHONDRION AS THE LOCATION OF RESPIRATORY ACTIVITY

The idea that respiration was a property of particulate parts of the cell initially arose before the First World War. As noted in the previous chapter, Otto Warburg in 1913 working on liver cells noted that respiration was located in cell granules.[2] He made a similar observation on sea-urchin eggs, and his colleague Otto Meyerhof commented on the association of respiration with particulate material in cells in his lectures published in 1924.

> The cause of the reaction velocity of vital oxidations [which he regarded as high], and the transformation of the oxidation energy into work—are closely connected with the structural properties of the cell.[3]

However, these observations do not appear to have significantly influenced other workers in the field.

More influential studies came from the cell biologists Robert Bensley (1867–1956) and Normand Hoerr (1902–1958) working in the University of Chicago. Bensley, originally from Ontario in Canada and educated in the University of Toronto, had moved to the department of anatomy at the University of Chicago in 1901. Here, as a coworker commented,

> He [Bensley] became intrigued by the problems of mitochondria. In the previous decade Altmann and others had described these small particles in cells . . . . His interest in mitochondria . . . finally led in 1934 to his separation of these particles from macerated liver cells by a method which has enabled us to analyze them directly.[4]

It was this separation of mitochondria by the method of differential centrifugation, described in a series of papers in the early 1930s, that opened the way for the biochemical study of mitochondria. It enabled Bensley and Hoerr to examine the chemistry of these particles and to conclude thus:

> Mitochondria in pure form have been separated from other cell constituents of the liver cell of the guinea pig and examined chemically.[5]

---

[2] Warburg 1913.
[3] Meyerhof 1924, p.10.
[4] Hoerr 1957, pp.8–9.
[5] Bensley and Hoerr 1934, p.455.

Although the purity of these mitochondrial preparations was questionable, there were clearly considerable possibilities inherent in this work. These were exploited in the mid-1940s at the Rockefeller Institute in New York by Albert Claude (1899–1983) and coworkers. Claude[6] was born and educated in Belgium and arrived at the Rockefeller Institute in 1929 where, in due course, he developed the methods associated with differential centrifugation. These enabled him and his colleagues to prepare and analyze the properties of the granular components of both normal and tumor cells by using high-speed centrifugation that he coupled with the simple electron microscopy available at that time. In 1946, together with George Hogeboom and Rollin Hotchkiss, he showed that the large granular fraction (mitochondria) possessed respiratory activities (succinate oxidase and cytochrome oxidase).[7] However, these mitochondria were different in appearance from those seen in cells and did not stain with Janus green. The Rockefeller group then developed improved methods of preparation by using sucrose solutions to suspend the particle and this produced normal-looking mitochondria.[8] As Claude commented in a lecture in 1948:

Together the preceding observations offer conclusive evidence to support the view that most, if not all, of cytochrome oxidase, succinoxidase and cytochrome c, three of the most important members of the respiratory system, are segregated in mitochondria or in granules of the same type. These results seem to be in agreement with the early findings of Warburg

. . . . These observations indicate that the power of respiration is not to be found in a diffused state in the cytoplasm, and that mitochondria may possibly be considered as the real power plants of the cell.[9]

Albert L. Lehninger (1917–1986) was attracted to biochemical work when he became aware of the discoveries of Warburg and Krebs. After being employed on blood plasma research in the Plasma Fractionation program during the Second World War, he joined the University of Chicago, where he worked on fatty acid metabolism. In 1948, using the methods for mitochondrial isolation developed by the Rockefeller group but with a cooled centrifuge (which produced much-improved mitochondria), he demonstrated the metabolic significance of mitochondrial particles:

The complex enzyme systems responsible for the oxidation of fatty acids and Krebs cycle intermediates and esterification of phosphate coupled to these oxidations are localized in that fraction of rat liver which consists of morphologically intact mitochondria or "large granules," almost completely free of other formed elements.[10]

[6] Claude shared the Nobel Prize in Physiology or Medicine for 1974 with Christian De Duve and George Palade "for their discoveries concerning the structural and functional organization of the cell."
[7] Hogeboom et al. 1946.
[8] Hogeboom et al. 1948.
[9] Claude 1949, p.137.
[10] Kennedy and Lehninger 1949, p.970.

The mitochondrion now became a biochemical entity as well as a cytological particle. This was, in fact, a step of critical importance in the development of cell bioenergetics.[11] By 1952, the main activities of mitochondria had been established—pyruvate oxidation, citric acid (Krebs) cycle, fatty acid oxidation, respiration, and oxidative phosphorylation.

During most of the 1950s, workers in bioenergetics seemed to view the significance of the mitochondrion primarily in terms of it being the location of oxidative phosphorylation. It provided the scaffolding on which the various proteins involved in the process were mounted. Isolation of mitochondria provided a basis for experimental procedures to study the process of oxidative phosphorylation separated from other cellular material. Nevertheless there was a perception that understanding mitochondria themselves might be necessary for appreciating respiration biochemistry. Britton Chance later commented that he had realized in the early 1950s that

> The Keilin and Hartree preparation[12] was significantly denatured and that an understanding of the respiratory chain would only come from studying mitochondria directly.[13]

In 1952–1953, electron microscopy had developed far enough to provide the first high-resolution pictures of mitochondria, which showed the presence of two membranes known as inner and outer (see Fig. 4.1).[14] By the end of the 1950s, many bioenergeticists believed that circumstantial evidence indicated that the inner membrane was the location of the respiratory system and its associated phosphorylation system. Firm evidence for this belief was obtained in the 1960s when studies of the mitochondrial membranes were substantially extended.

4.3 FURTHER ELUCIDATION OF THE RESPIRATORY CHAIN

4.3.1 Keilin and the Molteno Institute in the 1950s

Keilin continued to work on the respiratory chain through the 1950s at the Molteno Institute of which he was director until 1952. Bill (Edward C) Slater joined him in 1946 and continued until 1955; Ernest Hartree, who had worked with Keilin from 1934, remained until sometime after Keilin's formal retirement. Keilin's work in the late 1930s and early 1940s had established the principle of the respiratory chain.[15] However, it was

---

[11] See Matlin 2016 for a detailed discussion of this point, which is explored further in Chapter 9.

[12] The Keilin–Hartree preparation of finely divided heart muscle was used by Keilin and by others from the late 1930s onward as the experimental material of choice for their cytochrome studies (see Section 3.6).

[13] Chance 1991, see p.6.

[14] Palade 1952; Sjöstrand 1953.

[15] The respiratory chain also became known as the electron-transport chain, the terminology often attributed to David Green. The two terms are synonymous, although it should be noted that not all electron-transport chains are respiratory chains, for example, the photosynthetic

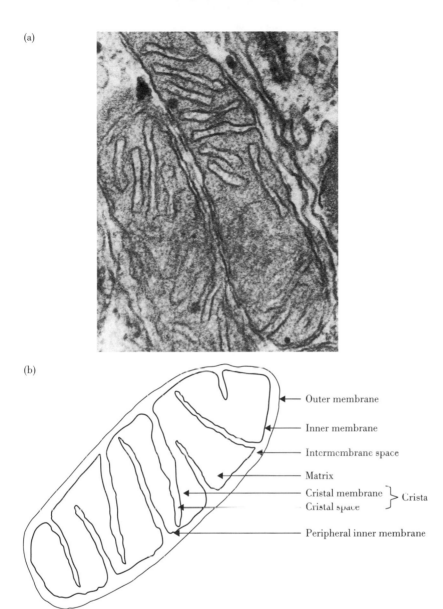

(a)

(b)

Outer membrane

Inner membrane

Intermembrane space

Matrix

Cristal membrane ⎫
                  ⎬ Crista
Cristal space   ⎭

Peripheral inner membrane

Figure 4.1 Electron Micrograph of the Mitochondrion
(a) An electron micrograph of a part of a mitochondrion showing the inner and outer membranes, the inner being invaginated to form the cristae. (Courtesy of the late Professor Stanley Bullivant, University of Auckland, New Zealand.) (b) Diagram of the structure of the mitochondrion. The diagram is reproduced with permission from Prebble 1981, *Mitochondria, Chloroplasts & Bacterial Membranes*, Pearson Education Ltd. (©Longman Group Ltd. 1981), fig. 4.1b, p.96.

electron-transport chain. The term is used here because the work discussed is a consequence of the studies on animal respiration.

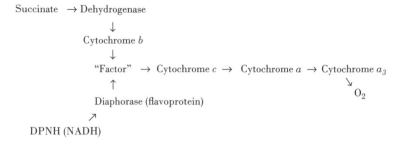

Succinate → Dehydrogenase
↓
Cytochrome *b*
↓
"Factor" → Cytochrome *c* → Cytochrome *a* → Cytochrome a_3
↑   ↘
Diaphorase (flavoprotein) O_2
↗
DPNH (NADH)

Figure 4.2 The Respiratory Chain as Understood in 1953
Either succinate or DPNH was seen as reducing "Slater's factor" by their respective dehydrogenases (a diaphorase for
DPNH) that passed electrons to the cytochrome chain, ultimately reducing oxygen to water. At this stage the enigmatic
cytochrome *b* was seen by Slater as associated with succinate oxidation. Based on Keilin and Slater 1953.

clear to those working in bioenergetics in the early 1950s that there was a great deal
still to learn about the respiratory chain. Although much had been achieved in the first
half of the twentieth century, understanding remained rudimentary although the basic
concepts were now widely accepted. It could be represented as a sequence of electron
carriers, including the cytochromes, from succinate and NADH to O_2. The respiratory
chain was now clearly seen as related to the web of metabolic pathways (see Fig. 3.1),
which showed how foodstuffs (carbohydrates, fats, proteins) were broken down through
the citric acid cycle. It was this cycle that was the primary source of reduction for the
respiratory chain.[16]

The chain as envisaged by Keilin and Slater in 1953 is shown in Figure 4.2, which
shows the oxidation of succinate and reduced DPN (diphosphopyridine nucleotide or
reduced coenzyme I, later known as NADH). DPN was oxidized by the diaphorase, the
nature of which remained a subject for debate. The "factor" was an unidentified compound detected by Slater during the late 1940s, which appeared to act immediately before cytochrome *c*.[17] It became known as Slater's factor until its identity was determined
a decade or two later.

This version of the respiratory system was soon shown to be incomplete in several
respects. For example, Japanese workers in 1940 had described an absorption band at
552 nm, very close to the 550-nm band of cytochrome *c* and not readily distinguished
from it.[18] They attributed this band to a new cytochrome, c_1. Although Slater, working
in Keilin's Molteno Institute, had later concluded that "evidence for the existence of cytochrome c_1 is . . . unsatisfactory,"[19] Keilin and Hartree redeemed the situation in 1955
when they identified cytochrome c_1 of the Japanese workers with a widely distributed

---

[16] Krebs discussed the integration of metabolism in an essay in 1953 showing a coherent set of
pathways for the breakdown of carbohydrates, fats, and proteins to the citric acid cycle in which
oxidations were linked to the respiratory chain and the energy conserved in the synthesis of ATP
(Krebs 1953).
[17] Keilin and Slater 1953.
[18] Yakushiji and Okunuki 1940.
[19] Slater 1949; Slater 1985, see pp.216–218 for a personal view on this error.

cytochrome previously described by Keilin and Hartree in 1949 as cytochrome *e*.[20] They concluded that their

> Cytochrome c_1 (= cytochrome *e*) is a widely distributed component forming an integral part of the intracellular electron transport chain between activated molecules of substrate and oxygen.[21]

In view of the priority of the Japanese work, the pigment described as cytochrome *e* became known as cytochrome c_1. Important in these studies was the examination of spectra at low temperatures (around $-190°$ C) when the absorption bands were sharpened and intensified, a technique applied to cytochromes by Keilin and Hartree in 1949.[22]

4.3.2 A New Approach to the Respiratory Chain: Britton Chance

Keilin's heir in the study of the respiratory chain was Britton Chance (1913–2010), a biophysicist at the Johnson Foundation at the University of Pennsylvania and also an Olympic gold medalist in sailing at the 1952 Olympics. He started work on spectrophotometry just before the Second World War. During the war he had worked on precision circuitry for radar at the MIT radiation laboratory, and it was the application of this expertise to spectrophotometry that transformed the methodology applied to the study of the respiratory chain.[23] He also developed stop-flow methods for rapid kinetic measurements of reactions. In 1947 Chance had visited the Molteno Institute in Cambridge, England, in order to do some experiments with Keilin on the heme-containing enzyme catalase. Chance assembled his electronic spectroscopic equipment, including rather bulky amplifiers and voltage stabilizers, in the Cambridge laboratory. Slater recalled Keilin's reaction to the appearance of the new approach: "I can well remember Keilin standing in the doorway with a look of astonishment at what was replacing his microspectroscope."[24]

Chance returned to the United States to apply his more advanced spectroscopic methods initially to the Keilin–Hartree heart-muscle preparations used in the Molteno Institute that Slater had demonstrated to him. He confirmed and extended Keilin's findings including the existence of cytochrome c_1 but concluded that these preparations were significantly denatured, leading him to study suspensions of mitochondria. Because both the muscle and the mitochondrial suspensions are turbid, Chance developed sensitive difference spectroscopy, measuring the difference between oxidized and reduced preparations. These studies gave a view of the kinetics of oxidation–reduction in the respiratory chain. His work enabled him to conclude that the various cytochromes were present in equimolar amounts, that is, cytochromes *a*, a_3, *b*, *c*, and c_1 were present in a

---

[20] Keilin and Hartree 1949, 1955.

[21] Keilin and Hartree 1955, p.206.

[22] Keilin and Hartree 1949.

[23] For an account of Chance's war work that influenced his subsequent development of spectroscopic methods for muscle and mitochondrial preparations, see Chance 1991.

[24] Slater 1997, see p.110.

ratio of 1:1:1:1:1; flavoproteins were present in higher quantity. It was possible to observe the sequence of oxidation of the components of the chain that confirmed the sequence determined by other means. He also concluded that cytochrome *b* participated in DPNH oxidation in phosphorylating mitochondria.[25] Further, he was able to show that the behavior he described for the respiratory chain in mitochondria could be confirmed in suspensions of intact mammalian cells.[26] At that time it was important to demonstrate that the observations in subcellular preparations were not experimental artefacts but reflected events occurring in intact cells.

These experiments confirmed and extended the knowledge of the respiratory chain conceived by Keilin and later supported by Slater. The chain could now be seen as a dynamic membrane-bound sequence of electron carriers in cell mitochondria, closely interacting with each other in the oxidation of succinate and NADH. Attention was directed to the thermodynamic aspects of the chain, but the failure to obtain satisfactory oxidation–reduction potentials for several components (for example, the flavoproteins and cytochrome *b*) limited the effectiveness of this approach. The new information encouraged visualization of the cytochromes as a spatial sequence on the mitochondrial membrane.[27] On a broader view, Chance was now able to view the important metabolic and regulatory functions of the chain as

(1) to transfer electrons or protons from substrates to oxygen and particularly to maintain the necessary level of oxidized DPN within the aerobic cell; (2) to act as a sequence of three or more energy conservation steps by which ADP is converted to ATP so that the latter is available as a common medium for energy expenditure throughout the cell; and (3) to regulate the metabolism in accordance with the levels of control substances, for example, of ADP itself or of a hormone upon the rate or efficiency of the energy conservation process.[28]

Such a view constitutes an update of Lipmann's views on bioenergetics expressed 15 years earlier. Chance's greatest contribution was his experimental exploration of the link between the respiratory chain and phosphorylation, to which we will return. This link had not been explored by Keilin, although it was pursued by Slater.

4.3.3 A Role for Quinones

One of the most significant achievements in the study of the respiratory chain during the 1950s was the discovery of ubiquinone (coenzyme Q). There was a growing interest in the lipids of the mitochondrion, although these were normally seen as having a structural

[25] Chance and Williams 1955.

[26] Chance and Hess 1959.

[27] Much of this is described in Chance and Williams 1956. Research with whole mammalian cells provided a means of checking that the information obtained with various preparations was relevant to the in vivo situation. For these studies see the work of Chance and Hess 1959, where ratios of individual cytochromes are also considered.

[28] Chance and Williams 1956, p.66.

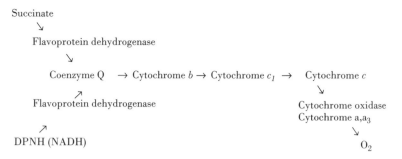

Figure 4.3 The respiratory chain after accommodating coenzyme Q in the early 1960s
See, for example, Lehninger 1964.

rather than a reactive role. In the lipid fraction, this period also saw a consideration of lipid-soluble quinones, such as α-tocopherol (vitamin E) and naphthoquinones (vitamin K), as possible significant participants in mitochondrial biochemistry. Frederick L. Crane (1925–), Youssef Hatefi (1929–) and coworkers at the Institute of Enzyme Research at the University of Wisconsin discovered a lipid substance that appeared to undergo oxidation–reduction in mitochondrial preparations, reduced by succinate and oxidized by O_2 through the cytochrome oxidase. The substance proved to be a quinone and was part of the respiratory chain in oxidative phosphorylation systems. Initially described as Q_{275} it became known as coenzyme Q.[29]

As the group commented,

> The data . . . provide ample evidence for considering coenzyme Q as a member of the mitochondrial electron transport system.[30]

Simultaneously, the same compound was isolated by Richard A. Morton (1899–1977) and his group at the University of Liverpool and its structure determined. It was found to have a wide distribution in nature and the group called it ubiquinone, which has become the preferred name.[31] Meanwhile the Wisconsin group had joined forces with a group led by Karl Folkers (1906–1997), an organic chemist at Merck Sharp & Dohme. This group also independently obtained the same structure for ubiquinone. This new compound was found to act in the region of cytochrome *b* and to be oxidized by cytochrome *c*, but its insolubility in water made experiments difficult. Clear conclusions from the early experiments were not easily drawn[32] and a fuller understanding of the role of ubiquinone had to wait until the 1970s when the Q-cycle was described.

By the end of the 1950s, the understanding of the respiratory chain could be summarized as in Figure 4.3. The position of cytochrome *b* remained uncertain. This is

[29] Crane et al. 1957; Hatefi et al. 1959; Hatefi 1959.
[30] Hatefi et al. 1959, p.500.
[31] For a description of the development of Morton's studies on ubiquinone, see Morton 1961.
[32] Hatefi 1963, see pp.305–306.

well illustrated by comparing comments of Chance writing to Slater in 1951 with those of Chance and Williams in 1956. In 1951 Chance wrote,

> I can understand your attachment to cytochrome b as a component of the main pathway for it is only with considerable reticence that I relegate it to a side track.[33]

But in 1956 cytochrome b was seen by Chance to be in the main pathway of DPNH oxidation. The difficulty in understanding the role of cytochrome b persisted in various forms until the acceptance of the Q-cycle in the 1970s or later.

Although the 1950s saw many achievements in understanding of the respiratory chain, the detailed situation at the end of the decade was still in considerable confusion. The function of cytochrome b remained unclear and disputed. There were questions about other components including nonheme iron and other quinones. More important, little was known about the proteins constituting the respiratory system; apart from cytochrome c almost all investigations were carried out using physical methods, primarily spectrophotometry. Biochemists of the 1950s were not in a position to seriously investigate membrane-bound systems because appropriate techniques were not available.

4.4 PHOSPHORYLATION

The early studies of respiration had proceeded without reference to phosphorylation. The respiratory system was viewed primarily as an oxidizing, not an energy-conserving, one. Slater later commented about the situation in the late 1940s: "quite a lot was known about the electron transfer chain, practically nothing was known about oxidative phosphorylation."[34] The elucidation of the respiratory chain had been a project on its own account, and Keilin, for example, had not seriously considered the question of phosphorylation. Slater wrote in relation to oxidative phosphorylation in the Molteno Institute around 1950,

> I do not recall that it was ever a topic for discussion in the Molteno Institute. The mechanism of the esterification of phosphate in glycolysis was, of course, well known . . . but the direct demonstration of oxidative phosphorylation by Engelhardt and Kalckar and the strong but indirect evidence by Ochoa and Belitzer and Tsybakova of the existence of what later became known as respiratory chain phosphorylation made very little impact on us in the Molteno Institute.[35]

However, during the 1950s the study of the phosphorylation, including Slater's work at the Molteno, began to become an integral part of investigations of the respiratory chain.

---

[33] Letter, B. Chance to E. C. Slater, December 24, 1951. EC Slater archive (379) Noord-hollands archief, Haarlem, The Netherlands.

[34] A recollection quoted in Slater 1997, see p.122.

[35] Ibid.

Earlier, Ochoa had obtained a P/O ratio of 3.0 for the complete oxidation of pyruvate to CO_2. Now the relationship between phosphorylation and respiration was explored in several laboratories by measuring the P/O ratio associated with the oxidation of citric acid cycle intermediates and related substances, in effect extending the original experiments of Krebs and others. But as Lehninger commented, the experiments were sometimes difficult to interpret:

> Attempts were made to determine which of the five oxidations involved in the complete oxidation of pyruvate[36] were accompanied by phosphorylation . . . une-quivocal evidence has been difficult to obtain. One of the experimental difficulties is the fact that when any individual cycle intermediate is added to a suspension of mitochondria . . . the oxidation which ensues may not be attributed solely to the one-step oxidation of the particular intermediate to its immediate oxidation product . . . there is a certain amount of oxidation in succeeding oxidation steps of the cycle.[37]

Despite the problems it became clear that all five oxidation steps in pyruvate oxidation were linked to phosphorylation and that the measured P/O ratio was in most cases a little higher than 2. Exceptions were succinate, which gave values significantly below 2, and α-ketoglutarate, which gave a relatively high value of around 3 or above. Measurements were made in several laboratories, and although results were variable, Lehninger felt able to conclude as follows:

> It appears reasonably certain that the average P:O ratio of the Krebs cycle oxidations is 3.0, the succinoxidase step being accompanied by only two and the keto acid oxidation steps perhaps by as many as four phosphorylations.[38]

It should be noted first that although this appears to have been the prevalent view, not everyone agreed, and second that it was assumed that the P/O ratio would be an integer on theoretical grounds.

A high P/O, significantly above 3, associated with ketoglutarate oxidation was difficult to explain until a phosphorylation associated with the enzyme responsible for conversion of ketoglutarate to succinate (the ketoglutarate dehydrogenase, part of the citric acid cycle; see Fig. 3.3) was shown to involve a phosphorylation.[39] Thus the ketoglutarate dehydrogenase was associated with one (substrate-level) phosphorylation, and the other three phosphorylations were associated with the respiratory chain.

The understanding of metabolic phosphorylation based on glycolytic phosphorylation (for example, the glyceraldehyde 3-phosphate oxidation and phosphorylation; Figs. 3.2 and 3.14) led to the view that the P/O ratio would be an integer and therefore

---

[36] One oxidation in the conversion of pyruvate to acetyl coenzyme A (CoA) and the other four in the citric acid cycle.
[37] Lehninger 1955, pp.187–188.
[38] Lehninger 1951a, p.344.
[39] Kaufman 1951.

the value obtained experimentally would need to be rounded up to the next-highest integer, a triumph of theory over experimentation! Slater felt that there was an equally strong argument for rounding down.[40] Even in 1966, the faith in integers was still sufficiently strong for Slater, when considering the problems of experiments, to sound a note of caution: "P:O ratios differing from whole numbers are to be expected." [41] Yet he still believed an integer to be appropriate for the theoretical value of the P/O ratio. In a letter to Slater, Chance stated that his ratio (ADP/O, equivalent to P/O) was less than 3:

> After a large number of controls, some of which have not been done as completely as we wish, our ADP/O value runs about 2.7 for beta-hydroxybutyrate [used as a means of generating reduced NAD], and for succinate the value is two thirds of this. We are considering in some detail the errors that may arise in this system.[42]

Such a view persisted, although rounding up the P/O ratio to the next-highest integer remained the majority view.

In the early 1950s, Lehninger devised an experiment to demonstrate phosphorylation when DPNH itself was oxidized in order to demonstrate phosphorylation associated directly with the respiratory chain.[43] This confirmed the view that the major sites of phosphorylation were not in the citric acid cycle but in the respiratory chain. Although this was influential at the time, the experimental results were subsequently found to be incorrectly interpreted because DPNH could not penetrate the mitochondrial membrane.

By 1953, an overall coherent view of ATP synthesis in metabolism had emerged. Krebs, considering the main reactions concerned with ATP synthesis in glycolysis, the citric acid cycle, and respiration and basing his comments in part on Lipmann's view about energy, summarized the situation as follows:

> Of the seven energy-yielding reactions proper, three occur when reduced pyridine nucleotide is oxidized by molecular oxygen . . . . A fourth is the oxidation of succinate by cytochrome c. Two more occur in anaerobic glycolysis and the seventh during the oxidative decarboxylation of α-ketonic acids [e.g., α-ketoglutarate, pyruvate]. The crucial feature of these reactions from the biological point of view is not the liberation of energy but the coupling of energy production with energy utilization, i.e., with the synthesis of the pyrophosphate bonds of adenosine triphosphate.[44]

[40] See Slater 1997, p.136; Slater and Holton 1954.
[41] Slater 1966, see p.334.
[42] Letter, Chance to Slater, May 25, 1954, North Holland archive.
[43] Lehninger 1951b. The uncorrected P/O ratio obtained was 1.89.
[44] Krebs 1953, p.101.

4.5 SITES FOR PHOSPHORYLATION

The studies of phosphorylation now needed to draw substantially on the long investigation of the respiratory chain in which thinking had concentrated on an endeavor to establish the sequence of the electron carriers. If the P/O ratio was in fact 3, did this imply three separate sites of phosphorylation? And if there were separate sites, in which part of the respiratory chain did each occur? A first step in understanding the mechanism of oxidative phosphorylation would be to determine which components of the respiratory chain were actually associated with the synthesis of ATP. Although attempts to measure phosphorylation in segments of the chain dated back to the beginning of the 1950s, because of the limited knowledge of the respiratory chain and the difficulty of measuring ATP synthesized, many of the experiments produced inconsistent results or were open to criticism and alternative explanations. For example, in 1952, J. H. Copenhaver Jr. and Henry Lardy noted that both Slater and also Morris Friedkin and Lehninger had failed to demonstrate phosphorylation in cytochrome c oxidation whereas J. D. Judah obtained evidence for such phosphorylation.[45]

The study of sections of the respiratory chain could proceed with added cytochrome c, which became a standard reagent in the 1950s.[46] The outcome of these experiments was to conclude that cytochrome c reduction by NADH (DPNH)[47] gave a P/O ratio of up to 1.8, and this persuaded Lehninger and also Chance and Ronald Williams to conclude that there were two sites of phosphorylation between NADH and cytochrome c.[48] In both Lardy's and later in Lehninger's laboratory, the oxidation of cytochrome c by cytochrome oxidase gave a P/O ratio of around 0.7, indicating that a third phosphorylation site was associated with the cytochrome oxidase,[49] the last segment of the respiratory chain, although this was later disputed.[50]

Experiments to separate the two phosphorylation sites prior to cytochrome c were much more difficult to design because suitable electron acceptors were not available. However, phosphorylation linked to the oxidation of NADH by mitochondria in the presence of ferricyanide and with antimycin A (which inhibited the transfer of electrons prior to cytochrome c) showed that there was a site in the early region of the respiratory chain associated with the oxidation of NADH and reduction of quinone (see Fig. 4.4).[51] This view was supported by the demonstration that reduction of cytochrome c by NADH produced significantly more phosphorylation than reduction of the cytochrome by succinate. If there was one site closely associated with NADH oxidation, then there would

[45] Copenhaver and Lardy 1952.
[46] Cytochrome c was the only cytochrome that could be readily detached from cell particulate material (mitochondria) and obtained in solution. It was prepared to a high degree of purity by Keilin and Hartree 1937 and by Theorell 1936. In the 1950s it became a normal reagent in bioenergetic studies. See Section 3.6.
[47] Note that DPNH (NADH) did not permeate the mitochondria, and β-hydroxybutyrate (which readily entered mitochondria) was used to reduce intramitochondrial DPN (NAD).
[48] Borgström, Sudduth, and Lehninger 1955; Lehninger 1955; Chance and Williams 1956.
[49] Maley and Lardy 1954; Nielsen and Lehninger 1955.
[50] See note 55.
[51] Copenhaver and Lardy 1952.

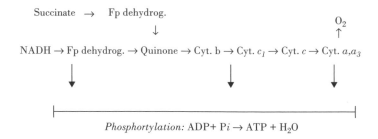

Succinate → Fp dehydrog.

NADH → Fp dehydrog. → Quinone → Cyt. b → Cyt. c_1 → Cyt. c → Cyt. a,a_3

Phosphortylation: ADP+ Pi → ATP + H_2O

Figure 4.4 The Three Sites of Phosphorylation in the Respiratory Chain

Based on Racker 1961 with the addition of ubiquinone ("quinone"), about which there was some uncertainty at the time.

also need to be a site for phosphorylation common to both NADH and succinate oxidation before cytochrome c (see Fig. 4.4). The experimental basis for such conclusions was not, however, undisputed.

Nevertheless, by the late 1950s it was possible to identify the three sites in the respiratory chain where oxidation was coupled to ATP synthesis (see Fig. 4.4). The first lay between the point of NADH oxidation and cytochrome b, the second between cytochromes b and c, and the last associated with the cytochrome oxidase.

Were the reactions at the three sites of phosphorylation basically the same? The compounds involved in the first site, which lacked cytochromes, were clearly very different from those at the other two. Racker concluded that the observed differences did not necessarily imply fundamental differences in reaction mechanisms.[52] Such a conclusion had been drawn by some workers much earlier.

A different and influential approach to the problem came from Chance, who by the mid-1950s had developed a powerful and sensitive spectrophotometric system for analysis of the respiratory chain. Initially he examined the effects of inhibitors on the oxidation–reduction state of the respiratory electron carriers. The addition of an inhibitor rendered the carriers before the site of inhibition (the substrate side such as NADH) more reduced and those on the O_2 side more oxidized. The point between greater reduction and greater oxidation he termed the crossover point, and this defined the site of action of the inhibitor. Antimycin A treatment rendered cytochrome b more reduced and cytochrome c more oxidized, indicating that the inhibitor acted between these cytochromes. Indeed, the period saw the identification of a number of inhibitors of respiration whose site of action was determined.

Another phenomenon to emerge in the mid-1950s was respiratory control. From work by Lardy and Harlene Wellman and by others it emerged that the rate of respiration in coupled mitochondria was regulated by phosphate acceptors, particularly by ADP levels.[53] In due course the ATP:ADP ratio was seen as the controlling factor. This property of the system was exploited by Chance in defining the relationship between phosphorylation and the respiratory chain.

[52] Racker 1961 summarizes the position.
[53] Lardy and Wellman 1952.

In phosphorylating mitochondria, a lack of ADP (the substrate phosphorylated to form ATP) substantially depressed the rate of respiration. Lack of ADP appeared to inhibit respiration. By sensitive adjustment of his experimental conditions, Chance detected the three crossover points for phosphorylations at which lack of ADP inhibited oxidation–reduction. This method led to the view that the phosphorylation sites lay between cytochromes c and a (in cytochrome oxidase), between cytochromes b and c, and between NADH and the flavoprotein. Such results were broadly consistent with those obtained by methods previously discussed but had the added advantages that they were noninvasive and did not involve significant manipulation of the mitochondrion.[54] However, locating the sites of phosphorylation remained one of the preoccupations of bioenergeticists during the 1950s, and although much agreement was achieved, the question of whether cytochrome oxidase was a site for phosphorylation was still being debated in the 1980s.[55]

During the 1950s, the study of respiration and phosphorylation was complicated by fundamental uncertainties about the nature of respiration itself. Keilin's heart-muscle preparations were uncoupled and carried out oxidation but not phosphorylation. This enabled the early experiments on cell respiration to be pursued by ignoring phosphorylation. This led to a view explored in the 1950s but ultimately rejected, that perhaps cells possessed two types of respiratory chain, only one being coupled to phosphorylation. As Lehninger commented,

> One of the major dilemmas is posed by the question of whether there are two separate sets of respiratory carrier enzymes, one of which is phosphorylating and the other non-phosphorylating. This dilemma arises because some of the electron carriers have been isolated in highly purified form and their interaction reconstructed *in vitro*, but these substances have shown no properties indicating potential participation in phosphorylation reactions.[56]

Further, the position of cytochrome b was not always seen as the same in each chain.

4.6 SEEKING TO UNDERSTAND THE MECHANISM OF PHOSPHORYLATION

The link between respiration and phosphorylation was a challenging puzzle for bioenergeticists in the 1950s and was arguably their central concern. How was the energy of oxidation in the respiratory chain used for the synthesis of ATP? Unlike phosphorylation linked to the glycolytic pathway, the coupling between respiration and ATP

---

[54] Chance and Williams 1956. They gave a general account of this work on crossover points for inhibitors and phosphorylation in this review on the respiratory chain and phosphorylation.

[55] The 1980s saw a dispute between Peter Mitchell and Mårten Wikström over whether the oxidase was linked to phosphorylation. Agreement was finally reached in 1986 that the oxidase was a phosphorylation site. See Prebble and Weber 2003, pp.222ff.

[56] Lehninger 1955, p.201.

synthesis was fragile and easily broken. The earliest comment on this was probably by Belitzer and Tsybakova:

> The coupling between respiration and phosphorylation is apparently accomplished only under definite conditions and is not necessarily mutually compulsive. Respiratory oxidation-reductions can also proceed as an "idle run," i.e. without esterification [phosphorylation]. Glycolytic oxidation-reduction, on the contrary, is coupled mutually with phosphorylation. Based on these facts it should be possible to distinguish two types of oxidation-reductions linked with phosphorylation: unconditionally coupled and conditionally coupled.[57]

Explaining coupling and uncoupling became a challenging problem for bioenergeticists.

Hunter, reviewing the field in 1951, noted that in oxidative phosphorylation, there were two means whereby the coupling might be broken experimentally, one simply by allowing the mitochondrial preparation to age and the other by treating the preparation with certain chemicals described as "uncouplers" or "uncoupling agents."[58] The initial compound found to act as an uncoupler was 2,4-dinitrophenol whose general effects on metabolism had been known since the nineteenth century.[59] However, it was not until 1948 that Loomis and Lipmann demonstrated the uncoupling effect of this compound on crude mitochondrial oxidative phosphorylation, showing that it blocks phosphorylation but not respiration. This observation was confirmed and extended by Judah who clearly demonstrated that the uncoupling effect was specific to oxidative phosphorylation and not to phosphorylation associated with glycolysis or citric acid cycle reactions such as ketoglutarate oxidation.[60] Other substances were found to uncouple oxidative phosphorylation, and the list of uncouplers was extended through the 1950s. The phenomenon of coupling continued to mystify, and clues to its nature were sought. Coupling required structural integrity, and damage to the mitochondria was associated with uncoupling. Dinitrophenol not only uncoupled but increased the rate of respiration and promoted mitochondrial ATPase activity.[61] Any mechanism for oxidative phosphorylation would need to take these properties of the system into account.

In 1946, Fritz Lipmann had suggested a simple mechanism for oxidative phosphorylation based on the notion that some members of the respiratory chain would be phosphorylated during oxidation. This phosphate would then be transferred from the respiratory intermediate to ADP to form ATP.[62]

---

[57] Belitzer and Tsybakova 1939. See Kalckar 1969, pp.211–227 for the English translation quoted here. (p.226).

[58] Hunter 1951.

[59] Parascandola 1975 traces the history of the effect of dinitrophenol and its derivatives on metabolism back to 1885, noting the early use of such compounds for food coloring and in munitions manufacture by the French. The toxicity of these compounds was soon realized.

[60] Loomis and Lipman 1948; Judah 1951.

[61] Lardy and Wellman 1953.

[62] Lipmann 1946.

$$AH_2 + C \quad \leftrightarrow \quad AH_2C$$

$$AH_2C + B \quad \leftrightarrow \quad A{\sim}C + BH_2$$

$$A{\sim}C + Pi \quad \rightarrow \quad A + P{\sim}C$$

$$P{\sim}C + ADP \quad \rightarrow \quad ATP + C$$

Figure 4.5 The Proposed Chemical Theory of Oxidative Phosphorylation

A and B are consecutive members of the respiratory chain at a coupling site, and they undertake oxidation reduction, reduced A being oxidized and B being reduced to BH_2 (as shown on line 2). The unspecified intermediate, C, is involved in the oxidation– reduction, and energy is stored in A~C. The squiggle, ~, represents the high-energy bond of Lipmann. A~C then reacts to form a high-energy phosphate, P~C, and the phosphate is finally transferred to ADP to form ATP. The proposal predicts a phosphorylated compound P~C, and it was this hypothetical compound that was extensively sought in the 1950s and 1960s using radioactive phosphorus ^{32}P. Based on Slater 1953, with the final equation expanded as proposed by Lehninger.

Slater revised this approach for several reasons, including the absence of any requirement for phosphate in respiration itself and also observations on the effect of uncouplers. The understanding of phosphorylation during this period was strongly influenced by the elucidation of the oxidation of triose phosphate in the glycolytic pathway where Needham and Pillai had stressed the link between oxidation and phosphorylation.[63] Subsequently a mechanism for the enzyme had been provided from Warburg's laboratory,[64] and this had formed the basis for the biological understanding of ATP synthesis linked to oxidation. Slater's hypothesis (Fig. 4.5) was modeled on the phosphorylation in the glycolytic pathway (see Fig. 3.10), the glyceraldehyde phosphate dehydrogenase reaction (with influence from some other phosphorylations, particularly the ketoglutarate dehydrogenase reaction) and reflects thinking of the period.[65] Its key proposal was that respiration would generate an unspecified high-energy compound that could then be used for the synthesis of ATP. The mechanism was expressed in a group of equations that underwent many revisions over the following 20 years. The proposal served to separate the utilization of inorganic phosphate in ATP synthesis from the operation of the respiratory chain itself.

The proposal was welcomed, and Slater's paper became one of the most widely cited papers in biochemistry. Its prediction that there was an unidentified high-energy intermediate at the core of oxidative phosphorylation set off an extensive hunt by bioenergeticists from many laboratories for the identity of this compound. Such was the significance attached to the possible discovery of this compound that numerous identifications were reported in the literature including at least one fraudulent one. Douglas Allchin lists sixteen isolations of the high-energy intermediate, some of which led to useful advances in other areas of biochemistry but not to the mechanism of oxidative phosphorylation.[66] Indeed Slater's proposal inspired a massive amount of research in bioenergetics, and the proposal was not formally abandoned until the early 1970s. The hypothesis was, however,

[63] Needham and Pillai 1937a, 1937b.

[64] Warburg and Christian 1939; Negelein and Brömel 1939. This was later revised by Racker and Krimsky 1952.

[65] Slater 1953.

[66] Allchin 1997.

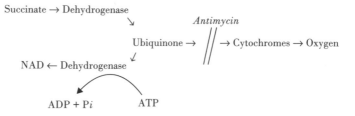

Figure 4.6 The Reversibility of Oxidative Phosphorylation

Chance showed that succinate would reduce NAD in coupled mitochondria at the expense of ATP. Normal oxidation through the cytochromes was blocked by antimycin. Thus the usual functioning of the NADH dehydrogenase, which would oxidize NAD and synthesize ATP, was reversed.

subjected to a variety of modifications as experimental results necessitated refining the equations.[67] A specific development of Slater's proposal was made by Lehninger's hypothetical mechanism for phosphorylation in the first coupling site associated with DPNH oxidation.[68] Indeed, when reviewing the various developments of the chemical theory in 1958, Slater commented thus:

> The first impression obtained from a study of the current literature on oxidative phosphorylation might be that there are about as many theories for the mechanism as there are workers in the field.[69]

In general, he divided them into those based on his original 1953 proposals (the majority) and a few based on Lipmann's proposal that involved phosphate in the oxidation–reduction process itself.

An important element in understanding the mechanism of oxidative phosphorylation was the realization that the process was reversible. Work in Chance's laboratory demonstrated that whereas respiration could drive ATP synthesis, ATP hydrolysis could drive electron flow back down the chain in coupled, but not uncoupled, mitochondria. The clearest demonstration of this was the reduction of NAD by electrons coming from succinate. Such a reaction was known to require energy. This could be obtained by passing the electrons back through the first coupling site to NAD at the expense of ATP (see Fig. 4.6). Hence the whole mechanism of coupling respiration to phosphorylation was a reversible process.[70]

---

[67] An example of this can be seen in Chance and William's consideration of the mechanism (Chance and Williams 1956), and one of the last amendments was proposed in Racker 1970a, see p.136.

[68] Lehninger 1955, see p.207.

[69] Slater 1958, p.286.

[70] See Chance and Hollunger 1960; Chance 1961.

4.7 THE PHOSPHORYLATING ENZYME

The developing knowledge of oxidative phosphorylation implied the existence of a phosphorylating enzyme. For a long time, mitochondria had been known to possess the ability to hydrolyze ATP to ADP and inorganic phosphate. A persistent question that proved difficult to resolve was whether this ATPase activity was due to the same enzyme that carried out the phosphorylation of ADP to ATP in intact mitochondria. In 1951, Kielley and Kielley, in a study of mitochondrial phosphorylation, suggested that the latent ATPase of freshly prepared mitochondria that became active as mitochondria aged and lost phosphorylation ability was the enzyme acting in ATP synthesis:[71]

$$ATP + H_2O \leftrightarrow ADP + P_i.$$

Such a view was reinforced through the demonstration by Mildred Cohn with isotopes that the phosphorylating enzyme of oxidative phosphorylation appeared to be readily reversible under suitable conditions.[72] The fact that Slater's chemical hypothesis for the mechanism of phosphorylation required transfer of high-energy phosphate (from the high-energy intermediate to ADP) rather than straight synthesis of ATP from inorganic phosphate seemed to raise relatively minor problems. More significant was the possibility that there were in fact several ATPases associated with the respiratory chain, one associated with each phosphorylating site. As Slater wrote in 1957,

> There are three dinitrophenol sensitive ATPases in liver and heart mitochondria, and we are now certain that these are related to the three phosphorylation steps in the respiratory chain.[73]

In contrast, Racker felt able to dismiss such possibilities in a review in 1961 but such ideas nevertheless lived on.[74] All in all the research carried out in the 1950s led to little progress with the predicted enzyme for phosphorylation until toward the very end of the decade, when Racker reported that he had successfully isolated such an enzyme. This initiated a new era of experimental research into the phosphorylation enzyme in the 1960s.

4.8 FRAGMENTING MITOCHONDRIA

Biochemists believed much might be achieved if the active parts of the oxidative phosphorylation system could be isolated for investigation. Many of the problems experienced

---

[71] Kielley and Kielley 1951.

[72] Cohn 1953, using the heavy isotope of O_2 ^{18}O, demonstrated an exchange between water and inorganic phosphate in coupled mitochondria, implying the reversibility of the ATP synthesizing enzyme. The experiments were repeated by Paul Boyer and extended by using the radioactive isotope ^{32}P.

[73] Letter, Slater to Cleland, March 26, 1957. EC Slater archive (379) Noord-hollands archief, Haarlem, The Netherlands.

[74] Racker 1961.

by investigators in the 1950s arose from the difficulty in separating out from other activities the systems they wished to study. The complex materials used (such as mitochondria) often led to experimental results that were capable of more than one interpretation. There was therefore a need to prepare the component parts of the system in isolation.

The location of the oxidative phosphorylation system in the mitochondrial membranes presented bioenergeticists with a severe challenge. Prior to the 1950s, almost all the major achievements in metabolic biochemistry were from studies with soluble systems. The initial studies of cell metabolism, although progressing only slowly during the early part of the twentieth century, had shown remarkable success during the 1930s and 1940s in elucidating the main pathway of carbohydrate breakdown through the glycolytic pathway and citric acid cycle.[75] Such achievements were to a large extent due to the application of chemistry, particularly organic chemistry, to the problems of metabolism supplemented by the growing knowledge of the enzymology of soluble systems. Thus traditional and successful biochemical methodology did not provide a good basis for the study of processes closely associated with mitochondrial membranes. As Slater has pointed out, "it was a period in which soluble enzymes held sway among biochemists."[76] Hence it became apparent that rather different skills would be necessary to master terminal oxidation and energy conservation, although workers still sought soluble systems. This, together with the frustration it induced, is reflected in comments of Lehninger in 1960:

> Since the first efforts of Warburg and Keilin many years ago, the respiratory enzymes in these [mitochondrial] membranes have been found to be remarkably refractory to isolation in soluble, homogeneous form. The relatively limited information we now have available on the respiratory carriers and coupling enzymes, as molecular entities, has figuratively had to be carved out of solid rock.[77]

The difficulties associated with working out methods for studying membrane-bound systems is probably a major reason why progress was so slow in the 1950s. As Green commented,

> The lack of progress during the past decade in our understanding of oxidative phosphorylation is in large measure attributable to the fragmentary information available about the structure of the electron transport system.[78]

Racker expressed a rather similar view:

[75] Krebs (1953) was able to take a unitary view (in three phases) of the whole of catabolic metabolism in some detail when looking at the way living organisms obtained their energy from the breakdown of foodstuffs.

[76] Slater 1997, see p.112.

[77] Lehninger 1961, p.31.

[78] Green 1959, p.123.

At the present time we are challenged by the complexity of the particulate systems which pose many experimental difficulties . . . .We have learned a great deal from studies of soluble enzyme systems and from simple chemical models. It is important to bear in mind that none of these systems describe what actually happens in oxidative phosphorylation.[79]

The quest for soluble preparations began to have some success during the late 1950s. Several claims to have solubilized the oxidative phosphorylation system were made, but these were subsequently discredited. At the end of the decade, one of the most successful isolations came from Racker's laboratory, where an enzyme capable of hydrolyzing ATP was isolated in solution. Further, this enzyme could be restored to the membrane from which it had been taken.[80] In due course this enzyme was shown to be directly involved in oxidative phosphorylation.

At the end of the decade, detergents began to be used for making soluble preparations of parts of the oxidative phosphorylation system drawn from the inner mitochondrial membrane. Although early preparations failed to live up to expectation, by 1962 successful isolations of complexes of the chain were made and opened up a new approach to the problems of oxidative phosphorylation.[81] David E. Green (1910–1983), an American who had trained in biochemistry in Hopkins' Cambridge Department in the 1930s and subsequently led a major research group at the Institute for Enzyme Research at Madison, University of Wisconsin, isolated small electron-transport particles (ETPs) from mitochondria. Initially these oxidized DPNH and succinate through a respiratory chain apparently identical with that in mitochondria. Improved preparations were also able to carry out oxidative phosphorylation.[82]

Toward the end of the 1950s there was a growing confidence that separate particles containing segments of the chain could be isolated. Detergents such as cholate were found to enable the isolation of a green particle having the properties of the cytochrome oxidase. These preparations of the oxidase left behind a red material capable of oxidizing DPNH and succinate and reducing added cytochrome c, effectively dividing the respiratory chain into two parts. Numerous attempts progressively produced better preparations,[83] and eventually four complexes of the respiratory chain were isolated in the early 1960s.[84]

---

[79] Racker 1961, see p.388.

[80] The first report of this work appeared in Pullman et al. 1958, but the full account appeared in Pullman et al. 1960 and Penefsky et al. 1960. This work is discussed further in the next chapter.

[81] See for example the preparation of the NADH dehydrogenase by Ziegler et al. 1959, which, however, did not have the correct flavoprotein, A more satisfactory isolation was made a little later by Hatefi's group; see Hatefi, Harvik, and Griffiths 1962.

[82] Green 1959 describes four methods for the isolation of ETPs, only one of which results in particles capable of oxidative phosphorylation.

[83] See for example Smith and Stotz 1954, who solubilized the enzyme so that it could be purified by conventional protein separation methods of the time. These authors trace the use of detergents for such preparations back to the very late 1930s.

[84] See Hatefi 1966 and the next chapter.

This was not the only approach to the isolation of the oxidative phosphorylation machinery from the mitochondrion. During the 1950s it was shown that small particles could be separated from mitochondrial preparations by either detergent treatment or the use of sonic vibrations. In most cases these proved to be vesicles derived from the inner mitochondrial membrane and possessed respiratory chain components; some could carry out phosphorylation. Green carried out an extensive study of mitochondrial fragments that he referred to as phosphorylating ETPs (PETPs) and that appeared to be mini-mitochondria. Using a sonic treatment of mitochondria, he obtained his ETP, which appeared to be primarily an inverted vesicle in which the inside of the inner membrane was exposed.[85] Numerous approaches produced various types of particles, from beef-heart mitochondria, particularly the digitonin particle (produced using digitonin) in which the membrane was "right side out," and sonic particles in which the membrane was inverted. The latter was important experimentally in providing access to the inside surface of the inner membrane. This methodology became particularly important in the 1960s.[86]

4.9 PHYSIOLOGICAL ASPECTS OF MITOCHONDRIA

During most of the 1950s, workers in bioenergetics seemed to view the significance of the mitochondrion primarily in terms of it being the location of oxidative phosphorylation. It provided the scaffolding on which the various proteins involved in the process were mounted. Thus the components of the respiratory chain could be viewed as a linear sequence in space held upon or within the membrane. More especially, isolation of mitochondria provided a basis for experimental procedures to study the process of oxidative phosphorylation. However, as time went on, the intimate relationship between phosphorylation and the mitochondrion raised questions about the bioenergetics of the particle itself and the relation of other mitochondrial processes to oxidative phosphorylation. A more comprehensive appreciation of the relationship of particle and process was required when mitochondria were observed to swell in the presence of thyroxine, phosphate, or calcium, etc., and when ATP was found to promote contraction. A number of such observations suggested that a close relationship existed between mitochondrial size and metabolism.[87] Explanations of these events included the presence of a contractile protein (comparable to that involved in muscular contraction) in the mitochondrial membranes. Such speculations led to suggestions of mechanisms for oxidative phosphorylation based on contractile membrane proteins.

A related aspect of mitochondrial physiology was the uptake of ions such as phosphate, sodium, and potassium, and also fumarate. Mitochondrial ion accumulation was shown to be linked with metabolic energy. "High-energy phosphate esters (such as ATP)

---

[85] See Green 1959.

[86] It should be noted that later work showed that beef-heart sonic particles were not 100% inverted and digitonin particles were not 100% right side out. Further, particles preparations from different sources produced different configurations.

[87] Packer 1961. Lehninger 1959 had shown swelling/contraction of mitochondria and ATP are linked.

can be used by mitochondria to move water and ions."[88] They could also be seen as an integral part of the cell's transport systems. As Walter Bartley, Robert Davies, and Krebs concluded,

> Mitochondria are not only the site where most of the available energy—the pyrophosphate bonds of ATP—are generated, they are also able to use this energy for the transport of solutes; they are basic units of secretory and absorptive activities.[89]

There were only modest investigations of such questions until the 1960s, but these and other findings began to make some bioenergeticists realize that the mitochondrial processes might have a more intimate role in oxidative phosphorylation than simply providing a framework on which the key proteins were mounted.

Cellular energy-dependent transport processes were yet another field of study that was about to cross-fertilize the field of oxidative phosphorylation. Physiologists had studied for some time a form of transport across membranes that required energy and that they had termed active transport.[90] This field was reviewed in 1960 by the Australian biologist Sir Rutherford Ness Robertson (1913–2001), who had a strong interest in bioenergetics. Robertson considered the research that had shown the dependence of transport in plants and acid secretion in the stomach on respiration and the bioenergetics implications. Robertson concluded that

> The active transport mechanism may be intimately connected with one of the steps in oxidative phosphorylation . . . . Understanding of the mechanism of oxidative phosphorylation might lead to understanding active transport, but knowledge about active transport may also contribute to understanding oxidative phosphorylation.[91]

Indeed, earlier, in 1952, Davies and Krebs had summarized a hypothetical mechanism for linking active transport to respiration and phosphorylation, a proposal that Davies felt anticipated the chemiosmotic hypothesis.[92] Those working in active transport in the 1950s were generally not in close contact with those in oxidative phosphorylation, a situation that changed after Mitchell brought the two together in the chemiosmotic hypothesis for oxidative phosphorylation proposed in 1961 but accepted only gradually.

---

[88] Bartley, Davies, and Krebs 1954.

[89] Ibid.

[90] Active transport is a mechanism by which metabolic energy is used to move substances or ions against a concentration gradient.

[91] Robertson 1960, p.258. Note that this review was influential in the formulation of the chemiosmotic hypothesis by Peter Mitchell.

[92] Davies and Krebs 1952. I have discussed the position of this hypothesis in Prebble 1996 and in Prebble and Weber 2003, see pp.88–89.

Thus by 1960, it was becoming clear to biochemists working on oxidative phosphorylation that they might need to take a broader view of their subject and that other properties of mitochondrial membranes might be crucial to understanding oxidative phosphorylation. The phosphorylation problem was much more a broad bioenergetics problem than a simple chemical one as initially summarized in Slater's chemical theory.

5

Defining the Mechanism
1960–1977

What was achieved in the 1950s was a partial definition of the core problem of bioenergetics—the nature of the mechanism of oxidative phosphorylation. This was realized by bringing together various hitherto independent strands of research into a coherent field of study. Although the two central strands were the respiratory chain and phosphorylation, other important ones came from cell biology through the identification of the mitochondrion as the site of respiration, phosphorylation, and energy conservation. This convergence with elements of cell biology proceeded further during the 1960s as the significance of an active role for the membrane in bioenergetic processes became apparent, particularly in the movements of ions into and out of the mitochondrion.

The key experimental developments concerned the ability to analyze membrane proteins, an issue that had delayed the growth of understanding in bioenergetics. Techniques for the isolation and partial purification of mitochondrial protein complexes concerned with electron transport gave a much better understanding of the operation of the respiratory chain. Analysis of proteins associated with ATP synthesis itself began to clarify the mysteries of the mitochondrial ATPase/ATP synthase enzyme. Reconstituting respiratory and phosphorylating systems from separated components confirmed the validity of the analytical techniques and some of the proposals for mechanisms.

The period also saw the introduction of newer spectroscopic techniques that led to the discovery of the last major component of the respiratory chain, the nonheme iron proteins. It was also a time when bioenergeticists fully realized that the process of oxidative phosphorylation could not be understood without an appreciation of the systems that transported both inorganic and organic ions across the mitochondrial membranes. Although these are the main topics arising from laboratory work that were addressed in bioenergetics over the period covered by this chapter, more fundamental was the theoretical question of the mechanism of oxidative phosphorylation. The 1960s through the early 1970s was a period when this became a very pressing question for bioenergeticists.

5.1 WHAT MECHANISM?

During the 1960s and much of the 1970s there was no shortage of proposed mechanisms for oxidative phosphorylation. The chemical theory that had reigned supreme in the 1950s was now challenged by several very disparate hypotheses. Indeed the field moved into near crisis during this period, brought about in large measure by the failure to find any agreed basic mechanism for mitochondrial bioenergetics. The ensuing frustration appears to have engendered disputes, sometimes acrimonious, and to have created divisions among laboratories and research groups.[1] The lack of any agreed paradigm for bioenergetics in the 1960s and much of the 1970s was probably the major reason why individual bioenergeticists championed their own pet theory of oxidative phosphorylation, tending to ignore alternatives. Resolving the problem of a mechanism became almost as much a social issue as it was a scientific issue.

Slater's chemical theory that had remained essentially unchallenged in the previous decade was founded on strong biochemical principles of phosphorylation, particularly the synthesis of ATP in the glycolytic oxidation of glyceraldehyde phosphate and the oxidation of ketoglutarate to succinate in the citric acid cycle. This had led to the search for the predicted phosphorylated intermediate of oxidative phosphorylation that was greatly aided by the availability of radioactive phosphate because a labeled intermediate should have been readily detected. In due course a number of phosphorylated intermediates were proposed but subsequently found to lack the required properties of the predicted intermediate. As already noted in Chapter 4, Douglas Allchin identified at least sixteen different claims to have isolated the elusive intermediate between 1956 and 1970 but none survived critical examination.[2] However, by 1960 there was already some uneasiness that the considerable effort being put into this search was not producing results and that other possibilities might be considered. Mitchell had met with Slater, Chance, Lehninger, and Daniel Arnon (a leader in photophosphorylation) to discuss the state of the chemical theory in Stockholm in 1960 and concluded,

> I was very impressed by the general admission of the weakness of the orthodox conceptions involving the hypothetical "energy rich" intermediates, X~I etc., none of which has been identified, although they have been assumed to exist for many years. The general climate of opinion during our discussions was quite favorably disposed toward considering alternative hypotheses, if such hypotheses could be formulated—especially if they would include a closer connection between the supramolecular structure and chemical function.[3]

At that date the failure to separate experimentally the process of oxidative phosphorylation from mitochondrial membranes was proving of some concern because the chemical theory did not require the involvement of a membrane. During the 1960s, several alternative hypotheses for oxidative phosphorylation, for which the membrane was essential, were proposed.

---

[1] This period in bioenergetics history has been referred to as the "Ox phos wars," a phrase that captures something of the strains in the field at this time. See for example Prebble 2002.

[2] Allchin 1997.

[3] Letter Mitchell to the editor of *Nature* (London), April 28, 1961. See Prebble and Weber 2003, pp.79–80.

5.2 THE FIRST PROTON THEORY: ROBERT J. P. WILLIAMS

It had not been lost on bioenergeticists that the normal formulation of the respiratory chain involved hydrogens (protons and electrons) for reduction of the NAD, flavin, and quinone, whereas the later part, the cytochromes, required only electrons for reduction. However, the final cytochrome oxidase, in which electrons reduced O_2 to water, also required protons. There appeared to be a flow of protons from the substrate end of the chain toward the oxidase accompanying the movement of electrons. This situation was exploited by Robert J. P. Williams (1926–2015) who proposed a proton-based hypothesis for the mechanism of oxidative phosphorylation. Williams, an internationally honored inorganic chemist in Oxford University, where he became the Royal Society Napier research professor, specialized in bioinorganic chemistry. During the 1950s he was concerned with the functions of metal ions in enzymes and particularly in reactions of metal complexes of the heme proteins of the membrane-bound respiratory chain. A concurrent interest was in the possible role of protons in membranes and in electrical potential differences in membranes. These strands came together in his proposal for oxidative phosphorylation.

The underlying principle of Williams's theory was that the respiratory chain, particularly the NADH dehydrogenase reaction, would produce protons in the membrane and that they moved to create an environment in which ATP could be synthesized from ADP and inorganic phosphate. The theory was based on the chemical fact that polyphosphates could be synthesized in the laboratory at high proton concentration *in the absence of water* and ATP was a polyphosphate. The protons could act as the sought-after intermediate coupling respiration to phosphorylation. At a high localized concentration, in an anhydric region of the membrane, this high-proton environment could drive ATP synthesis at the ATPase.[4] So this hypothesis required a lipid membrane to provide the anhydric environment.

Williams's proposal never really achieved a central position in the thinking of bioenergeticists. Nevertheless, it remained in contention at least until the basic issue of the nature of the link between respiration and ATP synthesis was settled toward the end of the 1970s, and some of its implications remained influential for the remainder of the twentieth century. During this period it was not easy for anyone to find ways of experimentally testing the hypothesis, and, as Williams himself commented to the author, an inorganic chemistry laboratory was not an appropriate place to carry out experiments on mitochondrial oxidative phosphorylation.

---

[4] Williams 1961. The basic idea was previewed in an earlier review, Williams 1959. Some revision of the hypothesis appeared in Williams 1962. Williams modified his views in the 1970s but the proton remained central to his position.

5.3 THE CHEMIOSMOTIC HYPOTHESIS
OF PETER MITCHELL

5.3.1 Connecting Transport and Metabolism

A very different hypothesis also based on protons and known as the chemiosmotic hypothesis was put forward by Peter D. Mitchell (1920–1992).[5] Mitchell, initially a biochemist at the University of Cambridge with strong microbiological interests, had developed an interest in biological membranes and the transport of molecules across them. The term active transport was used to describe energy-dependent transport across biological membranes, and it was this area, together with that of membrane-bound enzymes, that was of particular interest to Mitchell during the late 1950s when he was at Edinburgh University working on the bacterial membrane. Although active transport was energy-dependent, the link to metabolic processes was quite unclear. As Robertson noted,

> All living cells are capable of regulating their internal environment and of maintaining differences between inside and outside the cells, both in ionic composition and concentration . . . mechanisms dependent upon the metabolism of the cell are responsible. These mechanisms result in salts being moved against a concentration gradient and so energy expenditure by the cell is necessary. The link between the respiration process, which provides the energy and the movement of these ions, is still a matter for investigation. This type of ion movement, which is called *active transport*, is of general biological importance and probably occurs in all living cells.[6]

Mitchell concluded that transport must be integral with the reactions of membrane-bound enzymes and these must react vectorially (directionally in space) across membranes. He suggested systems for phosphate, succinate, and glutamate uptake into cells dependent on membrane-bound enzymes that metabolized these substrates:[7]

> We proposed the general hypothesis that the permeability of bacteria to many of their nutrients and waste products is a specific one, dependent upon the presence of the enzymes which adsorb the nutrients and desorb the products of metabolism in the surface of the plasma membrane.[8]

This was further developed to consider the possibility that some ions (such as hydroxyl and protons, the components of water) might be extracted from the membrane to drive chemical reactions such as the synthesis of glucose 6-phosphate from glucose and water:

---

[5] The relationship between the two proton-based theories was considered by Weber and Prebble 2006, particularly in relationship to the issue of priority and the possibility that Mitchell had drawn on Williams's work.

[6] Robertson 1960, pp.231–232.

[7] Many of Mitchell's ideas on these matters are summarized in Mitchell 1959 where the use of the word chemi-osmotic first occurs and its meaning is described rather than defined.

[8] Mitchell 1959, p.88.

$$\text{glucose} + \text{phosphate} \rightarrow \text{glucose 6-phosphate} + OH^- + H^+$$

Such considerations and particularly the chemiosmotic theory subsequently considered began to make mechanistic connections between the concept of active transport and metabolism.[9] Indeed, Mitchell almost felt a mission to close the gap between transport and metabolic studies. As he commented in a lecture in Edinburgh,

> The belief in the necessity for coupling separate membrane-transport systems and metabolic systems has partly been a reflection of the necessity for coupling physiologists to biochemists.[10]

5.3.2 The Original Chemiosmotic Theory

In his chemiosmotic proposal Mitchell considered both oxidative and photosynthetic phosphorylation. He assumed that respiration (or photosynthetic electron transport) would create a proton gradient (pH gradient)[11] across the mitochondrial (or photosynthetic) membrane. This would be achieved by the release of protons on one side of the membrane and hydroxyl ions on the other forming the gradient (see Fig. 5.1). The energy of oxidation would be conserved in the gradient, which could be dissipated by synthesizing ATP through extracting the elements of water (protons and hydroxyl ions) to opposite sides of the membrane:

$$P_i + ADP \rightarrow ATP + H^+ + OH^-.$$

In Mitchell's proposal the "high energy intermediate" of the chemical theory now became a pH gradient across the membrane.

The chemiosmotic theory was published in *Nature* in 1961.[12] Initially it was reviewed by Lehninger who felt it "demands the closest scrutiny" but saw it primarily as contributing to the understanding of the movement of ions across membranes more than to the core mechanism of oxidative phosphorylation.[13] After seeking Mitchell's comments on his draft, Slater reviewed the theory in 1966, describing it as

> beautifully simple, and as we have seen it explains all the experimental findings concerning oxidative phosphorylation which we have discussed to date. We shall

---

[9] Note that the reaction synthesizing glucose 6-phosphate from glucose and inorganic phosphate requires energy. The origins of Mitchell's theory (of which this reaction is a part) include philosophical elements, and these have been explored in Prebble 2001; aspects of this story are also told in Prebble and Weber 2003.

[10] Mitchell and Moyle 1959, p. 24.

[11] Mitchell's proposal of a gradient across the membrane is concerned primarily with pH. In the pH scale, high concentrations of protons (low hydroxyl ions) give low values whereas high concentrations of hydroxyl ions (low protons) give high values.

[12] Mitchell 1961.

[13] Lehninger and Wadkins 1962, pp.51–52.

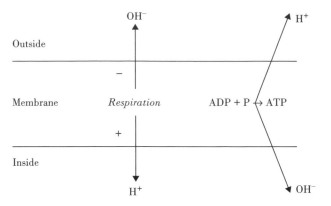

Figure 5.1 Mitchell's Chemiosmotic Theory of 1961
Respiration drives hydroxyl ions outward and protons inward, creating a pH gradient across the membrane. This gradient leads to the extraction of the components of water (H^+ and OH^-) to drive the synthesis of ATP.

see below it also explains in a simple way many other findings . . . . In general oxidative phosphorylation appears to be more complex than envisaged by the Mitchell theory.[14]

A number of others found Mitchell's ideas difficult to follow. Efraim Racker commented when meeting him in 1965 that he had not "really understood most of what Mitchell said . . . but I opened my mind to a new way of thinking."[15] Others also had difficulties, including Andre Jagendorf, who wrote

> Peter Mitchell's terminology and writing were difficult to understand. I was fortunate in having a brilliant postdoctoral from England, Geoffrey Hind, who could explain the concept to me.[16]

5.3.3 The Revised Theory

Subsequently the polarity of the membrane and the stoichiometry of the process (one proton transferred/ATP synthesized) were found to be wrong and the chemistry was criticized. After a period of illness, Mitchell resigned from Edinburgh University. Together with his research associate, Jennifer Moyle, he then set up his own private research institute Glynn, at Bodmin, Cornwall, in the southwest of England, using part of his family fortune. Here he revised his theory that retained the pH gradient across the membrane but the polarity was reversed (low pH, high proton concentration outside)

---

[14] Slater 1966, p.355.
[15] Racker 1976, p.43.
[16] Jagendorf 2002, p.236.

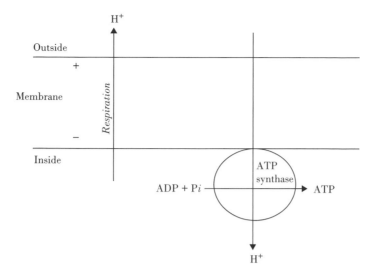

Figure 5.2 Mitchell's Revised 1966 Version of the Chemiosmotic Theory

The membrane potential (Δψ) of the membrane is shown as + and −. The operation of the respiratory chain moves protons outward, and the resulting proton gradient and membrane potential move protons back through the ATPase, synthesizing ATP.

and the hydroxyl ion disappeared from the mechanism. The protons were now seen as traversing the membrane, and two protons were associated with each ATP synthesized (see Fig. 5.2).[17] As with the first version of the theory, the movement of protons across the membrane not only created a pH gradient but also an electrical membrane potential. The energy of oxidation in the respiratory chain was stored in what Mitchell referred to as the proton motive force (PMF), being the sum of the pH gradient and the electrical potential·

$$PMF = \Delta\Psi - Z\Delta pH,$$

where $\Delta\Psi$ is the membrane potential and $Z\Delta pH$ is the proton gradient expressed in volts.

The mechanism whereby chemical uncouplers break the link between respiration and ATP synthesis was here seen as dissipating the PMF by freely transporting protons across the membrane. Mitchell's extensive description of his ideas, set out in privately

---

[17] Some of the early experimental work coming from Mitchell's laboratory is summarized in Mitchell and Moyle1965. This made an important contribution to the credibility of the theory, but more independent support was required. In part this was provided by Jagendorf but also by Guy Greville (1907–1969) working at the Agricultural Research Council Institute Babraham, Cambridge, England. Initially he provided supporting evidence but published a review of the theory in 1969 that was influential in making the theory more acceptable (Greville 1969).

published books, "the Grey Books," were now much more readily absorbed by fellow bioenergeticists, particularly the younger ones.[18]

Experimental support for the theory came only slowly. Mitchell showed that mitochondria carrying out respiration acidified the medium in which they were suspended, providing the first positive evidence together with the observation that the mitochondrial membrane itself was impermeable to protons. Of particular significance was the demonstration in Andre Jagendorf's laboratory that an artificially created proton gradient in a chloroplast preparation led to ATP synthesis in the dark (see Chapter 7). This impressive result in support of the chemiosmotic theory was somewhat muted by the failure to repeat this observation in mitochondrial preparations for several years. The 10 years following the publication of the second version of the theory saw an increasing effort to test the theory frequently but not always with positive results. The matter became more complex when other plausible theories were also proposed for the mechanism of oxidative phosphorylation.

The impact of the chemisomotic theory on workers in the field was somewhat mixed. Generally it was the younger workers who were inspired by its possibilities. Geoffrey Hind who drew Jagendorf's attention to the way in which the theory would explain their results on chloroplast phosphorylation was one such. Franklin Harold, a microbial physiologist and biochemist, was another, who subsequently wrote

> Mitchell's hypothesis revitalized and transformed bioenergetics; I and many others built our professional careers upon it. But it has a much wider significance. The chemiosmotic theory explicitly introduced spatial direction into biochemistry at a time when the metaphor of the bag of enzymes was still prevalent.[19]

5.4 REVISING THE RESPIRATORY CHAIN

Inadequate knowledge of the respiratory chain[20] in the early 1960s was seen as a major issue holding up progress in understanding oxidative phosphorylation. As Lehninger and Wadkins observed in 1962, "lack of detailed knowledge of the respiratory chain continues to hamper the study of oxidative phosphorylation."[21] During the 1960s the respiratory chain came to be seen as a series of lipoprotein complexes. Further, despite the

---

[18] Mitchell 1966a. Mitchell also published from his Institute, the Glynn Research Institute, a fuller version of his ideas, which became known as the first Grey Book (Mitchell 1966b). A second Grey Book was published 2 years later. Many workers came to understand Mitchell's ideas from these books, particularly the first, which were widely circulated among those laboratories devoted to bioenergetics research.

[19] Harold 2001, p.86.

[20] The respiratory chain began to be called the electron-transport chain during this period and in due course became the preferred term, the oxidation process being seen as the transfer of electrons from substrate to O_2.

[21] Lehninger and Wadkins 1962, p.47.

fact that work on the biochemistry of the respiratory system had proceeded for almost 40 years, it emerged that an important group of components had not been detected.

5.4.1 Iron–Sulfur Centers

Earlier work had drawn attention to the importance of iron in mitochondria, and there was a strong suspicion that not all the iron could be accounted for by association with the heme groups of the cytochromes. One of several new techniques introduced into biochemical work during the 1960s was electron paramagnetic resonance (EPR) spectroscopy in which an unpaired electron (in a reduced iron or semiquinone for example) placed in a variable magnetic field absorbed energy. In reducing conditions this technique detected a signal attributable to iron; this disappeared in oxidizing conditions.

Although probably not the first to detect nonheme iron, Helmut Beinert (1913–2007) did more than any other biochemist to apply EPR techniques to biochemical problems.[22] Beinert was born in Baden and educated in Heidelberg in Germany where he initially became a professional actor prior to being called up into the army. He obtained leave from the service in order to study chemistry at Heidelberg and Leipzig, where he obtained his doctoral degree. After the Second World War, he was one of a group of medics and chemists moved by the Control Commission of the Allied Forces to Texas. In 1950, he joined the Institute for Enzyme Research at the University of Wisconsin, where he carried out his major work on iron–sulfur proteins.

Initially, because EPR spectroscopy is carried out at low temperatures, the kinetics of iron oxidation and reduction could not be reliably studied. Nevertheless, by the late 1960s, bioenergeticists were reasonably convinced that iron, now known to be linked to sulfur, performed a role in the respiratory chain, particularly in the NADH and succinate dehydrogenases. As Beinert and his colleagues commented in relation to the EPR signal obtained from the NADH dehydrogenase enzyme, their work led them "to conclude that the component has the characteristics of an intermediate electron carrier in the enzyme."[23] In addition, in 1964, John S. Rieske (1923–2012) and coworkers, also at the Wisconsin Institute of Enzyme Research, described the partial separation of a nonheme iron protein (the Rieske protein) that they believed to be active in the ubiquinone–cytochrome c segment of the chain.[24] Such work was enhanced by finding iron–sulfur centers in other oxidizing enzymes such as xanthine oxidase and also, surprisingly, the citric acid cycle enzyme aconitase. Rather more important, soluble iron–sulfur proteins, ferredoxins capable of oxidation– reduction, were found in bacteria and chloroplasts (see Chapter 7). These latter iron–sulfur proteins were relatively stable, unlike those in the respiratory chain that could not be isolated. These discoveries initiated a decade of studies demonstrating the variety and importance of iron–sulfur centers in the respiratory chain.

[22] Kresge, Simoni, and Hill 2009.
[23] Beinert et al. 1965, p.475.
[24] Rieske et al. 1964.

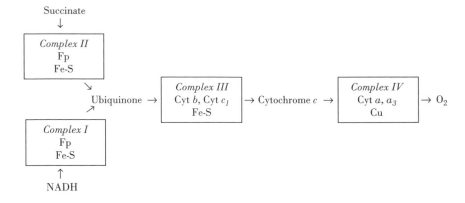

Figure 5.3 Complexes of the Respiratory Chain as Understood in the Late 1960s

Ubiquinone (coenzyme Q) is a mobile carrier between complexes I and III and between II and III. Cytochrome c links complexes III and IV. Fp is the flavoprotein involved in the NADH dehydrogenase (complex I) and succinate dehydrogenase (complex II). Fe-S denotes the iron–sulfur centers in complexes I, II, and III. Cu refers to the copper atoms in complex IV. See for example Green and Maclennan 1967.

5.4.2 Respiratory Chain Complexes

Perhaps the greatest barrier to progressing research in the electron-transport chain was the inability to study isolated parts of the system. At the end of the 1950s, particularly in David Green's laboratory in Wisconsin, work had begun on separating out complexes from the mitochondrial membranes. An important precondition for this type of mito-chondrial work was the ability to prepare substantial amounts of stable mitochondria. Referring to the mitochondrial preparations in use during much of the 1950s, Green commented,

> These suspensions had to be prepared daily or twice daily because of their limited stability which is a matter of some hours at 0°. These crippling exper-imental difficulties were in large measure responsible for the limited progress in and scope of many biochemical studies aimed at elucidating mitochondrial function . . . . The study of the mitochondrion of beef heart muscle showed that it had a stability which was of a different order of magnitude compared to that of liver. Suspensions could be frozen and stored without loss of the capacity for oxidative phosphorylation.[25]

Green's group devised methods for the large-scale production of mitochondrial material.

Initially isolations of electron-transport chain complexes were rather crude, but during the 1960s there was substantial success in making usable preparations of four multienzyme complexes that together made up the electron-transport chain. Each com-plex was capable of carrying out the oxidation reaction appropriate to that segment of the chain. The relationships of these complexes are shown in Figure 5.3, which includes their

---

[25] Green 1959, see p.78.

principle components. Complex I oxidized NADH and reduced ubiquinone, complex II oxidized succinate and also reduced ubiquinone, complex III oxidized the ubiquinone and reduced cytochrome c, and complex IV oxidized cytochrome c and reduced O_2 to water. Complexes I, III, and IV covered sections of the chain known to be coupled to ATP synthesis as sites 1, 2, and 3, respectively. In the early preparations the complexes contained about 20% lipid and were of substantially different sizes, complex I being the largest with a molecular weight in excess of a million. The isolated complexes could be recombined to give a functional respiratory chain.[26]

In fact, these studies brought about a fundamental change in the way the electron-transport chain was viewed. From the time of Keilin's earlier studies until the 1960s, the respiratory system was seen as an essentially linear chain of components. This linear system was understood as oxidizing NADH (DPNH) or succinate and reducing O_2 through the cytochrome oxidase, albeit with uncertainties about the role of cytochrome b. However, the ability to isolate individual complexes initiated a new view of the chain and a new program of research into the proteins that make up the respiratory chain. Hitherto the knowledge of the chain had depended almost exclusively on various spectroscopic techniques and had concerned the main groups undergoing oxidation–reduction. It was now possible to begin to analyze the constituents of the complexes and to look at the proteins involved, even if with rather limited and uncertain results at first. Analysis of the polypeptides that made up the complexes and also those of the ATP synthase were undertaken in many laboratories but not always with consistent results. Such an approach became very fruitful in the 1980s.

The analysis of iron–sulfur centers in complex I (the NADH dehydrogenase that by 1976 was seen as having a molecular weight of around 800,000) produced a range of results not always consistent, but by the mid-1970s it could be argued that there were six or seven centers. Nevertheless, in reviewing the situation, Ragan noted that "it is fair to say that everyone disagrees to some extent with everyone else."[27] The same review found between five and fifteen protein subunits in complex I. It must be admitted that complex I was much the most difficult to analyze and that the problems remained for many years. However, such results are indicative of the considerable challenges faced by those in the field in which the study of membrane proteins was still in its infancy.

By contrast the understanding of complex III, although still limited, was much clearer partly because of its smaller size (ca. 250,000). For a while the possibility of three b cytochromes was considered by some workers, but eventually the number was agreed on as two. Various laboratories concluded that there were about eight subunits in the complex and broadly agreed on their molecular sizes. Where disagreement arose was in assigning the cytochromes (b and c_1) and the single iron–sulfur center to the subunits.[28] A clearer view of complex III was achieved by the late 1970s.

However, the nature of cytochrome b remained a problem. Chance had identified three peaks in the cytochrome b region of the spectrum and these he attributed to two

[26] Hatefi et al. 1962.
[27] Ragan 1976.
[28] Rieske 1976.

cytochromes, one the Keilin pigment (designated b_K) with a peak at 562 nm and the other two peaks to a second b cytochrome (b_T).[29] Later his group showed that the cytochrome b_T underwent energy-linked changes,[30] which suggested that it could be involved with ATP synthesis in this complex. Similar observations to those of Chance's group were made in Slater's laboratory. This led to a proposal of a further mechanism for oxidative phosphorylation, at least for this complex, in which oxidation would result in the formation of the high-energy form of cytochrome b_T This would then return to its low-energy form, releasing energy for ATP synthesis.[31] This interpretation was disputed by Wikström[32] and others. The problem of cytochrome b is considered further in Chapter 8.

Complex IV, the cytochrome oxidase, had two cytochromes proposed by Keilin, a and a_3, together with two copper atoms, Cu_A and Cu_B. The complex was estimated to have a size of about 140,000 and to be composed of seven subunits.[33] In general, its structure was somewhat better understood at this stage than that of the other complexes. Extensive studies were made of the spectral changes that occurred during the oxidase reaction that progressively contributed to understanding the mechanism. It is discussed further in Chapter 8.

Complex II, the succinate dehydrogenase, was found to have a molecular weight of around 100,000 and was composed of only two main subunits, one containing flavin adenine dinucleotide (FAD) and the other iron–sulfur centers.[34]

The work in the 1960s and 1970s on the electron-transport chain itself transformed the way in which the system was seen. The chain could now be seen as interacting complexes and each complex could be understood in terms of its subunits, although this understanding was very much at a preliminary stage. The details of the manner in which the chain functioned became much clearer and information on the proteins involved began to be gathered, but in several areas this was subject to considerable disagreement. From this point on, however, the concept of the electron-transport chain became intimately involved with the question of its link to phosphorylation.

5.5 EXPLORING THE ATP SYNTHASE

One of the major advances in bioenergetics made in the 1960s was the isolation of the enzyme responsible for ATP synthesis. As already noted, in the 1950s, the mitochondrial enzyme responsible for hydrolyzing ATP was suggested, on rather limited evidence, to be the enzyme responsible for ATP synthesis in oxidative phosphorylation. The isolation of this ATP hydrolyzing enzyme in soluble form and its reattachment to mitochondrial

---

[29] Chance 1958.

[30] Treatment of mitochondria with ATP shifted the midpoint potential 250 mV to the positive, suggesting that two states of this cytochrome could be the mechanism for ATP synthesis during respiration. See Chance et al. 1970.

[31] See Slater et al. 1970.

[32] Wikström concluded "no evidence in favour of 'high energy' derivatives of cytochrome b" was found (Wikström 1971).

[33] Malmström 1979.

[34] Davis and Hatefi 1971.

membranes were demonstrated around 1960 by the group led by Efraim Racker. It was then shown to be the enzyme capable of synthesizing ATP in oxidative phosphorylation.

Racker had initially planned to make fine arts his life's work, and he remained an accomplished artist for the rest of his life; indeed many of his distinguished colleagues received Racker paintings from the artist. However, after graduating in medicine in Vienna, where he had become interested in the biochemical aspects of psychiatry, he left Austria after Hitler invaded in 1938. He moved to Cardiff, Wales, and then in 1941 migrated to the United States, where he became a major figure in the bioenergetics community. In 1943 he joined New York University. Later, after work on carbohydrate metabolism and moving to the Public Health Research Institute, Racker took up work on mitochondrial oxidative phosphorylation by using an approach based on resolution of the system into different fractions and reconstitution of the material to achieve the original activity.[35]

Using methods for large-scale production of beef-heart mitochondria as developed in Green's laboratory in Wisconsin, Racker broke up mitochondria until phosphorylation activity was lost and then added back the supernatant protein fraction to restore activity. Using this approach, he isolated a protein fraction capable of ATP hydrolysis that, when added to the membranous material, restored oxidative phosphorylation. The protein was referred to as F_1 and Racker's group showed that the mitochondrial ATP-hydrolyzing enzyme and the enzyme responsible for ATP synthesis in oxidative phosphorylation were the same.[36] A similar protein (named CF_1) was isolated from chloroplasts and was shown to restore light-driven photophosphorylation under appropriate conditions (see Chapter 7).[37]

It could be assumed that F_1 would be attached to a membrane protein. The clue to the identification of this protein came from the study of oligomycin sensitivity. Oligomycin had been shown by Lardy and others to inhibit the ATPase in mitochondria and submitochondrial particles.[38] The failure to get such inhibition with F_1 particles raised questions about the relationship of the particle to the ATPase. A protein (named F_o) was isolated from mitochondrial fragments that bound F_1 and also conferred oligomycin sensitivity.

Work on the electron microscopy of mitochondria now added to the understanding of the ATP synthesizing enzyme. Humberto Fernández-Morán (1924–1999) an electron microscopist then working at the Massachusetts General Hospital, Boston, and later a Venezualan politician, observed small particles (about 80–90 Å in diameter) attached to mitochondrial membranes (see Fig. 5.4b).[39] It eventually became clear that they were relatively densely packed on the inner membrane. This discovery created considerable interest among bioenergeticists, and in Green's laboratory it was argued that the particles were the site of the electron-transport chain.[40] However, Chance and colleagues showed

---

[35] For autobiographical accounts of Racker's work, see Racker and Racker 1981 and Racker 1976. See also Schatz 1997.

[36] Pullman et al. 1960; Penefsky et al. 1960. A preliminary account of this work appeared 2 years earlier; Pullman, Penefsky, and Racker 1958).

[37] Vambutas and Racker 1965.

[38] Lardy, Johnson, and McMurray 1958.

[39] Fernández-Morán 1962; Fernández-Morán et al. 1964.

[40] Green, Blair, and Oda 1963.

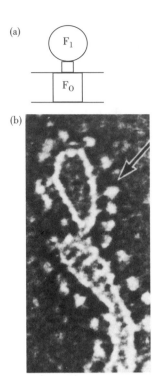

(a)

F_1

F_0

(b)

Figure 5.4 The Visualization of the ATPase (ATP Synthase) Enzyme

(a) the relationship of the components F_0 and F_1 to the membrane. (b) an electron micrograph showing the small spheres (F_1) attached to inner membrane vesicles (which are inverted) on which the model is based. Photo courtesy of Professor Humberto Fernández-Morán; reproduced with permission from Fernández-Morán et al. 1964, "A Macromolecular Repeating Unit of Mitochondrial Structure and Function" (*Journal of Cell Biology* 22: 63–100), Rockefeller University Press, fig. 5 (part), p.71.

that stripping membranes of these particles enhanced the concentration of respiratory components in the membrane.[41] Eventually Yasuo Kagawa and Racker showed that these membrane-bound particles were the F_1 ATPase that had earlier been isolated in Racker's laboratory.[42] Although there were arguments about whether the particles were an artefact of preparation of the material for electron microscopy, the consensus was that the ATPase was as shown in Figure 5.4.

Work in Racker's laboratory continued to isolate many more protein factors associated with the activity of the ATPase enzyme, which he referred to as "coupling factors." However, this work produced increasingly confusing results, and in retrospect Racker concluded that there were only four fairly well-characterized protein coupling factors, F_1, F_2, F_6, and the oligomycin-sensitivity-conferring factor.[43] Several other laboratories

[41] Chance, Parsons, and Williams 1964.

[42] Kagawa and Racker 1966. The significance of this work is explored by Bechtel and Abrahamsen 2007 (see pp.21ff.) who saw these structural–functional relationships as the outcome of the impact of cell biology on biochemistry.

[43] Racker 1976.

engaged in similar analyses, Sanadi and coworkers obtaining rather similar but less extensive results.[44]

5.6 RECONSTITUTING OXIDATIVE PHOSPHORYLATION

In the autumn of 1966, Racker moved to Cornell University and shortly after recruited Peter Hinkle, who spent the year 1968–1969 working in Mitchell's research institute, Glynn, in Bodmin, Cornwall, England. He returned to the United States convinced of the basic validity of the chemiosmotic mechanism, which he shared with Racker. The latter had, however, been brought up with the chemical theory and his comments gave an indication of the thinking of many of the older bioenergeticists in 1970:

> Having been raised by the music of substrate-level phosphorylation my own prejudices induce me to lean toward some aspects of the chemical hypothesis, e.g. the phosphorylated high energy intermediate. In its favor is the tendency of nature to repeat itself with respect to mechanisms. It also permits a brighter experimental outlook than the concerted reaction formulated by Mitchell. On the other hand, the experimental evidence in favor of an intimate relation between oxidative phosphorylation and the formation of a membrane potential and proton movements is mounting.[45]

Racker's conversion to the chemiosmotic mechanism was beginning, and these ideas gave him a rational approach to reconstitution experiments that recognized the possible importance of ion transport that could be measured only in intact vesicles.

In 1971, Hinkle's group showed that submitochondrial particles lacking F_1 had good proton permeability that was reduced by the addition of F_1 or by the ATPase inhibitor, oligomycin.[46] This led to the view that the membrane-bound F_o, a protein earlier detected by Racker in mitochondrial membranes was possibly a proton channel through the membrane and that the addition of F_1 or oligomycin blocked the channel. This view was strengthened by further research, particularly in Kagawa's laboratory in the Jichi Medical School in Japan.[47]

In Racker's laboratory extensive further work showed that the incorporation of the ATPase into artificial vesicles enabled the ^{32}P/ATP exchange showing that the enzyme was acting in a coupled manner.[48] A complete phosphorylation system was demonstrated

---

[44] See Andreoli, Lam, and Sanadi 1965. Factor A from this group was probably the same as F_1 whereas Racker's group regarded Factor B of Sanadi's group as equivalent to their factor F_2.

[45] Racker 1970a, p.137.

[46] Hinkle and Horstman 1971.

[47] In 1977 Kagawa's group used a very stable ATPase from a thermophilic bacterium to demonstrate the proton permeation properties of a purified CF_o (known as TF_o); see Okamoto et al. 1977.

[48] Kagawa and Racker 1971. The exchange experiments were initially carried out in the 1950s by Cohn and by Boyer. They are a means of demonstrating that the enzyme is coupled to the energy-conserving system; they are discussed in Chapter 4.

by incorporating into vesicles prepared with purified phospholipids, the ATPase coupling factors, the isolated cytochrome oxidase, and cytochrome c with an electron donor. Oxidation by molecular O_2 was observed to be coupled to ATP synthesis.[49] The first complex of the respiratory chain could also be incorporated into vesicles that oxidized NADH and reduced ubiquinone and with the ATPase could synthesize ATP.[50] Hence Racker's laboratory showed that isolated complexes of the respiratory chain could synthesize ATP when they were inserted into artificial vesicles together with the ATPase (ATP synthase).

Subsequently Mårten Wikström (1945–),[51] a Finnish bioenergeticist who had worked in Slater's and Chance's laboratories and who became professor of physical biochemistry in Helsinki, worked extensively on the cytochrome c oxidase. He incorporated this enzyme into vesicles and showed that it translocated protons at a stoichiometry of one per electron when reducing O_2.[52] Although these experiments strongly supported the chemiosmotic mechanism, at least the earlier ones were also open to other interpretations, which still left the issue of the mechanism not fully resolved. This work also provided an answer to another question that had concerned biochemists: Was the membrane symmetrical or were the two sides different? Many of these experiments required a design that recognized that the two sides were different.

5.7 BACTERIORHODOPSIN

At the end of the 1960s, work in the laboratory of Walther Stoeckenius (1921–2013)[53] began to reveal the unusual properties of the membranes of the archaebacterium, *Halobacterium*. These proved to be of special interest to bioenergeticists because of their system for ATP synthesis. This small specialized group of halophilic bacteria had evolved to occupy an ecological niche in the splash pools above the ocean high-tide mark where the salinity was high (2M sodium chloride). In their native environment these organisms were purple because of a single protein occurring in patches in the membrane. Together with Dieter Oesterhelt, Stoeckenius explored the purple protein, bacteriorhodopsin, which could readily be isolated in a relatively pure state. It had much in common with the mammalian eye pigment, rhodopsin, and, like the eye protein, was found to have retinene as its prosthetic group.[54] When illuminated, the protein was shown to pump protons outward across the bacterial membrane, and the resulting proton gradient was used for ATP synthesis.[55] Hence the organism's bioenergetics were based on the use of

---

[49] Racker and Kandrach 1973.

[50] Ragan and Racker 1973.

[51] For an autobiographical note on Wikström, see Wikström 2008.

[52] Krab and Wikström 1978.

[53] Stoeckenius was educated in Germany at the University of Hamburg, where he went on to work on membranes. He left Germany in 1959 to work at the Rockefeller University, New York, moving to the University of California in 1967. Here he carried out his studies of bacteriorhodopsin. See Kresge et al. 2011.

[54] Oesterhelt and Stoeckenius 1971.

[55] Oesterhelt and Stoeckenius 1973. See also Grote 2013 for a consideration of this work.

light energy for ATP synthesis, a chlorophyll-less form of photosynthesis. Clearly such a relatively simple system had relevance to the chemiosmotic theory.

Subsequently extensive studies explored the conformational mechanism by which the protein pumped protons as a result of illumination (see Chapter 8). Stoeckenius collaborated with Efraim Racker to place the purple protein into phospholipid vesicles, where uptake of protons into the vesicle in the light could be demonstrated. They then added the oligomycin-sensitive ATPase to the vesicles and showed that, in the light, the vesicles synthesized ATP.[56] Because there was no electron-transport chain involved, the synthesis of ATP was driven by the proton gradient created by the bacteriorhodopsin.

In general the Racker–Stoeckenius experiment demonstrated the operation of the chemiosmotic mechanism, providing convincing evidence for the proposals of Peter Mitchell. This seems to have been true for biochemists generally but bioenergeticists were more skeptical. The situation was summarized in a letter from Peter Mitchell to Walther Stoeckenius:

> I feel your work will do much to help promote interest in the chemiosmotic type of coupling mechanism outside the classical arena of oxidative and photosynthetic phosphorylation where the simple scientific realities still tend to be rather obscured by the dust and smoke of ancient battles.[57]

Although the experiment should have settled the broad issue of the mechanism of oxidative phosphorylation, for many of the older bioenergeticists particularly, it did not.

5.8 CONFORMATIONAL THEORIES

With the failure to agree a possible mechanism for oxidative phosphorylation, the field was open for new proposals. In the 1960s and early 1970s, none of the theories proposed by Slater, Williams, and Mitchell commanded broad support across the field. Several other proposals were made, more especially a group of theories based on conformational changes in the whole organelle, in membranes, or in individual proteins.

5.8.1 Mitochondrial Conformational Changes

To understand puzzling situations, biochemists often resort to analogy. Ever since the myosin ATPase was described by Engelhardt and Ljubimova,[58] the possibility that bioenergeticists might learn something from the biochemistry of muscular contraction relevant to their own problems had existed. As ATP hydrolysis in muscle is concerned with a change of conformation (contraction) of proteins, mitochondrial ATPase activity was seen as linked to observable changes in particle shape and structure; under suitable

---

[56] Racker and Stoeckenius 1974.
[57] Letter, Mitchell to Stoeckenius, June 11, 1973. P. D. Mitchell archive, Cambridge University Library, quoted in Prebble and Weber 2003, p.166.
[58] Engelhardt and Ljubimova 1939.

conditions mitochondria treated with ATP contract. Hence Lehninger found a number of similarities between "the contractile systems of the mitochondrion and of the myofibril." Claims were also made about the isolation of a contractile protein from mitochondria. In 1964, Lehninger concluded thus:

> The respiratory assemblies in the mitochondrial membrane that are responsible for coupling ATP synthesis to electron transport are polymodal energy-transducing systems that can convert oxidation-reduction into chemical or phosphate bond energy, into the osmotic energy of active transport, and into the mechanical energy of contraction.[59]

These ideas were pursued further, particularly by Charles Hackenbrock, a graduate student working at Columbia University, New York, who identified two states of mitochondria by electron microscopy. The orthodox conformation is that seen normally while in the condensed conformation, the inner membrane of the mitochondrion is randomly folded, the density of the matrix increased, and the volume decreased (see Fig. 5.5). Respiration limited by lack of ADP for phosphorylation gives an orthodox mode whereas the addition of ADP produces the condensed mode.[60] A detailed study including the use inhibitors and uncouplers led Hackenbrock to this conclusion:

> These and more recent observations with inhibitors of respiration and phosphorylation suggested that the ultrastructural transformations were manifestations of the transduction of oxidation-reduction energy into mechanical or conformational energy. It was postulated that conformational energy generated by electron transport is the immediate source of energy for the synthesis of ATP during oxidative phosphorylation.[61]

Hence conformational energy could be the "high-energy" intermediate in oxidative phosphorylation rather than a chemical intermediate (chemical theory) or an electrochemical proton gradient.

A further version of the mechanism based on mitochondrial morphology was proposed by David Green. It was influenced by Green's view of Fernández-Morán's observation of membrane particles and identified three morphological forms of mitochondrial membranes, nonenergized, energized, and energized twisted. Either electron transport or ATP hydrolysis could convert the nonenergized to the energized. The membrane

---

[59] See Lehninger 1964, p.202.
[60] Hackenbrock 1968a.
[61] Hackenbrock 1968b, p.598.

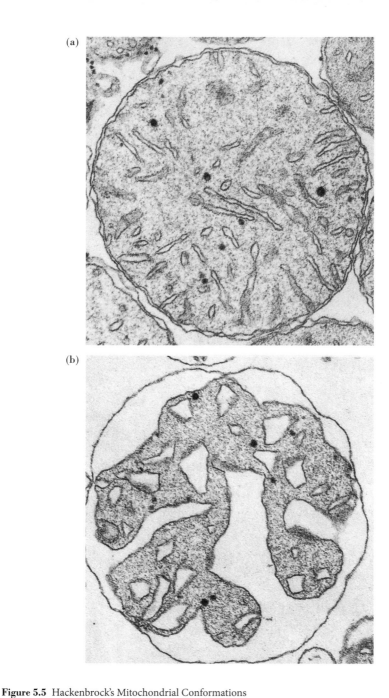

(a)

(b)

Figure 5.5 Hackenbrock's Mitochondrial Conformations

(a) orthodox or high-energy state; (b) condensed or low-energy state.

Reproduced with permission from Hackenbrock 1968, "Ultrastructural Bases for Metabolically Linked Mechanical Activity in Mitochondria" (*Journal of Cell Biology* 37: 345–369), Rockefeller University Press, figs. 19 and 20, pp.364 and 365.

particles attached to the membrane by stalks also underwent conformational change.[62] This hypothesis had only limited influence.

The value attached to organelle conformational studies declined in the early 1970s when it was realized that the observed changes seemed too slow for other than a secondary effect of the phosphorylation mechanism. The morphological changes seen in mitochondria appeared to be the result of osmotic affects associated with ion movements (see next section).

5.8.2 Protein Conformational Changes

A different approach came from the enzymologist Paul Boyer (1918–2018) and stemmed from his endeavor to find the high-energy intermediate predicted by the chemical hypothesis. After a first degree in chemistry, Boyer went to the University of Wisconsin where he worked on enzymology and became interested in bioenergetics stimulated in part by the notable "Symposium on Respiratory Enzymes" of 1942. There he listened to several of the leaders in contemporary biochemistry including Otto Meyerhof, Herman Kalckar, and Fritz Lipmann.[63] Later, at the University of Minnesota, in the mid-1950s he carried out experiments on ATP synthesis using an O_2 isotope (see Section 4.7). In June 1963 he joined the chemistry department of UCLA where he continued work on oxidative phosphorylation, subsequently becoming the director of the new Molecular Biology Institute.[64]

In the mid-1960s, as part of the search for the elusive high-energy intermediate of the chemical theory,[65] he proposed that the energy of oxidation might be conserved by the formation of an acyl-S bond in a protein. Such a bond might provide the energy to phosphorylate a histidine group in the protein where the phosphate might then be used for ATP synthesis. In order to couple the formation of the acyl-S bond with the energy available from the oxidation in the respiratory chain, Boyer thought that changes in protein conformation of respiratory intermediates would be necessary.[66] The hypothesis also drew inspiration from the fact that a protein conformational change was associated with ATP hydrolysis in muscle contraction.[67] Such a view saw the source of energy as the conformational changes in the respiratory chain during oxidation but found no place for the high-energy intermediates of the conventional chemical theory (other than the acyl-S group), membrane protons (Williams theory), or transmembrane proton potential gradients (chemiosmotic theory). Later Boyer made proposals to accommodate a proton electrochemical gradient.

---

[62] See Green and Baum 1970 for a full version of Green's views. These were further developed as the electromechanical hypothesis; see Green 1974.

[63] University of Wisconsin 1942.

[64] For autobiographical notes, see Boyer 1981, 1998.

[65] See Allchin 2002 for a discussion of Boyer's development of conformational ideas.

[66] Different protein conformations were seen as having different energy levels.

[67] The original version of this theory was proposed in Boyer 1965, which treats the glycolytic phosphorylation as basic to thinking on oxidative phosphorylation.

These views were not developed further until the early 1970s, when it occurred to Boyer that his earlier experiments measuring O_2 exchange between phosphate and water were open to a new interpretation. They could be explained in terms of the use of energy from respiration to *release* ATP from the ATP synthase (rather than to synthesize the co-valent bond in ATP). Such a view implied a conformational change in the enzyme itself as the means of supplying energy for ATP release. This conceptual advance, published in 1973,[68] was really concerned only with the ATP synthase enzyme. So Boyer's proposals were first for a theory of the mechanism of oxidative phosphorylation based on conforma-tional change. Proteins in the respiratory chain would undergo conformational changes that would induce conformational changes in other proteins including the ATP synthase. A second and separate proposal was limited to a conformational mechanism for the ATP synthase enzyme itself. The existence of two separate conformational proposals created a situation that engendered some confusion. It was the second proposal that saw the ATP synthase as synthesizing ATP on the surface of the enzyme that then underwent an energy-dependent conformational change bringing about ATP release.[69] Both theories continued to be canvassed for several years but the development of the mechanism of the ATP synthase itself eventually emerged as a monumental biochemical achievement

5.9 MITOCHONDRIAL MEMBRANES

The 1960s saw a much greater understanding of the mitochondrial membranes. Not only was it possible to separate and partially purify active complexes from mitochondria but it now became possible to understand many of the properties of the membranes them-selves. The permeability of the membranes was determined by experiments on osmotic swelling and contraction. The osmotically sensitive region of the mitochondrion was re stricted to the inner part (inner membrane and matrix), and it became clear that the inner membrane was a classical semipermeable membrane.[70] The outer membrane was found to be permeable to molecules up to a molecular weight of around 10,000, that is, it was seen to be permeable to metabolites but not to almost all proteins.

In the mid-1960s, methods for controlled disruption of mitochondria and separation of membranes were developed so that the long-believed location of the oxidative phos-phorylation system in the inner mitochondrial membrane could be confirmed among other metabolic–structural questions.[71] Understanding of the membrane structure

---

[68] Boyer, Cross, and Momsen1973. There was only limited evidence for Boyer's proposals, and they were rejected by the *Journal of Biological Chemistry* but subsequently accepted by the *Proceedings of the National Academy of Sciences.*

[69] See Boyer 1981, pp.234–235, for a personal comment on these events.

[70] Permeable to water but not to sugars, metabolites, and inorganic ions. Permeation by metabolites and ions, etc., is then dependent on specific systems present in the membrane such as proteins that translocate substances across the membrane.

[71] Methods initiated between Parsons and Chance (Parsons et al. 1966) were developed in Ernster's laboratory (Sottocasa et al. 1967).

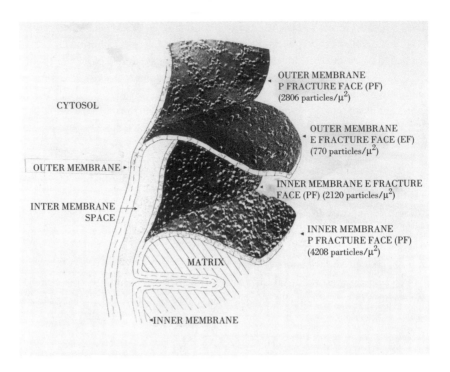

CYTOSOL

OUTER MEMBRANE
P FRACTURE FACE (PF)
(2806 particles/μ²)

OUTER MEMBRANE
E FRACTURE FACE (EF)
(770 particles/μ²)

INNER MEMBRANE E FRACTURE
FACE (PF) (2120 particles/μ²)

OUTER MEMBRANE

INTER MEMBRANE
SPACE

INNER MEMBRANE
P FRACTURE FACE (PF)
(4208 particles/μ²)

MATRIX

INNER MEMBRANE

Figure 5.6 Montage Showing the Distribution of Protein Particles in the Inner and Outer Membranes of the Mitochondrion by Freeze-Fracturing Techniques

Freeze fracturing splits the membrane along the middle, revealing the inner faces of the inner and outer halves. This shows that the inner half of the inner membrane has a high protein density whereas the outer half has fewer protein particles. The outer membrane has a lower protein concentration. Reproduced with permission from Packer 173, *Membrane Particles of Mitochondria* (Mechanisms in Bioenergetics, Academic), Elsevier, fig. 2, p.36.

developed during the 1960s when it was appreciated that the lipid bilayer was the only plausible model.[72]

This period saw the introduction by electron microscopists of the freeze-fracturing and freeze-etching techniques that showed the distribution of protein particles within the membranes (Fig. 5.6).[73] Such developments were conceptually enhanced by the appreciation of the fluid nature of natural membranes. The introduction of the fluid mosaic model in the early 1970s gave a new way of looking at membrane proteins that were now free to translate laterally in the membrane.[74] The model probably overemphasized the fluidity aspect. Such developments made it easier to understand the migration of ubiquinone in the lipid phase between membrane-bound proteins complexes. They also

[72] For a history of models and ideas on membrane structure starting in the nineteenth century with the views of Schwann and continuing through the twentieth century, see Robertson 1981.

[73] These techniques were applied to mitochondria by Packer 1973 and discussed in Prebble 1981, pp.99ff. They showed an inner membrane with large numbers of protein particles on the inner half (around 4200/μ²) and a smaller number on the outer half (ca. 2100/ μ²).

[74] Singer and Nicolson 1972.

enabled bioenergeticists to visualize the interactions of the protein complexes of the respiratory chain.

5.10 ION MOVEMENTS ACROSS THE MITOCHONDRIAL MEMBRANE

5.10.1 Inorganic Cations

Around 1960 the understanding of oxidative phosphorylation began to undergo a conceptual change when bioenergeticists realized that ion movements into and out of the mitochondrion were an integral part of the mechanism and could no longer be treated as side reactions. Questions about uptake of inorganic ions initially arose in the 1950s when it was shown that isolated mitochondria retain ions such as potassium and magnesium.[75] The binding of potassium was shown to be strongly influenced by oxidative phosphorylation.[76] Robertson's view on the likely links between research in oxidative phosphorylation and research in "active transport" were noted earlier.[77] Lehninger commented that

> Much of the recent interest in this field [mitochondrial ion transport] attaches to the relationship between active ion transport and oxidative phosphorylation, particularly to the hypothesis that energy-dependent ion movements may be the driving force for ATP synthesis.[78]

Early in the 1960s, the bioenergetics world was amazed by the ability of mitochondria to take up calcium. Respiring mammalian mitochondria could take up large amounts of calcium very rapidly.[79] Under appropriate conditions (in the presence of phosphate for example), crystals of calcium phosphate could be found in the matrix and the uptake could proceed to the point of destroying the structure of the particles. Uptake was dependent on either respiration or ATP hydrolysis as an energy source and was blocked by uncouplers such as dinitrophenol. Uptakes of strontium and manganese were rather similar to calcium uptake although magnesium uptake was different and could be shown under only some conditions. The interpretation of these results led to the conclusion that calcium uptake was driven by the high energy intermediate of the chemical theory or by the proton motive force of Mitchell's chemiosmotic hypothesis.[80] It was, however, noted that the uptake of calcium was accompanied by an ejection of protons (or an uptake of hydroxyl ions). There was in fact a clear stoichiometric relationship between calcium uptake and respiration of up to two calcium ions (Ca^{++}) taken up for each pair of electrons passing through a coupling site or up to six calcium ions per O_2 reduced, again measured under suitable conditions. Correspondingly, a little fewer than two calcium ions

---

[75] See for example Bartley and Davies 1954.
[76] Gamble 1957.
[77] Robertson 1960, quoted at the end of the previous chapter.
[78] Lehninger, Carafoli, and Rossi 1967, p.261.
[79] Vasington and Murphy 1962
[80] Lehninger, Carafoli, and Rossi1967.

were taken up for each ATP hydrolyzed when ATP was the energy source. This research directed attention to the role of ion transport in oxidative phosphorylation.

Experiments on potassium uptake by mitochondria tended to use the antibiotic ionophores,[81] valinomycin and gramicidin, which rendered membranes permeable to this ion; these compounds also uncoupled oxidative phosphorylation. In general, rather similar results to those with calcium were obtained;[82] approximately four atoms of potassium[83] were accumulated per pair of electrons passing through a coupling site, up to almost twelve per O_2 atom taken up in respiration.[84] A significant difference from calcium uptake was that potassium influx was balanced by a significant efflux. Such studies were valuable in elucidating the relationship of ion transport to respiration and phosphorylation. Other experiments in the absence of an ionophore showed a slow energy-dependent accumulation of potassium.

While adding to the knowledge of the physiology of mitochondria, the essential contribution of these experiments on cation accumulation was to throw a new light on the mechanism of oxidative phosphorylation. Whereas previously the link between respiration and phosphorylation was an intermediate state whether a high-energy chemical substance or an electrochemical proton gradient (or other physical or chemical states as seen by Williams or Boyer, for example), the intermediate state now had to link with cation transport. Energy from ATP hydrolysis or from respiration was able to create a cation gradient. The link with the chemiosmotic hypothesis was significant, as Azzone and Massari observed:

> The ion transport field has been markedly influenced by the chemiosmotic model, since the process of ion translocation has been thought of as the most suitable test for the idea that the primary energy-conserving step involves a membrane potential.[85]

5.10.2 Anion Movements

Since early work had pointed to the operation of the citric acid cycle, fatty acid oxidation, and some other key metabolic processes as being properties of the mitochondrial matrix, the question of transport across the inner membrane became a subject for study. A number of specific systems for the transport of dicarboxylic acids (malate, succinate, etc.), citrate and α-ketoglutarate among others, were identified during the 1960s.[86]

---

[81] An ionophore is a molecule that is able to transport an ion such as potassium, K^+, across a lipid membrane.

[82] Harris, Cockrell, and Pressman 1966.

[83] Note that calcium is a divalent ion and potassium is monovalent so that twice as many potassium ions as calcium ions would be expected to be moved.

[84] Azone and Massari 1973. This result was later confirmed in Lehninger's laboratory.

[85] Azzone and Massari 1973, p.221.

[86] The translocating systems in mitochondria were initially described by Chappell; see Chappell 1968. For an early review, see Klingenberg 1970.

However, several of these systems were found to have an important energetic aspect in that transport across the membrane was driven by the energy systems of the mitochondrion associated with oxidative phosphorylation, directly or indirectly. A particular example is phosphate, in which the internal concentration of phosphate was higher than that outside the mitochondrion and in which the phosphate carrier was linked to the proton gradient. Consequently it became clear during the early 1970s that the general process of metabolism was intimately linked with the respiratory phosphorylation system through the operation of membrane translocators.[87] These translocators were seen as proteins located in the inner membrane of the mitochondrion.

Of particular interest in oxidative phosphorylation was the protein responsible for the transport of adenine nucleotides across the inner mitochondrial membrane.[88] This was studied by several groups but particularly by the German bioenergeticist Martin Klingenberg (1928–).[89] The adenine nucleotides, ADP and ATP, were shown to enter the uncoupled mitochondrion without preference, but in coupled mitochondria, there was a preference for the entry of ADP. This situation, which complements the metabolic function of the mitochondrion where ADP enters, is phosphorylated and exits as ATP.[90] The protein responsible for the translocation was found to be present at high concentrations in the inner mitochondrial membrane and became a major subject for investigation in the 1980s.

In a review of the field in 1979, Kathryn LaNoue and Anton Schoolwerth noted the consistency of membrane transport studies with Mitchell's hypothesis:

> One striking feature of the findings is their essential harmony with important concepts of the chemiosmotic hypothesis of oxidative phosphorylation . . . The carriers appear to possess molecular mechanisms that make use of the electrochemical potential gradient of protons across the membrane to facilitate transport in either direction.[91]

It became clear that in order to properly understand the process of oxidative phosphorylation in the mitochondrion, it was also necessary to understand the role of inner membrane translocators. In the case of the adenine nucleotide and phosphate translocators, the ADP/ATP ratio and the phosphate concentration increased in the matrix, so lowering the energy required for ATP synthesis by the ATP synthase bound to the inner surface of the inner membrane. Thus these systems were both involved in the overall process of conservation of the energy of respiration in ATP synthesis, and they were also involved in understanding the heart of the mechanism of oxidative phosphorylation, the nature of the intermediate between the respiratory chain, and the ATP synthase enzyme.

---

[87] For a review of aspects of this issue as seen during the early 1970s, see Meijer and van Dam 1974.

[88] This transport system was first identified by Pfaff, Klingenberg, and Heldt 1965.

[89] See Klingenberg 1986 for an autobiographical note.

[90] Pfaff and Klingenberg 1968.

[91] LaNoue and Schoolwerth 1979, see pp.872–873.

5.11 RESOLVING THE MECHANISM

After some two decades of highly intensive research into oxidative phosphorylation in a number of laboratories, Albert Lehninger summarized the overall situation on the mechanism in 1971:

> Despite years of effort the mechanism by which ATP is generated during electron transport is still unknown. However, it is clear that the inner mitochondrial membrane is an essential element in the mechanism . . . . If this membrane is damaged or disrupted, then oxidative phosphorylation of ADP no longer occurs, although electron transport may still take place
> . . . . The *chemical coupling hypothesis* holds that oxidative phosphorylation is catalyzed by a sequence of enzymes acting consecutively through common chemical intermediates, similar to the consecutive reactions of glycolysis. The membrane according to this hypothesis must supply the framework for bringing together the enzymes
> . . . . The *chemiosmotic hypothesis*, on the other hand, holds that electron transport pumps protons across the membrane, and that the gradient of H^+ ions so produced is the immediate driving force for ATP formation. The third hypothesis, called the *conformational hypothesis*, holds that electron transport causes conformational changes in the membrane of the mitochondria, which are converted into phosphate-bond energy. The mechanism of ATP formation in the mitochondria is one of the most lively and challenging problems in biology today.[92]

The conformational changes referred to here are those affecting membranes such as discussed by Hackenbrock rather than the protein conformational changes of Boyer, which he appears to ignore.

The early 1970s saw the progressive demise of the chemical coupling hypothesis. Numerous proposals for the high-energy intermediate had been made but had not stood up to rigorous testing. In a lecture in 1974, Slater, who originally proposed the hypothesis in 1953, concluded,

> The gathering conviction that the high-energy intermediates in the sense envisaged by my 1953 hypothesis do not exist, and the intellectually satisfying structure built up with great brilliance and originality, by Peter Mitchell . . . have persuaded many of its fundamental correctness.[93]

Slater went on to consider that an alternative to Mitchell's hypothesis should be kept in view, namely the conformational theory of Paul Boyer. By the mid-1970s, the chemical

---

[92] Lehninger 1971, pp.95 and 98.
[93] Slater 1974, p.1153.

coupling hypothesis was effectively abandoned because its predicted intermediate could not be shown to exist.[94]

The conformational hypotheses as based on observable morphological changes linked to metabolic state of mitochondria were abandoned by the mid-1970s when it became evident that such changes could be the result of osmotic effects. In any case such changes were too slow to be a key element in the mechanism of oxidative phosphorylation. However, the proposals of Boyer involving a conformational change of a protein in the respiratory chain that would then induce a change in the conformation of the ATP synthase remained a live issue. The evidence for conformational changes lay essentially with the ATP synthase itself, although changes in the respiratory chain proteins had been observed, for example those described earlier by Chance and by Slater in cytochrome *b*. It was work on the ATP synthase enzyme, particularly in Boyer's laboratory, that had led to the conclusion that energy was used not for the synthesis of ATP itself but rather for the release of ATP from the surface of the enzyme. As Boyer became more fascinated by the mechanism of the enzyme, his enthusiasm for the conformational theory abated. In 1975, he accepted that the link between respiration and the enzyme could be as Mitchell had proposed so that the conformational changes in the ATP synthase would be induced by the electrochemical proton gradient. Such a suggestion was not acceptable to Mitchell, and Boyer still wished to keep his conformational theory as an option.[95]

Mitchell sought to persuade the bioenergetics community of the reasonableness of his hypothesis with almost a crusading zeal. However, this approach was modulated by a conviction, arising from his admiration for the philosopher Karl Popper, that every means possible should be found to falsify the theory. Obviously one means of doing so was to demonstrate oxidative phosphorylation in a soluble, nonmembranous system, and several claims were made that this had been achieved, although they were not confirmed.[96] The first impressive support for Mitchell's proposal had come from the laboratory of Andre Jagendorf, with chloroplasts showing that a proton gradient could induce ATP synthesis in the dark. A similar demonstration in mitochondria was unsuccessful in several laboratories, although Mitchell claimed to demonstrate small amounts of ATP synthesis in this way; the interpretation of his experiment was open to significant criticism.

Mitchell based his hypothesis on four postulates that can be summarized as (1) an ATP synthase that reversibly translocates protons; (2) the electron-transport chain (in mitochondria, bacteria or chloroplasts) translocates protons during oxidation; (3) exchange–diffusion systems couple the proton translocation to translocation of

[94] A form of chemical coupling hypothesis did emerge when Griffiths claimed that lipoic acid could be seen as a coupling factor between respiration and ATP synthesis (Griffiths 1976). The claim could not be confirmed.

[95] Although first suggested in a letter to Mitchell, Boyer's compromise proposal was published as Boyer 1975b. I have analyzed the correspondence between Boyer and Mitchell in which Mitchell's insistence that the proton was involved in catalytic activity at the active site of the enzyme maintained a permanent gulf between the two. Prebble 2013.

[96] See for example Green 1974, p.29, in which three laboratories are cited as demonstrating oxidative phosphorylation in preparations devoid of membranous vesicles, and the author therefore concludes that Mitchell's chemiosmotic proposal "is invalid."

anions and cations; (4) the coupling membrane is impermeable to ions.[97] By the early 1970s, experimental evidence in support of all these postulates had been obtained, though not necessarily widely accepted, so that the validity of the theory continued to be a subject for debate.

Some objected to the chemistry of Mitchell's schemes for ATP synthesis. In particular, Boyer could not see how the proton could bring about ATP synthesis and found Mitchell's detailed proposals unacceptable.[98]

> I wish to emphasize that my objections are not to the possibility that you have so elegantly raised that energy may be transmitted by proton or potential gradients but to your specific chemical suggestions as to how such proton or potential gradients might be used for ATP synthesis. I hope you will give serious consideration to the possibility that proton or potential gradients drive ATP synthesis through energy-requiring protein conformational changes . . . . To the student of enzyme mechanism, a molecular description of how ATP might be synthesized by a proton or potential gradient is a formidable problem that could take many years to solve.[99]

Among objections from several laboratories were those that came from Chance, who found inadequate proton movements in mitochondria.[100] Thus there were both experimental and theoretical objections to Mitchell's proposals. In fact, Mitchell accompanied his proposals by detailed schemes for the mechanism, and it was these schemes that were unconvincing to many and proved a major part of the problem, as in the case for the ATP synthase reaction.

However, not all Mitchell's schemes failed to convince his fellow bioenergeticists. In 1975 he published a scheme that both provided a rationale for the odd behavior of the *b* cytochromes and also showed in more convincing detail how protons might be translocated across the membrane. Named the protonmotive Q-cycle by Mitchell, it used a cyclical electron flow involving ubiquinone and cytochrome *b* to translocate two protons per electron in complex III. In the cycle, the quinone was both reduced by cytochrome *b* on the inside of the inner membrane and oxidized by cytochrome *b* on the outside. The proposal was met with some skepticism initially, but over several years the principles it embodied were confirmed (see Chapter 8).[101]

---

[97] Mitchell 1966a. See p.494 for a brief summary of the four postulates.

[98] Boyer published a detailed objection to the chemistry proposed by Mitchell to explain how the proton was involved at the active site of the ATP synthase (Boyer 1975a).

[99] Letter, Boyer to Mitchell, November 21, 1974 (note that this is one of two letters of this date). P. D. Mitchell archive, Cambridge University Library. As Boyer comments in his autobiographical essay, "I could not see a satisfactory way to use the protonmotive force to make ATP. Mitchell's suggestions for ATP formation were without experimental support and seemed chemically unattractive to me" (Boyer 1981, pp.236–237). Objections to Mitchell's chemistry also came from R. J. P. Williams (see Prebble 2013).

[100] Chance and Mela 1966.

[101] Mitchell 1975a,1975b. Some ideas underlying the role of the *b* cytochromes in the cycle can also be found in Wikström 1973, see particularly pp.176ff. Professor Wikström advises me that in

In the 1970s, the overall assessment of the chemiosmotic proposal began to change. Particularly because of his reconstitution experiments, Racker had begun to conclude that the chemiosmotic proposal was along the right lines. Slater had concluded that the chemical hypothesis was no longer viable. Boyer was prepared to consider a scheme in which the chemiosmotic electrochemical proton gradient was the link between the conformational changes in the ATP synthase and the respiratory chain while still wishing to keep his conformational coupling proposal on the table. Racker summarized the situation in a letter to Mitchell:

I feel that a new confrontation between the "conformational" and the chemiosmotic hypothesis is ahead of us. I thought we could forget about the original conformational hypothesis of 1965 without holding a wake and accept a new version which centers on the oligomycin-sensitive ATPase.[102]

Mitchell responded:

David's letter of 28 May (and a private letter that I received from Paul) shows that we would be unwise to forget about the original conformational hypothesis of 1965. The original conformational hypothesis of coupling between the redox (or photoredox) chain and the reversible ATPase system, and David's recent electromechanicochemical model that is an "outgrowth" from it, are both fundamentally different from the chemiosmotic coupling hypothesis.[103]

Probably what finally led to the resolution of the mechanism question started with a practical concern of Racker. He wrote to the leaders in the field:

Many investigators in bioenergetics have felt for some time that our field has not received the appreciation it deserves. This has led to problems with the funding of grants . . . . It is my impression that among the major reasons for the unfavorable image are the rather confusing controversies which have dominated some sessions at the Federation meetings as well as published records of contradictory data and conclusions. Although we all know that this happens in other areas of

---

Amsterdam with Berden in 1971–1972, he described to Mitchell his "idea that ubiquinol may be oxidised by bifurcated electron transfer, i.e. transfer of one of its electrons to the high-potential chain (cytochrome c) and the other to cytochrome b with a semiquinone intermediate." (Email to the author, July 14, 2001.) Although there is no reason to doubt the claim, no published evidence associated with the last point has been offered.

[102] Letter, Racker to Mitchell, May 28, 1974 (copied to ten others). P. D. Mitchell archive, University of Cambridge Library. The "original" conformational hypothesis is that which regarded protein conformational changes as constituting the high-energy intermediate between respiration and ATP synthesis. The second hypothesis, "new version," concerned the mechanism of the ATPase.

[103] Letter, Mitchell to Racker, June 13, 1974 (copied to numerous other workers). P.D. Mitchell archive, University of Cambridge Library. Those referred to are David Green and Paul Boyer.

science also, it is the consensus of our biochemical colleagues that we have had a rather excessive share of it.[104]

He went on to make suggestions including the proposal for a published note describing in broad terms the present state of our knowledge signed by an international group of bioenergeticists.

The letter received a somewhat discouraging response from Mitchell and a mixed response from the other recipients. However, Lars Ernster, who had made a similar suggestion 18 months earlier, suggested with Boyer, a member of the editorial board of the *Annual Review of Biochemistry*, that a review should be prepared for that periodical. There followed an extensive debate, mainly by correspondence, leading to a final agreement that the review should involve six independent articles by Boyer, Chance, Ernster, Mitchell, Racker, and Slater, covered by a short simple introduction.[105] This was published in 1977 and was seen by many people as the resolution of the conflict.

The major hypotheses discussed were Boyer's conformational and Mitchell's chemiosmotic but Williams's proposal was considered by most contributors. The chemical hypothesis was included by some for historical reasons, and the proposals of Green were largely ignored. Mitchell and Racker clearly supported the chemiosmotic proposal. Ernster felt that the current state of the evidence led him to conclude

> Energy transfer between energy-transducing units may occur by direct conformational interactions or intramembrane proton fluxes [Williams's hypothesis] in addition to bulk transmembrane electrochemical potentials.[106]

The views of Boyer and Slater were rather similar. Boyer stressed his proposals for the ATP synthase and indicated that they would be necessary for the chemiosmotic proposal to explain how ATP was synthesized. Slater wrote in summary:

> The feasibility of the essential features of the Mitchell chemisosmotic hypothesis is now firmly supported experimentally. The main question to be settled is whether, in addition to the link between electron transfer and ATP synthesis that is provided by the fact that both are coupled to proton translocation across the membrane in the same direction, a more direct functional and mechanistic link

---

[104] Letter, Racker to Mitchell, March 11, 1974. P. D. Mitchell archive, University of Cambridge Library. It was copied to Boyer, Chance, Ernster, Green, King, Lardy, Lehninger, Sanadi, and Slater. A more extensive quotation from this letter can be found in Prebble and Weber 2003, pp.196–197; where a fuller account of the debates and arguments behind the review can also be found. It should be noted that the sociologists Gilbert and Mulkay recorded a rather similar observation on the strains in the field where one of their interviewees had commented on the strong and often personal antagonisms evident in scientific meetings in the United States. (See Section 9.7 for a quotation from the interview.) See Gilbert and Mulkay 1984, p.36.

[105] Boyer et al. 1977. It was agreed that Lehninger's views would be represented by Chance. The other obvious omission was David Green.

[106] Boyer et al. 1977, p.992

exists between the two systems in mitochondria and chloroplasts. It is also important to establish the stoichiometry of the proton-translocating systems.[107]

It was this last point that particularly concerned Chance, who endeavored to avoid consideration

> of the controversial or fixed viewpoints of several hypotheses. At the same time it must be emphasized that the mechanism of membrane-bound energy-coupling processes remains one of the most challenging and as yet unsolved problems of modern biology, whose ultimate solution at a level of mechanistic detail acceptable to physical chemists demands information on structures, pathways and functional intermediates.[108]

Chance was critical of the numbers of protons per electron passing through the respiratory chain being transported outward in mitochondria whereas Mitchell's schemes were based on $H^+/e = 1$ at each coupling site. He found this unacceptable on thermodynamic grounds and also inconsistent with some experimental results.

The proton stoichiometry became a major issue for Mitchell in the early 1980s.

Overall it is possible to see some broad agreement (probably excepting Chance) that the chemiosmotic hypothesis constitutes the basis on which the mechanism of oxidative phosphorylation might be understood. There is, however, a significant current of opinion that the protein conformational theory for energy transmission from the respiratory chain to the ATP synthase cannot be ruled out, a view underlying the contributions of Boyer, Ernster, and Slater.

The following year, the Nobel Foundation awarded its prize in chemistry to Peter Mitchell "For his contribution to the understanding of biological energy transfer through the formulation of the chemiosmotic theory." The Nobel Committee for Chemistry had been chaired by Bo Malmström, who later commented that

> The committee was well aware of the fact that Mitchell was wrong about how the proton gradient is created (proton pumps rather than redox loops) and how it is utilized to make ATP (conformational coupling rather than mass action). This is why the academy's citation read rather vaguely.[109]

The award of the prize to Mitchell confirmed in the eyes of most scientists the rightness of the chemiosmotic approach. Although few were prepared to argue that Mitchell should not have been awarded the prize, there were several who felt that it should have been shared. Probably their principal candidate would have been Efraim Racker.[110]

---

[107] Boyer et al. 1977, p.1024.
[108] Ibid., p.968.
[109] Malmström 2000, p.356.
[110] See Schatz 1997.

6

Discovering Photosynthesis

Any discussion of the history of cellular bioenergetics would be incomplete without a consideration of the process of photosynthesis in plant chloroplasts and photosynthetic bacteria.[1] Indeed there are instances in which research in photosynthesis has made an essential contribution to understanding mitochondrial bioenergetics.

The word "photosynthesis" did not come into use until the end of the nineteenth century.[2] Although the early understanding of photosynthesis and respiration (eventually oxidative phosphorylation) initially developed from some of the same observations in the eighteenth century, their stories subsequently diverged. Because of the challenge of the role of light, research in the two fields followed rather different paths until the late 1950s when their interconnectedness began to be fully realized. Indeed, the first independent experimental support for the chemiosmotic hypothesis (discussed Section 5.3) came from chloroplast studies in the mid-1960s.

6.1 THE DEVELOPMENT OF IDEAS ON PHOTOSYNTHESIS

The story told in this chapter involves three periods of research in photosynthesis up to the end of the 1950s. The first period covers the initial studies in the eighteenth century and the early nineteenth century. This was concerned with the discovery of the fundamental properties of the process, namely the realization that green plants are capable of assimilation of CO_2 and water with the evolution of O_2. The process was found to require light. The second period, stretching over much of the nineteenth century and the early twentieth century, was marked by studies on chlorophyll, establishing its chemistry and role in the process. The period also included appreciation of photosynthesis in three small groups of photosynthetic bacteria that did not evolve O_2. However, although much experimentation led to only a very limited understanding of the process, the significant finding in the early twentieth century was that photosynthesis could be divided into a fast light reaction and a slow dark reaction. Furthermore, theoretical ideas derived from a study of photosynthetic bacteria, led to the view that the light reaction associated with

---

[1] Table 1.1 includes a summary time line for photosynthesis but a detailed time-line for the history of O_2 evolving photosynthesis can be found in Govindjee and Krogmann 2004.
[2] See Gest 2002.

chlorophyll might be an oxidation–reduction. This insight provided a valuable introduction to the modern study of photosynthesis.

The third period began with experiments that showed that the photochemical process was indeed an oxidation–reduction. This work introduced the first successful use of chloroplast preparations. Following these groundbreaking studies, the carbon reduction (Calvin) cycle was elucidated and photophosphorylation was demonstrated. The 1950s saw the relationships of the main features of photosynthesis fully realized, the fixation of CO_2, photosynthetic formation of ATP, and O_2 evolution. Although the importance of the chloroplast had become apparent earlier, its full significance awaited the understanding of these relationships. It was also in this period that studies of photosynthetic bacteria began to make a unique contribution to the understanding of the mechanism of photosynthesis.

6.2 INITIAL STUDIES OF PHOTOSYNTHESIS

Although others preceded him, the English natural scientist and perpetual curate of Teddington, England, Stephen Hales (1677–1761), was probably the first to show some understanding of the process of photosynthesis. In his book *Vegetable Staticks* in which he described a large number of experiments, he appears to have recognized a role of leaves in gas exchange and compares them with the lungs of animals: "Plants very probably drawing thro' their leaves some part of their nourishment from the air."[3] He also recognized a role for light, noting that a rather similar suggestion had been made by Isaac Newton in "Queries" in his *Opticks* (1704):

> And may not light also, by freely entring the expanded surfaces of leaves and flowers, contribute much to the ennobling the principles of vegetables? For Sir Isaac Newton puts it as a very probable query, "Are not gross bodies and light convertible into one another and may not bodies receive much of their activity from the particles of light, which enter their composition?"[4]

Hales's work was certainly influential and, as noted in Chapter 2, was carefully considered by Lavoisier and also by Ingen-Housz. There were other early experimentalists who noted the importance of light. Charles Bonnet (1720–1793), a Swiss naturalist from Geneva, had observed that shoots of grapevine immersed in water became covered with numerous air bubbles as long as they were in sunlight. The phenomenon disappeared after dark.

Joseph Priestley (1733–1804), an English experimental chemist, nonconformist minister, and theologian, was one of those influenced by Hales's work. Priestley can be regarded as taking the first significant steps in elucidating the biochemistry of photosynthesis. Among his many experiments, he showed unambiguously that green vegetation

[3] Hales 1727, p.325
[4] Ibid., p.327.

restored the quality of air in which candles had burnt out. In his classic paper to the Royal Society he recorded this:

> Accordingly on the 17th August, 1771, I put a sprig of mint into a quantity of air, in which a wax candle had burned out, and found that, on the 27th of the same month, another candle burned perfectly well in it.

In summary,

> Experiments made in the year 1772, abundantly confirmed my conclusion concerning the restoration of air, in which candles had burned out by plants growing in it.[5]

The experiment, which he repeated several times, was understood in terms of the phlogiston theory, an understanding he never abandoned.[6] But Priestley was primarily an experimentalist and seems not to have been overly concerned about interpreting the significance of his work. However, it was clear to his successors that Priestley had observed the evolution of O_2 in green plants and its use in combustion.

Some[7] have seen the Dutch physician Jan Ingen-Housz (1730–1799) as the discoverer of photosynthesis.[8] Apart from this distinction, Ingen-Housz achieved a reputation as an expert in the technique of inoculation against smallpox and became court physician to Austrian Empress Maria Theresa. In 1779, he spent the summer at a country house near London carrying out an intensive series of experiments on plants. The results were published in the same year in a book, *Experiments Upon Vegetables*. He confirmed many of Priestley's results, but probably his major contribution was to show that light was essential in Priestley's type of experiment, a point that the latter seems to have missed:

> I observed that plants not only have a faculty to correct bad air in six or ten days, by growing in it, as the experiments of Dr Priestley indicate, but that they perform this important office in a compleat manner in a few hours; that this wonderful operation is by no means owing to the vegetation of the plant, but to the influence of the light of the sun upon the plant.[9]

This achievement is summed up in a letter (dated October 17, 1779) Priestley wrote to a friend, Giovanni Fabroni, after reading *Experiments Upon Vegetables*:

---

[5] Priestley 1772, pp.168–169.

[6] It should be noted that later Priestley had some difficulties in repeating this experiment. The Swedish chemist Carl Wilhelm Scheele (1742–1786) was also unable to confirm Priestley's result. In both cases this was probably because of a failure to appreciate the importance of light.

[7] See Gest 2000.

[8] The discovery of plant respiration is also attributed to him.

[9] Ingen-Housz 1779, pp.xxxiii–xxxiv.

The things of most value that he [Ingen-Housz] hit upon and I missed are that leaves without the rest of the plants will produce pure air and that the difference between day and night is so considerable.[10]

Ingen-Housz also clearly demonstrated that green leaves, not other parts of the plant, are the loci of photochemical activity. Unlike Priestley, Ingen-Housz, who met Lavoisier in Paris, abandoned the phlogiston theory in his later paper published in 1796. He recognized that green parts of plants evolve O_2 in the light. This paper also suggests that carbonic acid is a source of carbon in plants. Ingen-Housz argues that the source of plant nourishment must be in the soil or the air or in both: "We ought to take it for granted that the soil or what exists in the soil is not the only food of plants." This leads him to argue that carbonic acid from the air is a source of plant nourishment.[11]

A Swiss pastor, Jean Senebier (1742–1809) from Geneva, a contemporary of Ingen-Housz, published several works on vegetable physiology.[12] In 1782, he published his book *Mémoires physico-chemiques sur l'influence de la lumière solaire pour modifier les être des trois règnes de la nature et surtout ceux du règne vegetal* [*Physicochemical Memoirs of the Influence of Sunlight on the Changes of the Three Kingdoms of Nature, Particularly the Plant Kingdom*], which confirmed many of Ingen-Housz's observations, particularly the role of light. Of special significance in Senebier's work was his observation that carbonic acid (CO_2) was essential for the evolution of O_2 in light. This tended to confirm Ingen-Housz's suggestion that the origin of plant carbon was CO_2. Senebier interpreted this as the dissociation of the O_2 from carbonic acid in sunlight. Thus he saw CO_2 as essential to plant growth.

These results of careful though simple experimentation by eighteenth-century workers provided a basic and elementary understanding of photosynthesis that formed the basis for the researches of nineteenth-century investigators.

6.3 THE IMPORTANCE OF WATER AND CO$_2$

In many ways more remarkable were the conclusions of another Swiss scientist from Geneva, Nicolas Théodore de Saussure (1767–1845),[13] who published his work in a book in 1804, *Recherches chimiques sur la vegetation* [*Chemical Research on Plants*]. He carried out many careful analyses of plant growth by weighing his experimental material. These studies confirmed the earlier proposals by Ingen-Housz and Senebier that plants find an adequate supply of carbon from the air. However, his results showed that the increase in dry weight associated with the assimilation of carbon from the atmosphere was considerably greater than that of the carbon itself. This he attributed to water taken up by the plant and not removable by drying. So de Saussure's work completed the basic understanding

---

[10] Extract quoted in Gest 2000. Gest also makes it clear that later Priestley disputed that he had failed to appreciate the role of light.
[11] See Gest 1997.
[12] For a brief note on Senebier, see Bay 1931.
[13] See Hart 1930 for a biographical note on de Saussure.

of photosynthesis showing the importance of CO_2 and particularly water so that the process could be represented as

$$light$$
$$\text{Carbon dioxide + water} \rightarrow \text{plant carbon + oxygen}^{\cdot}$$

By the beginning of the nineteenth century, the basic overall mechanism of photosynthesis was established.

Although Justus Liebig was somewhat critical of the work of many plant scientists, he admired the work of de Saussure. He approached the problem of carbon assimilation from a more theoretical position. From the argument that respiration and ordinary combustion continually add to the CO_2 of the atmosphere and the fact that the level of CO_2 remains more or less constant (using de Saussure's measurement of CO_2 in the atmosphere), he concluded that

> The facts which we have stated in the preceding pages prove that the carbon of plants must be derived exclusively from the atmosphere. Now, carbon exists in the atmosphere only in the form of carbonic acid and therefore in a state of combination with oxygen.[14]

Such an approach confirmed the role of photosynthesis in maintaining the atmospheric CO_2 concentration.

6.4 ENERGY

Early in the second period of photosynthetic research, an important contribution came from the physicists. The Role of light in the process had been clearly demonstrated in the eighteenth century but its significance remained to be appreciated. This began to be clarified by a group of mid-nineteenth-century scientists interested in energy, including James Joule, Hermann von Helmholtz, particularly Robert Mayer, and later Ludwig Boltzmann (1844–1906). Of great relevance to the process of photosynthesis from the 1840s onward was the appreciation of the conservation of energy (or force) by Mayer[15] and others. Consideration of these issues included the importance of sunlight in providing energy for the planet. In 1845 Mayer wrote,

> The task nature has set itself is to catch the earth's influx of light in flight, the most mobile of all forces, and to convert it into rigid form, storing it. To achieve this, it has attracted over the earth's crust organisms that absorb sunlight and generate continuous chemical processes using this power. These organisms are plants, the plants form a reservoir, in which fleeting solar rays are fixed for beneficial use, the

---

[14] Liebig 1843, p.15.
[15] See Chapter 2.

economical care to which the physical existence of the human race is inseparably tied. The plants take a force, light, and bring forth a force, the chemical process.[16]

Such views were certainly not unique to Mayer. In 1847 Hermann von Helmholtz also took the same approach to the importance of plant photosynthesis in understanding energetics:

> In plants the processes are mainly chemical, although we also find at least in some cases, a slight development of heat occurring. In particular there is a vast quantity of chemical tensional forces stored up in plants, the equivalent of which we obtain as heat when they are burned. The only *vis viva* [living energy] which we know to be absorbed during the growth of plants, is that of the chemical rays of sunlight.[17]

Hence, from the middle of the nineteenth century, photosynthesis could be considered as a process that converted light energy into chemical energy, although the appreciation of what this might mean in molecular terms had to wait until the middle of the next century.

Another physicist to be concerned with the physical aspects of photosynthesis was the Austrian, Ludwig Boltzmann. In his 1886 paper concerned with entropy, he wrote

> To make the most use of this transition [the energy transition from the hot sun to the cold earth] green plants spread the enormous surface of their leaves and in a still unknown way, force the energy of the sun to carry out chemical syntheses . . . . The chemical syntheses are to us in our laboratories complete mysteries.[18]

The principles of photosynthesis were clear but the mechanism remained a mystery. However, appreciation of the significance of energy would be essential in clarifying some of the questions that bioenergeticists would need to ask in order to elucidate the mechanism of photosynthesis.

6.5 DISCOVERING CHLOROPHYLL AND CHLOROPLASTS

Probably the major concern in this field during the nineteenth century was the nature and role of the ubiquitous green substance in leaves, and these constituted a second phase of photosynthesis research. Ingen-Housz had commented, referring to the ability of plants to make the atmosphere more fit for animal life,

---

[16] Mayer 1845. See Rabinowitch 1945, p.25. Note that Mayer used the term force where later energy would be used.

[17] Helmholtz 1847; see Kahl 1971, p.48.

[18] Boltzmann 1886, p.41.

that this office is not performed by the whole plant but only by the leaves and the green stalks that support them.[19]

The nature of the green pigments of plants that Ingen-Housz had identified as essential to photosynthesis was pursued by two French chemists working in Paris. Joseph Pelletier (1788–1842) and Joseph Caventou (1795–1877) appear to have been the first to isolate the green pigment. They had previously isolated a number of substances, particularly alkaloids, from plant materials. In 1817 they extracted the green substance from leaves and gave it the name chlorophyll. Further noteworthy work on the chemistry of chlorophyll was not carried out until Richard Willstätter's studies nearly 100 years later.

The nineteenth century saw the development of ideas about the chloroplast, the organelle for photosynthesis. Because of its much larger size than that of the mitochondrion, it was much more accessible to microscopic observation. The German botanist Hugo von Mohl (1805–1872) at the University of Tübingen was a good light microscopist, and, in 1837, he noted the presence of chlorophyll in cell organelles, which he called granules. Mohl also noted the association of starch with his chlorophyll granules,[20] an observation confirmed by others including the Swiss botanist, Carl Wilhelm Nägeli (1817–1891), who noted in 1858 that starch granules were frequently embedded in chlorophyll granules. In 1883, the term plastid was introduced by Andreas F. W. Schimper (1856–1901) for the chlorophyll granule, but at the beginning of the twentieth century this particle became known as the chloroplast. Arthur Meyer (1850–1922), a cell biologist at the University of Marburg, pioneered early work on chloroplast structure and described the green chlorophyll-containing bodies within chloroplasts, and visible under the light microscope, as grana (sing. granum).[21]

The German plant physiologist Julius Sachs (1832–1887), professor of botany at the University of Würzburg from 1868, brought together some of the earlier ideas about photosynthesis in his experiments described in the 1860s. His main contribution was to identify a role for sunlight in the formation of starch in chlorophyll granules (chloroplasts). Unilluminated parts of the plant did not contain starch, as judged from the iodine test that gives a blue color with starch. As he wrote in 1862,

> Through experiments . . . it can now be shown . . . *that for the production of starch in the chlorophyll bodies, light of higher intensity is necessary,* that for instance light in the interior of an ordinary living room is not sufficient . . . . I do not wish with these words to say that the starch in the chlorophyll so arises that from carbonic acid and water by elimination of oxygen immediately ready starch is formed. Rather the possibility remains open that here, inside the chlorophyll bodies themselves a longer series of chemical reactions sets in. With this assumption the single characteristic circumstance must be emphasized that here the process begins with inorganic substances and ends with the production of starch, so that

[19] Ingen Housz 1779, p.xxxv.
[20] Mohl 1837.
[21] Meyer 1883.

one can consider the starch produced here to be the first formed from inorganic substances [carbonic acid, water, mineral salts].[22]

Thus Sachs demonstrated the requirement for light in the synthesis of starch. Significantly he saw the possibility that the chloroplast could be a metabolic unit for starch synthesis from CO_2 and water, an activity involving a chain of metabolic events and for which an adequate level of illumination was necessary.

This understanding of the role of the chloroplast in photosynthesis was taken a step further by the German physiologist Theodor Wilhelm Engelmann (1843–1909), professor of physiology in Utrecht, later in Berlin, and a friend of the composer Johannes Brahms, particularly during his time in Holland.[23] Apart from major contributions to muscle physiology, Engelmann developed the use of motile bacteria that in suspension moved to regions of higher O_2 concentration. Using these bacteria, he found that O_2 production was associated with the green parts of algae. From his observations, he concluded that

> The chlorophyll-containing cells give off oxygen in the light and it is this which allows the bacteria to move and assemble at the source of oxygen. In the dark liberation of oxygen ceases and the oxygen lack resulting from the rapid oxygen consumption by the bacteria now brings the movements to an end.[24]

He employed these bacteria to show where O_2 was being evolved in a filamentous alga illuminated with the visible spectrum. The bacteria congregated at the red and blue parts of the spectrum (see Fig. 6.1), regions where chlorophyll was known to absorb light. Thus he used this early action spectrum to show the role of chlorophyll in photosynthetic O_2 production.[25]

That the chloroplast possessed a number of pigments began to be recognized in the second half of the nineteenth century. The Cambridge professor of mathematics, George Gabriel Stokes (1819–1903), had interests that extended well beyond pure mathematics into a number of physical issues. In 1852 he published an extensive paper on fluorescence (a term he invented) that described the fluorescence of many substances including chlorophyll and particularly of alcoholic extracts of green leaves, leaf green.[26] Fluorescence assumed major importance in studies of chloroplast photochemistry from around the middle of the twentieth century. Later he showed that the green extracts of leaves contained several pigments. He commented on these:

[22] Sachs 1862, pp.370–371; see Teich 1992, pp.88–89.
[23] See Kamen 1986 for a biographical note on Engelmann, including his love of music. This also shows that Brahms dedicated a string quartet (Opus 67) to his friend Professor Th. W. Engelmann.
[24] Engelmann 1881, p.444; see Teich 1992, p.90.
[25] Note that in the 1870s, the decade before Engelmann's ingenious experiment, C. A. Timiriazeff (1843–1920), a Russian physiologist, established a red absorption maximum for chlorophyll and showed that red light absorbed by chlorophyll is efficient in photosynthesis.
[26] Stokes 1852, in particular see pp.486–492.

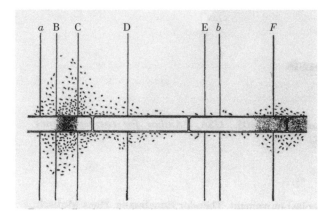

Figure 6.1 Illumination of a Filamentous Alga With the Visible Spectrum in a Suspension of Motile Bacteria (Engelmann 1882)

The diagram shows the accumulation of the oxygen-seeking bacteria in the red and blue parts of the spectrum. The unicellular alga is shown across the middle of the diagram, the red region of the spectrum lies between B and C, and the blue region is at F. Reproduced with permission from Hangarter and Gest 2004, "Pictorial Demonstrations of Photosynthesis" (*Photosynthesis Research* 80: 421–425), Springer, fig. 1, p.422.

> I find the chlorophyll of land-plants to be a mixture of four substances, two green and two yellow, all possessing highly distinctive optical properties. The green substances yield solutions exhibiting a strong red fluorescence; the yellow substances do not.[27]

The following years saw a number of other investigations of chlorophyll composition but it was the Russian botanist Mikhail Semenovich Tswett (1872–1919) who, at the beginning of the twentieth century, introduced the method of column chromatography. This technique enabled him to demonstrate the existence of two chlorophyll pigments, chlorophyllin α and chlorophyllin β together with yellow pigments, xanthophylls:

> The green pigment of leaves, chlorophyll, is recognized as a pigment mixture of varying complexity as estimated by different investigators. Chromatographic analysis is qualified to finally determine the degree of this complexity.[28]

The composition of the pigments of photosynthesizing systems was to a large extent resolved into two chlorophylls and two yellow carotenoids, although this subsequently

---

[27] Stokes 1863, p.144.

[28] Tswett 1906, p.388. See Strain and Sherma 1967, p.240. Tswett's work was significant not just for the plant pigments but in terms of the introduction of the chromatographic method to the chemistry of natural products and particularly in terms of its development in biochemical methodology. The underlying principles were exploited in a variety of ways in the second half of the twentieth century.

proved too simple a view. However, it did not carry total conviction without knowledge of the chemical structures.

A major contribution to the knowledge of chlorophyll was made by the German natural products chemist, Richard Martin Willstätter (1872–1942), who worked in Berlin, Zurich, and Munich. He showed from a wide range of analyses that plant chlorophyll was composed of two substances, chlorophyll a and chlorophyll b. Together with coworkers, he determined the composition of chlorophyll a and b[29] and also the yellow leaf pigments, carotene and xanthophyll. He was awarded the Nobel Prize in Chemistry for his work on plant pigments, especially chlorophyll.[30] This work was subsequently extended by studies of the structure of chlorophyll by the German chemist Hans Fischer (1881–1945), who was also awarded a Nobel Prize in Chemistry for his work on hemin and chlorophyll in 1930. The structure was not finally confirmed until 1960, when Robert Woodward's group achieved a total synthesis of chlorophyll a.[31]

6.6 UNDERSTANDING THE NATURE OF PHOTOSYNTHESIS

The early studies of photosynthesis successfully showed that the process was concerned with the assimilation of CO_2 and water in light, and the presence of chlorophyll and that O_2 was evolved. Starch could be seen as a product of the process. However, the mechanism for the process remained obscure although there were several attempts in the second half of the nineteenth century to propose a plausible mechanism.

One influential proposal was put forward by Adolf von Baeyer[32] (1837–1917), a German chemist who had taught in Berlin and Strasbourg and then succeeded Justus Liebig as professor of chemistry in Munich. Baeyer was interested in condensation reactions and considered that, in photosynthesis, several molecules of formaldehyde might be condensed to form sugar, a reaction that could be demonstrated in the laboratory. In 1870 he suggested that the photochemical reaction could be the formation of formaldehyde and O_2 from CO_2 and water (Fig. 6.2).[33]

Although supported by others, this proposal was only one of several suggestions for which solid evidence could not be obtained. When formaldehyde was found to be toxic to plants the proposal became more problematic but was not abandoned.

A number of mechanisms for the photosynthetic process emerged during the early part of the twentieth century. For example, Willstätter and Stoll proposed a mechanism

---

[29] Willstätter and colleagues determined the composition of chlorophyllin a, $C_{32}H_{30}ON_4Mg$ (COOH)$_2$, and chlorophyllin b, $C_{32}H_{28}ON_4Mg$ (COOH)$_2$. They noted that the chlorophylls are esters of methanol and the long-chain alcohol phytol. (See Willstätter and Stoll 1913.)

[30] The Nobel Prize for 1915 was awarded to Willstätter "for his researches on plant pigments especially chlorophyll."

[31] Woodward et al. 1960. Robert Woodward (1917–1979) was awarded the Nobel Prize in Chemistry in 1965 "For his outstanding achievements in the art of organic synthesis."

[32] Baeyer made many contributions to organic chemistry and was awarded the Nobel Prize for Chemistry in 1905 for "his researches on organic dyestuffs and hydroaromatic compounds."

[33] Baeyer 1870.

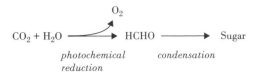

Figure 6.2 A Nineteenth-Century View of Carbon Fixation in Photosynthesis as Understood by Adolf von Baeyer in 1870

involving a reaction between the chlorophyll molecule and CO_2 to give a chlorophyll–carbonic acid compound. This compound would undergo a photochemical reaction, releasing O_2 and generating formaldehyde, which was seen to be the basis for carbohydrate synthesis.[34] This approach was strongly supported by the observation that colloidal solutions of chlorophyll in water absorb CO_2. Proposals that the basic mechanism of photosynthesis was the photochemical reduction of CO_2 to formaldehyde were probably the majority view in the early part of the twentieth century. Formaldehyde was seen as the precursor of sugars, and such a process could be demonstrated chemically in the absence of light. However, despite extensive experimentation, evidence that a process of this type was involved in plant photosynthesis failed to materialize.[35]

At the beginning of the twentieth century, an observation by the Czech-Austrian botanist Hans Molisch (1856–1937) proved influential some 30 or so years later. Working in Prague and Vienna (later in Japan), Molisch observed the slow evolution of O_2 from dried leaf powders when they were ground up in water and illuminated.[36] The significance of this work was later realized by Robin Hill.

It was Otto Warburg who also promoted the idea of a photochemical reaction involving primarily CO_2.[37] Warburg's contribution to the field was significant, partly because he introduced his manometric methods (the Warburg apparatus for measuring gas changes in metabolic reactions) into the study of photosynthesis.[38] He also initiated the use of a unicellular alga, *Chlorella*, for his investigations, an organism ideally suited to manometry as suspensions of cells were stable and easily handled.[39] He was convinced that the mechanism of photosynthesis was based on a reaction involving the photolysis of CO_2 that had been combined with chlorophyll. As his biographer Hans Krebs noted,

[34] Willstätter and Stoll 1918.
[35] See for example Dhar 1935 for a discussion of the problem as it then appeared in the early 1930s.
[36] Molisch first described his experiments in 1904 but returned to them in Molisch 1925. For a comment on Molisch, who also repeated Engelmann's experiments, see Gest 1991. See Section 6.9.
[37] Warburg's contribution to photosynthesis is set in context by Nickelsen 2009, who also considers the influence of his father, the physicist Emil Warburg.
[38] See Chapter 3 for a discussion of Warburg and his modification of the Haldane–Barcroft method, which became known as the Warburg apparatus.
[39] For a consideration of the importance of unicellular algae in earlier photosynthesis research, see Zallen 1993b.

Warburg remained convinced that the light reaction is a photolysis of CO_2 ... and he chose to ignore the powerful argument that photosynthetic bacteria, which can also convert CO_2 to carbohydrate in the light do not produce O_2 [see subsequent discussion].[40]

The general view of the mechanism for CO_2 assimilation was that carbonic acid (H_2CO_3) would form a complex with chlorophyll that would then rearrange to form a peroxide. This would break down to form O_2:

$$H_2CO_3 \rightarrow H_2CO.O_2 \rightarrow (CHOH) + O_2.$$

The resulting carbohydrate is shown as (CHOH).

To determine the number of photochemical events occurring in the fixation of a molecule of CO_2, in 1922 Warburg, together with Negelein, endeavored to measure the number of quanta[41] required. He obtained a value of about four.[42] This made photosynthesis an extremely efficient process and left a puzzle of how four photochemical events led to the fixation of one molecule of CO_2. Later, in the early 1940s Emerson and Lewis obtained values of approximately eight to twelve.

For 60 years, most workers in the field of photosynthesis believed that the basic reaction concerned the photolysis of CO_2. Few regarded the light reaction as an oxidation–reduction so that progress in the field was very slow. The situation began to change in the 1930s to one in which light-driven oxidation–reduction was the key to understanding the mechanism.

6.7 PHOTOSYNTHETIC BACTERIA AND AN OXIDATION–REDUCTION MECHANISM

In the early part of the twentieth century the understanding of biological oxidation had undergone a revolution led by Heinrich Wieland (see Chapter 3). Thorsten Thunberg had been a major contributor to the new understanding and had suggested (among other possibilities) that photosynthesis itself might involve a photochemical splitting of water, generating hydrogen that would reduce CO_2 to form formaldehyde.[43] In 1926 the Dutch microbiologist Albert Jan Kluyver (1888–1956) with H. J. L. Donker built on the ideas of Wieland and Thunberg. They proposed that water should be considered as the specific hydrogen donor for CO_2 reduction.[44] This view was given a much firmer basis from studies of photosynthetic bacteria carried out in the 1930s.

[40] Krebs 1972, p.655.
[41] Pigments absorb light energy in fixed amounts (quanta). Thus the *number* of quanta absorbed for a given activity (e.g., evolution of a molecule of O_2) will be the number of photochemical events necessary for the process.
[42] Warburg and Negelein 1922.
[43] Thunberg 1923.
[44] Kluyver and Donker 1926.

Toward the end of the nineteenth century, understanding the nature of photosynthesis was greatly aided by the study of a small group of bacteria which were discovered primarily by three microbiologists. Theodor Engelmann was one of the first to observe photosynthetic bacteria, although he thought they evolved O_2. He described a pigment from purple bacteria, bacteriopurpurin, as a true chlorophyll. Sergei Winogradsky (1856–1953), born in Kiev (then in the Russian Empire), was a leading microbiologist who described purple sulfur bacteria including *Chromatium*. He identified Engelmann's organism as a mixture of *Chromatium* species. Photosynthetic nonsulfur bacteria such as *Rhodopseudomonas* were described by Hans Molisch in 1907. A rather separate group, the green sulfur bacteria exemplified by *Chlorobium*, also oxidized hydrogen sulfide (H_2S) but were described somewhat later. Of the three classical groups of photosynthetic bacteria, it was the purple nonsulfur group exemplified by *Rhodospirillum* and *Rhodopseudomonas* that received the most attention initially. The photosynthetic bacteria, unlike green plants, did not evolve O_2 and were mostly anaerobic. Their chlorophyll was slightly different from plant chlorophylls and became known as bacteriochlorophyll.

The Dutch microbiologist Cornelius Bernardus van Niel[45] (1897–1985) was a student and an assistant of Kluyver, who asked him to culture iron and sulfur bacteria for teaching purposes. However, van Niel found the purple bacteria "aesthetically pleasing" and continued to work on them in Delft (Holland) and later after he had migrated to the United States in 1928 to work at the Johns Hopkins Marine Station. Van Niel was strongly influenced by Kluyver and particularly the latter's view of the importance of hydrogen transfer reactions in metabolism. He early concluded that purple sulfur bacteria carry out a novel form of photosynthesis in which O_2 is not produced and CO_2 is reduced by hydrogen derived from H_2S. These studies led him to conclude that an oxidation–reduction reaction must lie at the heart of the photochemical reaction.

Continuing his studies of the green and purple photosynthetic sulfur bacteria in the early 1930s, he concluded that the reaction for oxidation of H_2S in light might be represented as

$$CO_2 + 2H_2S \rightarrow (CH_2O) + H_2O + 2S.$$

The reaction could be generalized for both the sulfur bacteria and green plants carrying out photosynthesis in the light as follows:

$$CO_2 + 2H_2A \rightarrow CH_2O + H_2O + 2A,$$

where A is the O_2 of water in the case of green plants and the sulfur of H_2S in the case of the photosynthetic sulfur bacteria. In conclusion, van Niel commented thus:

> If one tries to understand the meaning of the generalized equation for photosynthesis it becomes clear that all those mechanisms proposed for the photosynthetic reaction which imply the formation of a carbonic acid-chlorophyll complex

---

[45] For a short biography of van Niel, see Barker and Hungate 1990.

which is subsequently transformed into a formaldehyde peroxide are not quite in accordance with the formulation of photosynthesis as an oxidation-reduction process. Such schemes fail to give a satisfactory explanation for the photosynthetic process carried out by the green and purple sulfur bacteria.

From a unified point of view, as laid down in the generalized equation, green plant photosynthesis should be considered as a reduction of CO_2 with hydrogen obtained from H_2O, and the oxygen produced during illumination as dehydrogenated H_2O.[46]

According to van Niel photosynthesis should be understood as having a light-driven oxidation–reduction reaction at its center and that the O_2 evolved is derived not from CO_2 but from water. He concluded that the fundamental reaction of photosynthesis was the photolysis of water in both green plants and bacteria.[47] Such a view brought about a major change in how the central photochemical reaction of photosynthesis should be seen, shifting the majority understanding of the core reaction from a photochemical reaction involving CO_2 and chlorophyll to the photochemical splitting of water.

This fundamental change in the understanding of photosynthesis was supported by the work of two American biochemists, Harland G. Wood (1907–1991) and Chester H Werkman (1893–1962), using heterotrophic bacteria. They initially found in 1935–1936 that propionic acid bacteria appeared to utilize CO_2 in forming 4-carbon dicarboxylic acids from 3-carbon substrates. Several reactions of this type were found, initially in microorganisms and later in mammalian systems.[48] In the 1940s, they became known as Wood–Werkman reactions. The significance of this work was that it demonstrated that CO_2 could be assimilated by simple nonphotochemical reactions. This led to the suggestion that CO_2 assimilation in photosynthesis might involve the formation of a carboxyl group by nonphotochemical means.[49] This proposal was consistent with the discovery at the beginning of the century that photosynthesis could be analyzed into light and dark reactions in which CO_2 assimilation might be a dark reaction.

6.8 LIGHT AND DARK REACTIONS

At the beginning of the twentieth century the Cambridge plant physiologist Frederick Blackman (1866–1947) examined the effect of temperature on photosynthesis. It was well known that the rate of photosynthesis increased with increasing light intensity. However, at saturating light intensities, the rate of photosynthesis could be determined by

[46] Van Niel 1935, pp.142–143.
[47] Van Niel 1941.
[48] Werkman and Wood 1942.
[49] Werkman and Wood 1942 pointed out that evidence for nonphotochemical CO_2 assimilation had been available from the end of the nineteenth century when chemoautotrophic bacteria had been described by Winogradsky. These organisms were able to synthesize their cellular constituents from CO_2 by using energy obtained from the oxidation of inorganic compounds such as inorganic sulfur compounds (*Thiobacillus*) and ammonia (*Nitrosomonas*). Despite this, the carboxylation reactions proposed by Wood and Werkman were not readily received.

temperature. Hence he was able to identify a light-dependent process and a temperature-dependent process. These he associated with the photochemical and chemical stages of the process.[50] Blackman's student George Edward Briggs referred to the temperature-dependent process as the dark reaction and Warburg referred to it as the Blackman reaction. Warburg confirmed the notion of two reactions, one the light or photochemical reaction and the other the Blackman or chemical reaction. He also used intermittent light, a technique exploited by Robert Emerson in the 1930s.

Following in the steps of Warburg, Robert Emerson[51] (1903–1959), an American plant physiologist, and his undergraduate student William Arnold (1904–2001) applied manometry to the measurement of photosynthesis in the alga *Chlorella* in 1932. Using flashing light they found that *efficient* photosynthesis required a temperature-dependent dark period between the flashes. They concluded thus:

> The experiments . . . show that photosynthesis involves a light reaction not affected by temperature, and capable of proceeding at great speed, and a dark reaction dependent on temperature, which requires a relatively long time to run its course. The light reaction can take place in about a hundred thousandth of a second. The dark reaction requires less than 0.04 second for completion at 25°C.[52]

Although in the 1930s there were some uncertainties about these results, they were important in setting the background for further research. Photosynthesis was now understood by the majority to involve a "light reaction" and a "dark reaction."

Perhaps more significant was a second contribution from Emerson and Arnold in 1932, again using flashes of light, now very brief, in which they calculated that only one chlorophyll in about 2500 was activated in the most efficient photosynthesis:

> Measurements of photosynthesis were made in continuous and flashing light of high intensity, using cells varying in chlorophyll content. The amount of chlorophyll present per molecule of carbon dioxide reduced per single flash of light was found to be about 2480 molecules.[53]

So it appeared that the vast majority of chlorophyll molecules were inactive in photosynthesis! The problem of the apparent inactivity of most chlorophyll molecules led to the notion of the photosynthetic unit in which a number of chlorophyll molecules collaborate to bring about a photochemical event. Such a situation implies the transfer of excitation energy between pigment molecules, as suggested by Hans Gaffron and K. Wohl in 1936.[54] The questions arising from Emerson and Arnold's work were pursued rather later,

---

[50] Blackman 1905.
[51] For a biographical memoir of Emerson, see Rabinowitch 1961.
[52] Emerson and Arnold 1932a, p.417.
[53] Emerson and Arnold 1932b, p.204.
[54] Gaffron and Wohl 1936.

particularly after the transfer of energy from a carotenoid pigment to chlorophyll was demonstrated in diatoms.[55]

6.9 O_2 EVOLUTION AND THE HILL REACTION

It is interesting to note a comment on research in photosynthesis made in 1938 and quoted by Hans Gaffron:

> The enormous amount of research upon the process of photosynthesis during the past half century has thrown little or no additional light upon the subject. The problem is evidently too complex for any specialists in any one field of science to solve.[56]

The comment is perhaps unduly negative, but it nevertheless expressed the fact that little progress had been made over many years and that hard experimental information lagged behind that elsewhere in bioenergetics. The situation did, however, begin to change in the late 1930s with the investigations of the Cambridge biochemist Robert Hill (1899–1991). His work can be seen as marking the commencement of the modern study of photosynthesis and the introduction of a simple filtering method for making chloroplast preparations for biochemical study. It initiates the third period in this history of photosynthesis.

Hill (known as Robin Hill) joined the young Cambridge department of biochemistry under the leadership of Gowland Hopkins in 1922.[57] Although already interested in plant pigments, Hopkins directed Hill into work on hemoglobin and related metalloporphyrins. However, in the 1930s, Hill moved into the photosynthetic field, introducing isolated chloroplasts as the subject of experimentation. Hill was influenced not so much by van Niel's work but especially by the earlier work of the botanist Hans Molisch, who had shown that wetted leaf powders could evolve O_2 in the light.[58]

Hill found that isolated chloroplasts would evolve significant amounts of O_2 when illuminated in the presence of ferric oxalate. The reaction did not require CO_2 and appeared as a photochemical splitting of water. Later other electron acceptors were used including ferricyanide, which was reduced to ferrocyanide. The chloroplast reaction could be represented as in Figure 6.3.

As Hill wrote,

> The photosynthesis of green plants is characterized by the production of molecular oxygen when the living cells containing chlorophyll are illuminated. The oxygen however is only produced when carbon dioxide is taken up, and no other substance at present known will cause the photosynthetic reaction on the living

---

[55] A group of unicellular algae.
[56] Quoted from E. C. Miller, *Plant Physiology,* 2nd ed., 1938, by Hans Gaffron (1946).
[57] For a biography of Hill, see Bendall 1994 and Walker 2002.
[58] See Section 6.6.

HILL REACTION

$$2H_2O + 4\,\text{ferricyanide} \xrightarrow[chloroplast]{light} O_2 + 4\,\text{ferrocyanide} + 4H^+$$

PHOTOLYSIS OF WATER

$$2H_2O \xrightarrow{light} 4H^+ + 4e$$

Figure 6.3 Hill's Chloroplast Reaction in the Light

Above, a representation of the Hill reaction using a suitable electron acceptor, ferricyanide. Below, the essential reaction, the photolysis of water generating oxygen and reducing power.

cell. The chloroplasts, which are organized subcellular units containing chlorophyll, will themselves evolve traces of oxygen in light after removal from cells. It was shown (Hill 1939) that, in light, chloroplasts would reduce ferric oxalate to ferrous oxalate and that an equivalent amount of oxygen was set free. This reaction, here referred to as 'the ferric oxalate reaction',[59] was capable of measurement and is probably the first indication of any measurable activity apart from the whole cell which has any bearing on photosynthesis. The chloroplasts after removal from the cells have the property of evolving oxygen from certain hydrogen acceptors, but carbon dioxide[60] will not act as a hydrogen acceptor.[61]

Hill further commented that

> The new conclusion that can be drawn from the work on isolated chloroplasts is that oxygen itself is formed in a photochemical reaction during which there is no reaction involving carbon dioxide.[62]

All this work pointed to the origin of evolved O_2 being water (as suggested by van Niel), rather than CO_2, as had been proposed by Warburg, Willstätter, and several others. Perhaps it is not surprising that there were those who wondered whether the Hill reaction was an artefact.

The conviction that O_2 in photosynthesis was derived from water was given very strong support from experiments that used the heavy isotope of O_2, ^{18}O. Sam Ruben (1913–1943) together with Martin Kamen (1913–2002) of the chemistry department at University of California Berkeley developed the early use of isotopes in biological research, including carbon-14.[63] Ruben and colleagues in 1941 suspended *Chlorella* cells in

---

[59] The reaction became known as the Hill reaction after it was so described by French and Anson 1941.

[60] Small amounts of CO_2 were later shown to be necessary for the Hill reaction, and this was used by Warburg to maintain his view that O_2 was derived from CO_2.

[61] Hill 1940, pp.238–239. See also Hill 1939.

[62] Hill 1940, p.254.

[63] Gest 2004. See also the next section for comments on the use of carbon-14.

heavy O_2 water in one experiment and in heavy O_2 carbonate/bicarbonate in the other. He examined the O_2 evolved during photosynthesis and found that the origin of the O_2 evolved was water, not carbonate:

> Using ^{18}O as a tracer we have found that the oxygen evolved in photosynthesis comes from water rather than from the carbon dioxide.[64]

From the beginning of the 1940s, it became clear that the central reaction of photosynthesis was likely to be a photolysis of water, although such a view was not universally accepted. This central reaction also appeared to be an oxidation–reduction.

If the Hill reaction represented a true photosynthetic reaction, what was the natural substance that accepted electrons rather than the ferric compounds used by Hill? Ochoa had considered the nature of heterotrophic carboxylations and noted that they were associated with reductions involving NAD and NADP. In cases such as carboxylation of pyruvate to malate by the malic enzyme, NADP would provide the reductant. In 1951 Wolf Vishniac and Ochoa and also L. J. Tolmach[65] found that pyridine nucleotides, NAD and NADP, would act as electron acceptors in photosynthesis, suggesting that they were the natural electron acceptors for the process.

In view of his friendship with David Keilin and his earlier interest in heme compounds, it is not surprising that Hill turned to looking at heme compounds in plants. Because of the interruption caused by the war, his next major contribution did not appear until 1951. He described a new c-type cytochrome associated with the green parts of the plant, which was similar to, but distinct from, cytochrome c.

> We also found the green parts of the plants contain a non-autoxidizable cytochrome component (α-band 555 mμ) resembling cytochrome c but with a remarkably sharp absorption spectrum. This cytochrome component which can be obtained in a soluble form is called here cytochrome f (Latin *frons*[66]), on account of the association with the green parts of plants.[67]

Hill concluded that the cytochrome was located in the chloroplasts. Another substance apparently involved in the Hill type of reaction was isolated by Davenport, Hill, and Whatley, who obtained a protein from green leaves that could be shown to be necessary for O_2 evolution when methaemoglobin[68] was the electron acceptor.[69] The authors gave the name of the "methaemoglobin reducing factor" to this substance. Its full identity was not determined for almost 10 years.

---

[64] Ruben et al.1941, p.87.
[65] Vishniac and Ochoa 1951; Tolmach 1951.
[66] A leaf.
[67] Hill and Scarisbrick 1951, p.99.
[68] Methemoglobin was found to be a useful electron acceptor for the Hill type of reaction, but it then emerged that a protein in chloroplasts was necessary for its reduction.
[69] Davenport, Hill, and Whatley 1952.

By the beginning of the 1950s, the possibility that the chloroplast might possess an electron-transport chain comparable to the respiratory chain in mitochondria was evident. Such an idea was prompted by the oxidation–reduction nature of the Hill reaction and the discovery of chloroplast cytochromes. Hill (with Davenport) noted

> that the processes of respiration and photosynthesis consist of a series of similar but not necessarily identical steps in hydrogen transport.[70]

In 1954 he described an autoxidizable b cytochrome from chloroplasts, b_6 (α-band 563 mμ).[71] Another prompt was the ability of the pyridine nucleotides, NAD and NADP, to act as electron acceptors in the Hill type of reaction. A comparison of chloroplast systems with those in mitochondria was now an obvious subject for debate, and many of the insights from mitochondrial studies became relevant to understanding photosynthesis.

6.10 CO$_2$ ASSIMILATION

The work of Hill and others that represented the central photochemical reaction of photosynthesis as an oxidation–reduction reaction that did not require CO_2 tended to suggest that carbon assimilation might be a straightforward metabolic process. Hill, considering his earlier work, wrote that he

> concluded from his experiments with chloroplasts that the reducing potential was in fact less than enough to reduce CO_2 directly. This led him to suggest that photosynthesis in green plants was essentially a chemosynthetic process . . . . This view was attractive, however, because it would represent photosynthesis in green plants as only a special case of CO_2 fixation which from 1936 was coming to be regarded as a general phenomenon in living cells.[72]

Indeed, Wood and Werkman's demonstration of CO_2 fixation in *Propionibacterium* in 1936 and later of CO_2 fixation in animals showed the plausibility of a straightforward nonphotochemical carbon fixation in photosynthesis.

The possibility that newly available radioisotopes might help solve the problem was early realized. However, experiments with the short-lived (half-life 20 min) isotope ^{11}C by Sam Ruben and his colleagues in 1939 proved inconclusive, although they supported the formation of a carboxyl group from CO_2.[73] The rapid decay of the isotope precluded further progress. Work then centered on obtaining a radioactive carbon isotope with a long life that could be used as a tracer to follow the path of carbon in photosynthesis. Ruben and Kamen discovered ^{14}C in 1940 and were able to obtain the isotope by bombarding graphite.[74] The death of Ruben in 1943 curtailed this work until 1945. New supplies of

---

[70] Davenport and Hill 1952, p.344.
[71] Hill 1954.
[72] Hill 1951, p.32.
[73] Ruben et al. 1939.
[74] See Benson 2002; Gest 2004.

the isotope began to become available after the Second World War as a result of atomic research associated with the war effort. The nuclear reactor rather than the cyclotron became the source of radioisotopes, especially ^{14}C, leading to a marked stimulation of biochemical research.

In 1945 the group of Melvin Calvin (1911–1997), a chemist in the University of California, and Andrew Benson (1917–2015), also from University of California, started work on the path followed by carbon in photosynthesis. Their work was made possible partly by the ready availability of a long-lived isotope of carbon that enabled them to follow the path of CO_2 fixation but also by the development of the paper form of chromatography.[75] This provided an effective method for separating the labeled products of photosynthesis with $^{14}CO_2$ that could be visualized by placing x-ray film against the chromatograms (see Fig. 6.4). By 1950 and using suitable solvent systems to develop their chromatograms in two dimensions, they were able to separate the products of CO_2 fixation after algae (*Chlorella* and *Scenedesmus*) were given a brief exposure to $^{14}CO_2$. Labeled products could be eluted from the chromatogram and analyzed. They identified 3-phosphoglyceric acid as the first formed stable compound with the label primarily in the carboxyl carbon.[76] This confirmed the earlier proposals that CO_2 was initially converted to a carboxyl group.

The chromatograms normally showed a number of ^{14}C-labeled substances. Detailed studies of these compounds and variation of the conditions for carbon fixation enabled the University of California group to formulate a cycle for the fixation of CO_2. The 3-carbon phosphoglyceric acid was formed by carboxylation of a 5-carbon sugar, ribulose bisphosphate, and the enzyme catalyzing this reaction[77] was found in a cell-free extract of algae.[78] Further extensive studies led to the elucidation of the carbon reduction cycle, more popularly known as the Calvin cycle or the C3 pathway (Fig. 6.5).[79] As formulated by Calvin and colleagues, the cycle required a supply of reduced NADP and ATP to maintain its activity. CO_2 fixation in higher plants was demonstrated to occur by a similar pathway to that in the algae used in initial studies.

However, some plants were found to have other forms of carbon assimilation. From around 1954, reports appeared that suggested that in plants such as sugarcane and maize there might be a different pattern of labeling from that obtained by Calvin and his colleagues. Malate and aspartate in particular became rapidly labeled. This led to the formulation of the dicarboxylic acid (C4) pathway by Marshall Hatch (1932–) and Roger Slack (1937–) in the mid-1960s. The carboxylation to form malate is then followed by the

---

[75] In the mid-1940s, the technique of paper chromatography had been described by Consden, Gordon, and Martin 1944.

[76] Benson et al. 1950.

[77] The enzyme was initially named carboxydismutase but later became known as rubisco and more formally as ribulose bisphosphate carboxylase. It was identical with fraction 1 protein originally isolated from chloroplasts by Sam Wildman in 1947; see Wildman 2002.

[78] Quayle et al. 1954.

[79] The first version of the cycle was published in 1954 (Bassham et al.) but this was amended (see Bassham and Calvin 1957). Calvin was awarded the Nobel Prize in Chemistry for "his research on the carbon dioxide assimilation in plants."

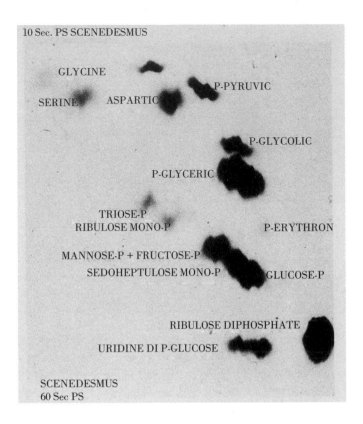

Figure 6.4 Two-Dimensional Radiochromatogram of the Labeled Products From 60-s Photosynthesis in *Scenedesmus*

A suspension of the unicellular alga *Scenedesmus* was exposed to $^{14}CO_2$ in the light for 60 s. The products were extracted and separated by two-dimensional paper chromatography. The main labeled products found were phosphoglyceric acid and sugar phosphates. Reproduced with permission from Benson 2002, "Following the Path of Carbon in Photosynthesis: A Personal Story" (*Photosynthesis Research* 73: 29–49), Springer, fig. 4, p.36.

release of CO_2 in special cells for a Calvin cycle type of carbon assimilation. It became clear that a number of plants fix CO_2 by this pathway known as the C4 pathway.[80] More specialized forms of carbon assimilation have also been discovered.[81] In photosynthetic bacteria evidence accumulated during the 1960s and 1970s that the carbon reduction cycle functioned in photosynthetic bacteria. But in the green photosynthetic bacterium, *Chlorobium*, Arnon's group identified a different cyclic pathway, a reductive carboxylic acid cycle.[82]

[80] Hatch and Slack (1966).

[81] In addition to the C3 and C4 pathways, a photosynthetic adaptation to drought conditions is known as Crassulacean acid metabolism and also concerns the fixation of CO_2 in malate. It was first noted by de Saussure in 1804 and was rediscovered by Meirion Thomas in 1946. See Black and Osmond 2003.

[82] Evans, Buchanan, and Arnon 1966.

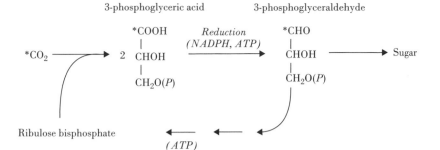

$$\text{3-phosphoglyceric acid} \qquad \text{3-phosphoglyceraldehyde}$$

*COOH $\qquad$ *CHO

*CO$_2$ $\longrightarrow$ 2 CHOH $\xrightarrow{\text{Reduction (NADPH, ATP)}}$ CHOH $\longrightarrow$ Sugar

CH$_2$O(P) $\qquad$ CH$_2$O(P)

Ribulose bisphosphate $\longleftarrow \longleftarrow \longleftarrow$

(ATP)

Figure 6.5 The Pathway for Carbon Dioxide Fixation in Summary as Elucidated by Calvin and Coworkers

Carbon dioxide (here shown labeled) reacts with the 5-carbon sugar phosphate, ribulose bisphosphate, to form two molecules of 3-phosphaglyceric acid, which are then reduced to 3-phosphoglyceraldehyde in a reaction requiring both reduced NADP and ATP. The phosphoglyceraldehyde is then either converted to sugar or used to regenerate the ribulose phosphate.

The work of Calvin and his colleagues finally resolved the problem of the mechanism of CO_2 fixation. Although in the period after the demonstration of the Hill reaction, there were various doubts about whether this represented the process in vivo, by the 1960s, most of that skepticism had evaporated. This provided final confirmation that carbon assimilation did not directly involve chlorophyll and that it did not involve a photochemical event but was the dark or Blackman reaction of Emerson, Arnold, Warburg, and others in the 1930s.

There was, however, a major dispute with the renowned biochemist Otto Warburg over the fundamental nature of photosynthesis. For Warburg, CO_2 assimilation was intimately involved with the basic light reaction of photosynthesis, and it was this process that was associated with O_2 evolution. This model of photosynthesis contrasted sharply with the work on the Hill reaction and Calvin's work on CO_2 fixation, which had effectively removed carbon assimilation from direct participation in the light reaction of photosynthesis.[83] Warburg's view was supported, for example, by the experimental demonstration that CO_2 was necessary for the Hill reaction, although the amount required turned out to be very small.[84] As Hill noted, Warburg's schemes led to new experimental discoveries even if they were basically unacceptable to most.

6.11 DISCOVERING PHOTOPHOSPHORYLATION

The idea that photosynthesis could be viewed as the conversion of light energy into chemical energy arose from the thinking of Julius Robert Mayer in the mid-nineteenth century. As Daniel Arnon (1910–1994),[85] one of the discoverers of photophosphorylation, commented,

The removal of CO_2 assimilation as the unique problem of photosynthesis has focused attention on what was already recognized a century ago as the key event

[83] Aspects of this long debate can be found in Krebs 1972 and Hill 1965.
[84] Warburg and Krippahl 1960.
[85] Daniel Israel Arnon was born in Warsaw, Poland, but, influenced by the famines following the First World War, he migrated at the age of 18 to the United States and eventually joined the

in photosynthesis, on which all life on our planet depends: the conversion of light into chemical energy. This was first expressed with great clarity by two founders of thermodynamics, J. R. Mayer (1845), and Ludwig Boltzmann (1886). Mayer, who three years earlier enunciated the Law of Conservation of Energy, now stressed its applicability to plant life: "Plants are able only to convert energy not to create it."[86]

This view had been given some substance by Ruben, who considered the conversion of light energy into chemical energy as involving the regeneration of a phosphate donor (possibly ATP).[87]

An initial attempt to elucidate the synthesis of ATP in photosynthesis considered the possibility that oxidative phosphorylation might have a role to play. Following the demonstration that the Hill reaction could involve reduction of pyridine nucleotides, the proposal was made that the reduced pyridine nucleotides could be oxidized in the mitochondria, giving ATP synthesis by oxidative phosphorylation. An experimental system was prepared that carried out phosphorylation by this means in the light.[88]

Although successful, there were doubts about whether this was the true mechanism for ATP synthesis.

The apparent inability of chloroplasts to assimilate CO_2 while carrying out the Hill reaction raised questions about whether photosynthesis required activities located outside the particles in the cytosol. However, as noted earlier, the similarity of the oxidative events in the chloroplast and mitochondrion became obvious after Hill had drawn attention to the presence of cytochromes in the chloroplast and had predicted some sort of electron-transport chain.[89] It followed from this that phosphorylation might also be a property of the chloroplast.

The initial question of whether complete photosynthesis depended on activities outside the chloroplast was resolved by Arnon in 1954, working at the University of California. He demonstrated that isolated spinach chloroplasts were capable of carrying out complete photosynthesis. His experiments showed isolated chloroplast preparations could carry out not only the photolysis of water but also CO_2 fixation and the synthesis of ATP.[90] So washed chloroplasts have the ability, without addition of other factors or enzymes, to use light energy for ATP synthesis and CO_2 assimilation. However, in these experiments the amounts of ATP synthesis were small.[91] Nevertheless such experiments demonstrated that chloroplasts alone were capable of carrying out photosynthesis.

University of California, Berkeley, where he remained for the whole of his professional career apart from war service in the Second World War.

[86] Arnon 1961, p.224. See earlier comments on the contributions of physicists in Section 6.4.
[87] Ruben 1943.
[88] Vishniac and Ochoa 1952.
[89] See Section 6.9.
[90] Arnon, Allen, and Whatley 1954.
[91] It should be noted that Arnon's group used adenylic acid (AMP), not ADP, as the substrate for ATP synthesis. Later experiments used ADP. (See Jagendorf 1998, p.219.)

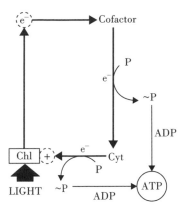

Figure 6.6 Arnon's Interpretation of His Experiments Demonstrating Cyclic Photophosphorylation
With Flavin mononucleotide and Vitamin K as Cofactors
The cofactors catalyze a cyclical electron flow from chlorophyll at the reaction center to cytochrome, which returns
the electron to the chlorophyll. The electron flow is seen as driving the synthesis of ATP at two points in the electron
pathway in a manner comparable to that in mitochondrial oxidative phosphorylation. Reproduced with permission
from Arnon 1961, *Photosynthetic Phosphorylation and the Energy-Conversion Process in Photosynthesis* (Biological
Structure and Function 2, Academic), Elsevier, fig. 1, p.354.

Although it is difficult to appreciate the significance of this experiment now, when it was
first performed it was seen as groundbreaking particularly by nonscientists. As *The Times
London* noted on New Year's Day 1955, the process in which plants and sunlight create
food out of CO_2 and water had been accomplished outside the living cell for the first
time: "This solar engine [the chloroplast], it was found, could do all the work of photosynthesis even after being removed from a living plant."[92]

In early work, the rates of ATP synthesis by chloroplast material were relatively low.
Further investigation established that a number of cofactors could stimulate the rate of
phosphorylation, including flavin mononucleotide and the vitamin K analogue, mena-
dione. In preparations with limited CO_2 fixation ability, the rate of phosphorylation could
be increased by over 100 times. Indeed the rates were higher than those obtained with
oxidative phosphorylation.[93]

Arnon's interpretation of these experiments in which photosynthesis was stimulated by
flavin mononucleotide or vitamin K was to regard the process as a cyclic flow of electrons
(see Fig. 6.6). On light activation of the chlorophyll, an electron would be ejected at low
potential and return to the chlorophyll by a route of increasing potential. The electron
flow would be coupled to phosphorylation in a manner similar to that in oxidative phosphorylation, however that occurred. Figure 6.6 can be compared with Figure 4.4 showing
oxidative phosphorylation linked to the mitochondrial respiratory chain.

Indeed, the developing understanding of photophosphorylation rested heavily on
the concepts developed in studies of mitochondrial phosphorylation that at this stage

[92] *The Times London*, January 1, 1955.
[93] Allen, Whatley, and Arnon 1958.

were guided by variations on the chemical hypothesis of Slater. Arnon's interpretation of events can be seen in the following extract:

> The simplest hypothesis to account for the formation of ATP in photosynthetic phosphorylation is to assume that, as in the dark phosphorylations of glycolysis and respiration, the formation of a pyrophosphate bond is also coupled with a release of free energy which occurs during electron transport, i.e., when an electron drops from the higher energy level that it has when it resides in the electron donor molecule to the lower energy level that it assumes on joining the electron acceptor molecule. But a mechanism for photosynthetic phosphorylation must also account for its unique features: ATP is formed without the consumption of an exogenous electron donor and electron acceptor. Unlike oxidative phosphorylation, photosynthetic phosphorylation consumes neither exogenous substrate nor molecular oxygen, only light energy.
>
> A mechanism for photosynthetic phosphorylation must, therefore, provide for the generation of both an electron donor and an electron acceptor in the primary photochemical act when radiant energy is absorbed by chlorophyll.[94]

The role of chlorophyll was now seen as carrying out an energy-dependent oxidation–reduction reaction.

Arnon's cyclic photophosphorylation (not associated with a Hill-type O_2 evolution), although adding significantly to the knowledge of photosynthesis, raised the question of the relationship of this photochemical reaction to the photochemical reaction in the Hill reaction. The answer lay in work from the laboratories of both Arnon and Andre Jagendorf[95] (1926–2017). The latter's group showed that ADP was phosphorylated in light by a chloroplast suspension evolving O_2 and reducing ferricyanide; ADP, phosphate, and magnesium increased the rate of ferricyanide reduction.[96] Arnon's group used ferricyanide or NADP as electron acceptors and also obtained ATP synthesis associated with O_2 evolution. The addition of catalysts of cyclic photophosphorylation increased ATP synthesis but reduced O_2 evolution.[97] Arnon described phosphorylation associated with the Hill reaction as noncyclic photophosphorylation, a process having much in common with oxidative phosphorylation.

The similarity between the mitochondrial and chloroplast systems was pursued further. Somewhat surprisingly, uncoupling agents used in the mitochondrial system were not particularly effective in chloroplasts, but it was shown in 1959 that ammonium chloride was an effective uncoupler in chloroplasts.[98]

[94] Arnon 1961, p.235.
[95] Andre T. Jagendorf was an American plant physiologist and biochemist at Johns Hopkins and later at Cornell University.
[96] Avron, Krogmann, and Jagendorf 1958.
[97] Arnon, Whatley, and Allen 1959.
[98] Krogmann, Jagendorf, and Avron 1959.

Concurrently with Arnon's work on photophosphorylation in chloroplasts, Albert W. Frenkel[99] (1919–2015) demonstrated light-induced phosphorylation by cell-free preparations of the photosynthetic nonsulfur bacterium, *Rhodospirillum*.[100] Further work showed that ADP was the specific substrate for phosphorylation. Later, Arnon, in a study of another photosynthetic bacterium, *Chromatium*, which had rather different carbon assimilation to that in green plants, concluded that

> Photosynthetic bacteria, which have greatly contributed to our understanding of photosynthesis by showing that this process can occur without oxygen evolution, now also provide us with evidence that photosynthesis can occur without carbon dioxide assimilation. What remains fundamental to all photosynthesis is the conversion of light energy to chemical energy. This conversion appears to be intimately bound to photosynthetic phosphorylation and its product adenosine triphosphate, which is the universal "energy currency" of all cells, whether they live in the light or in the dark.[101]

The core reaction of photosynthesis had now shifted from photolysis of water to photophosphorylation. The process could now be understood as the conversion of light energy into chemical energy.[102]

By around 1960 the overall process of photosynthesis had been clarified considerably and the major components of the process had been identified. The basic photochemical reaction was now seen as an oxidation– reduction driving an electron-transport system and, in green plants but not bacteria, evolving O_2. The electron-transport system, like that in the mitochondrion, was linked to the synthesis of ATP, which was now named photophosphorylation. The dark or Blackman reaction was seen as the fixation of CO_2 to form carbohydrate but this process was now not a photochemical one but involved a conventional metabolic cycle. The direct photochemical reduction of a chlorophyll–CO_2 complex had finally been abandoned.

Indeed it was possible to physically separate the light reaction associated with chloroplast grana from the dark reaction associated with the stroma and to show that together they carried out normal photosynthesis.[103] However, the nature of the light reaction itself had still to be worked out although the modern study of photosynthesis was now well underway.

---

[99] Albert W. Frenkel was born in Germany in 1919 but in 1937 migrated to the United States where he eventually joined University of California Berkeley Radiation Laboratory.

[100] Frenkel 1954, 1956.

[101] Losada et al. 1960, p.759.

[102] Arnon 1959.

[103] Trebst, Tsujimoto, and Arnon 1958. Note that later the stroma lamellae were shown to be capable of a light reaction.

7

Elucidating the Photosynthetic
Light Reaction

Studies of photosynthesis in the 1960s were primarily questions about the nature of the light reaction. There was a considerable amount of evidence that pointed the way to solve the problems posed by the light reaction. In particular, there was the demonstration that it was fundamentally an oxidation–reduction reaction, a view that had emerged experimentally from both Hill's study of O_2 evolution and Arnon's study of photophosphorylation as well as the earlier work of van Niel. This was supported by the discovery of chloroplast cytochromes and other redox components. The time was now ripe for major discoveries in the field, and this generated the fourth phase of the history.

7.1 THE FOURTH PERIOD OF
PHOTOSYNTHETIC HISTORY

This fourth period was concerned with creating a more or less coherent mechanism for the process. Studies during the third phase in the 1940s and 1950s had shown that CO_2 fixation could be equated with the dark or Blackman reaction; further reactions that could occur in "dark" periods following illumination began to be apparent. The light reaction appeared to be essentially an oxidation–reduction reaction involving chlorophyll, but little could be said with any certainty about its mechanism. In 1960, Robin Hill and Fay Bendall summarized the current experimental understanding of the light reaction of photosynthesis as follows:

> The conversion of light energy into a form of chemically available energy by chloroplasts *in vitro* can be observed in two ways: by the reduction of a hydrogen acceptor with the production of a stoichiometric equivalent of molecular oxygen or by the formation of adenosine triphosphate from inorganic phosphate in the presence of adenosine diphosphate. In the green plant, coenzyme II [TPN, NADP] can be regarded as the hydrogen acceptor for the complete photochemical system of the chloroplast.[1]

---

[1] Hill and Bendall 1960, p.136.

This phase of the history concerned itself first with understanding the way workers approached the question of numerous chlorophyll molecules associated with a single photochemical reaction—the problem of the photosynthetic unit and its reaction center. Second, a key issue would be to elucidate the meaning of Arnon's assertion that photosynthesis should be understood as the conversion of light energy into chemical energy, the synthesis of ATP. ATP synthesis might be seen as having a mechanism akin to that in mitochondria if that mechanism could be successfully elucidated. Resolving such an issue would draw on knowledge of mitochondrial oxidative phosphorylation to explain chloroplast photophosphorylation linked to a chloroplast electron-transport chain. All of this was complicated by a major surprise: There were two independent light reactions, not one! The period from around 1960 to the late 1970s was a time of substantial innovation in thinking about photosynthesis.

This phase was also marked by a particular burst of activity in the field of bacterial photosynthesis in which prokaryotes provided a simpler system. The isolation of a reaction center complex where the key photochemical reaction occurred led to major developments in research. This was supplemented by introduction of picosecond spectroscopy, enabling very fast reactions to be detected. Thus bacterial studies led to an important step forward in understanding the core process.

The light reaction presented a number of substantial challenges. Why were so many chlorophyll molecules apparently inactive, as suggested by Emerson and Arnold? If the answer lay in some sort of coordinated complex, what exactly was it? How did ten or so quanta function in the evolution of one molecule of O_2 (or assimilation of a molecule of CO_2)? One point had emerged about the function of the light reaction in relation to the Calvin carbon reduction cycle—the cycle as formulated required a supply of reduced NADP (or NAD) and ATP to maintain its activity. It could be presumed that this would be provided by the light reaction as understood from studies of the Hill type of reaction and photophosphorylation.

7.2 SEEKING A COHERENT MODEL OF PHOTOSYNTHESIS

The study of chloroplast structure had been initiated by the microscopists of the nineteenth century such as Hugo Mohl and Arthur Meyer who saw these particles and the internal chlorophyll-containing bodies that they contained. These bodies were termed grana (see Chapter 6). The apparently colorless matrix of the chloroplast was referred to as the stroma, which was much later associated with the dark carbon reduction cycle. The introduction of high-resolution electron microscopy more or less concurrently with similar work on mitochondria gave a very different view of the chloroplast, as shown in Figure 7.1.[2] The new structure became an important element in understanding the photosynthetic process taking place in the particle.

[2] See Steinmann and Sjostrand 1955 and also Sager and Palade 1957.

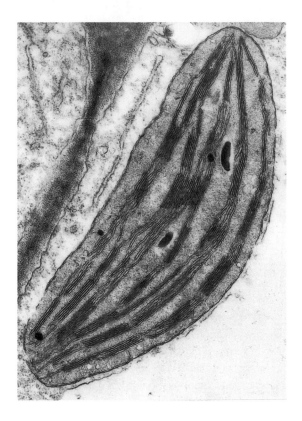

Figure 7.1 Electron Micrograph of the Chloroplast

The lens-shaped (in cross section) particle is surrounded by a double membrane or envelope. Internal membranes, thylakoid membranes that are basically flattened sacs with an internal space, the loculus, cross the particle occasionally, being part of a stacked group or granum. Space between the membranes is termed the matrix. Photograph courtesy of Dr. J. M. Whatley.

Reproduced with permission from Prebble 1981, *Mitochondria, Chloroplasts and Bacterial Membranes*, Longman, Pearsons Educational Ltd., fig. 12.1a, p.260.

As Staehelin remarked,

> In the late 1940s, photosynthesis researchers were still limited by the same 0.2 μm resolution of light microscopes encountered by the German botanist Hugo von Mohl, who, in 1837 provided the first definitive description of "Chlorophyllkörnern" (chlorophyll granules) in green plant cells. Between 1950 and 1960 the introduction of the electron microscope and of thin specimen preparation methods improved the 2-D resolution to <100 Å, which led to the discovery of thylakoids and to the first characterization of their 3-D architecture.[3]

In higher plants the chloroplast was described as a lens-shaped particle. Grana were now seen as tightly stacked membranes, and some of these membranes also crossed the

---

[3] Staehelin 2003.

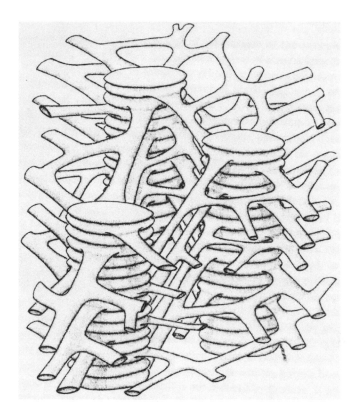

Figure 7.2 One of Several Models Produced During the 1960s Showing Possible Three-Dimensional Arrangement of the Thylakoid Membranes

The model shows the interconnections of the granum thylakoids, indicating that the stacked membranes in the grana link through the stroma to other grana. The internal space of the thylakoid, the loculus, is thus also connected. Reproduced with permission from Weier et al. 1963, "The Grana as Structural Units of Chloroplasts of Mesophyll of *Nicotiana*" (*Journal of Ultrastructure Research* 8: 122–143), Elsevier, fig.19, p.141.

stroma. The membranes (often referred to as lamellae) were seen as forming flattened vesicles named thylakoids by Wilhelm Menke.[4]

Interpreting two-dimensional electron micrographs in three dimensions produced a number of models (Fig. 7.2). The proposals of Werner Wehrmeyer in 1964 became the basis for a more sophisticated version by Dominick J. Paolillo at the end of the decade.[5] A key element of the models was that the granum thylakoids were connected by their membranes that crossed the stroma. This resulted in the loculi (internal spaces) bounded by these membranes also being at least partially connected. The chloroplast now had two membrane-bound internal spaces, the loculus of the thylakoid and the matrix space outside the thylakoid but bound by the double membrane of the envelope. The mechanism of photosynthesis would need to be understood within this architecture.

[4] Menke 1962.
[5] Paolillo 1970.

In the 1960s, the introduction of freeze-fracturing techniques similar to those used with mitochondria further advanced the knowledge of the thylakoid membranes. This development enabled workers to see the distribution of protein particles within the membrane and showed differences between the inner and outer halves of the thylakoid membranes.[6] The outer half possessed a rather dense distribution of protein particles whereas the inner half showed fewer but larger particles.

In the 1960s and 1970s, the structure of the chloroplast began to play a significant role in understanding the mechanism of photosynthesis. Initially the carbon reduction cycle was associated with the matrix and the light reaction with the grana. Indeed the light reaction of chloroplasts could be physically separated from the dark CO_2-assimilation part by separating grana and stroma, and the two could be recombined to give the complete process.[7] However, subsequently the light reaction was shown to be a property of the thylakoid membranes whether organized in grana or not. Later studies refined this view, especially distinguishing between stacked thylakoid membranes (adpressed membranes) and those traversing the stroma.

As with the mitochondrion, the permeability of the membranes and the properties they possessed were an important part of understanding photosynthesis—particularly in relation to the mechanism of phosphorylation. Integration of structure and biochemistry was a key part of elucidating the mechanism of photosynthesis.

7.3 THE PHOTOSYNTHETIC UNIT

The work of Emerson and Arnold in 1932, which showed that one CO_2 assimilated or one molecule of O_2 evolved was associated with 2400 chlorophyll molecules, presented a challenge. Hans Gaffron (1902–1979) was born in Peru of German parents, educated in Germany, and moved to the United States in 1937. In 1936, together with Kurt Wohl (1896–1962), he carried out experiments in weak light (and efficient photosynthesis) in a thick suspension of *Chlorella*, which also suggested an optical unit of 2400 chlorophyll molecules. He interpreted this as chlorophyll molecules collecting light energy that migrated to the reduction centers (later called reaction centers).[8] Over the following years the idea of a photosynthetic unit with a reduction center, or a reaction center, was much debated. A reaction center was considered to be a pigment absorbing at long wavelengths[9] to which light energy absorbed by a pigment molecule would migrate.

Fluorescence by green leaves and by chlorophyll activated by light had been observed in the nineteenth century, particularly by the Cambridge physicist and mathematician George Stokes.[10] Interest in chlorophyll fluorescence developed as an indicator of the excited (or energized) state of chlorophyll that emitted the light. However, precise measurements of fluorescence proved difficult. Following work by Hans Kautsky in 1931,

---

[6] Park and Pfeifhofer 1969.

[7] Trebst, Tsujimoto, and Arnon 1958.

[8] Gaffron and Wohl 1936.

[9] Pigments activated by light of longer wavelengths have less energy than those activated by shorter wavelengths. See note 22.

[10] Stokes made initial observation on fluorescence on "leaf green" (chlorophyll). Stokes 1852.

Edward McAlister and Jack Myers had observed an inverse relationship between carbon assimilation and fluorescence.[11] From a study of numerous experimental results, Evert Wassink felt that the relationship between fluorescence and the photosynthetic processes had not proved easy to interpret:

> At present, it seems hardly possible to give a clear description of the mutual relation between the rate of photosynthesis and the intensity of fluorescence that will cover all observations.[12]

Nevertheless fluorescence remained a valuable tool in measuring the excited state of chlorophyll formed by absorption of a quantum of light.

The migration of absorbed energy between pigment molecules proposed by Gaffron and Wohl was demonstrated in a diatom (unicellular alga). Leslie Dutton and coworkers detected the transfer of energy between the yellow carotenoid pigment, fucoxanthin, and chlorophyll. Activation of the system with wavelengths in which the carotenoid absorbed light but the chlorophyll did not resulted in fluorescence of the chlorophyll.[13] They were uncertain about the mechanism and suggested several possibilities.

The mechanism for energy transfer between pigment molecules that gained broad support among workers was resonance energy transfer. The theory behind this process[14] and its application to photosynthesis was examined by Theodor Förster. The probability of resonance energy transfer increases with the degree of overlap between the absorption band of the acceptor and the fluorescence band of the donor so that chlorophyll, where there is a good overlap, is well suited to energy transfer between pigment molecules.

The growing interest in resonance energy transfer as an important physical process in photosynthesis was expressed by Eugene Rabinowitch in 1952:

> Resonance interaction of photosynthetic pigments in living cells seems to be of a strength that makes possible the migration of excitation energy through a small number of pigment molecules. This can account for the observed sensitized fluorescence of chlorophyll *in vivo*, and makes feasible a two-step mechanism of sensitization of photosynthesis, in which one kind of pigment molecule absorbs the quanta and transfers them to another kind which interacts with the substrate. A similar energy transfer must be possible also between identical molecules.[15]

It was noted that resonance energy transfer between identical or similar molecules would be much more difficult to demonstrate experimentally.

---

[11] McAlister and Myers 1940.

[12] Wassink 1951, p.195.

[13] Dutton, Manning, and Duggar 1943.

[14] The probability of resonance energy transfer between molecules decreases with the sixth power of the distance between them. Tightly packed chlorophyll molecules in the grana appeared to be well located for the process.

[15] Rabinowitch 1952, p.244.

After Willstätter and Stoll's monographs on chlorophyll and photosynthesis, published early in the twentieth century,[16] chlorophyll was seen as the major pigment in photosynthesis, more particularly chlorophyll *a*. However, light absorbed by other pigments (carotenoids and the phycobilins found in many algae) was also shown to be effective in photosynthesis.[17] Whereas most of these pigments were less efficient than chlorophyll, phycoerythrin (a red phycobilin pigment in red algae) was apparently more efficient than chlorophyll. Because the general view was that chlorophyll *a* was the key pigment in photosynthesis, these other pigments became known as *accessory pigments*.

In 1951 Duysens used a red alga to demonstrate that light absorbed by the pigment phycoerythrin gave chlorophyll fluorescence, showing that energy was being transferred from this pigment to chlorophyll.[18] C. Stacy French (1907–1995)[19] and Young obtained similar results with red algae and concluded thus:

> It appears probable that chlorophyll plays a specific chemical role in photosynthesis in addition to acting as a light absorber.[20]

In fact, energy transfer to chlorophyll *a* was also observed in blue-green algae[21] and green algae and in photosynthetic bacteria to bacteriochlorophyll.

By the late 1950s, it was being realized that chlorophyll *a* itself might not be as simple a pigment as had been thought. For a long time there had been a suspicion that chlorophyll was bound to proteins in the chloroplast grana. Then workers, using improved spectrophotometric techniques, observed that chlorophyll *a* in vivo had several absorption peaks, suggesting that there were different forms of the pigment. Such might be the result of chlorophyll molecules being bound to proteins. In general chlorophyll *b* had a peak at around 653 nm, whereas the chlorophyll *a* forms had peaks at 673, 683, and 695 nm.[22]

The preceding suggestion by French and Young that chlorophyll *a* might be responsible for the photochemical reaction in photosynthesis developed progressively. The existence of different spectral forms of chlorophyll *a* began to give a more complex picture. From basic physical principles it was expected that a quantum absorbed by a pigment molecule somewhere in a mass of such molecules would be able to migrate by resonance energy transfer to a molecule of longer-wavelength absorption.[23] If there were to

---

[16] Willstätter and Stoll 1913, 1918. See Chapter 6, Sections 6.5 and 6.6.

[17] Haxo and Blinks 1950 in a survey of algae found that monochromatic light activated photosynthesis when it was being absorbed by chlorophyll or by other pigments.

[18] Duysens 1951.

[19] Stacy French, a plant scientist, was born in Massachusetts and studied at Harvard. Early on he worked with Otto Warburg and later was at the Carnegie Institution of Washington.

[20] French and Young 1952, p.889.

[21] Blue-green algae are, in fact, prokaryotes, otherwise known as cyanobacteria. They carry out O_2-evolving photosynthesis.

[22] See Brown and French 1959.

[23] From the work of Planck and Einstein at the beginning of the twentieth century, it was known that there was an inverse relationship between the energy of a quantum of light and wavelength. Pigments with absorption maxima at shorter wavelengths have more energy when activated

be, as suggested by Gaffron and Wohl, a reduction center or reaction center (French and Young's chlorophyll undertaking a chemical reaction) to which the quantum might migrate, it would be expected to be a long-wavelength absorbing pigment. In 1956, Bessel Kok (1918–1978) reported the bleaching of a pigment, present at low concentrations, with an absorption at approximately 705 nm. Kok was a Dutch biochemist who initially studied and worked at Utrecht but later moved to the United States. He considered that the bleaching of this pigment might be due to energy transfer from chlorophyll *a*. The unknown pigment occurred universally in the plant kingdom.[24] The photosynthetic unit with a reaction center, possibly with an absorption maximum around 700 nm, was becoming a meaningful concept. However, there were other major considerations that modified the concept significantly.

7.4 TWO LIGHT REACTIONS AND A REACTION CENTER

The idea that green-plant photosynthesis might involve not one but two separate light reactions began to emerge toward the end of the 1950s. The first hint of this situation was a consequence of work in 1943 by Robert Emerson and Charlton Lewis who measured the quantum yield for photosynthesis across the visible spectrum in the unicellular alga, *Chlorella*. They obtained a value of approximately 10–12 quanta per O_2 molecule,[25] but efficiency of the process fell unaccountably in the far-red end of the spectrum (above a wavelength 685 nm) in a region of the spectrum where chlorophyll still absorbed light. Thus those chlorophyll molecules that absorbed light at longer wavelengths appeared to be inactive in photosynthesis. The phenomenon, poor photosynthesis when illuminated with far-red light, was known as the "red-drop." The authors concluded that the quanta absorbed by these chlorophylls did not have sufficient energy "to raise chlorophyll to the excited state required for the photochemical primary process."[26]

An important discovery that provided an interpretation of the red-drop came from Emerson's group more than 10 years later. They found that far-red light would give efficient photosynthesis if supplemented by shorter-wavelength light. Photosynthesis appeared to require two wavelengths of monochromatic light. Emerson suggested somewhat tentatively "that photosynthesis requires excitation of two different pigments."[27] This observation became known as the Emerson enhancement effect.

It took a few years for workers in the field to be fully convinced that the Emerson enhancement effect was due to the existence of two photochemical events. From a detailed study of activation of photosynthesis at different wavelengths, Myers and French obtained evidence for a direct role of chlorophyll *b* in photosynthesis. They concluded,

by light than those of longer wavelengths. Consequently energy will transfer from pigments of shorter wavelength maxima to those of longer.

[24] Kok 1957.
[25] Warburg had obtained values of 4 and below that, if true, made photosynthesis an unusually efficient biological process.
[26] Emerson and Lewis 1943, p.178.
[27] Emerson, Chalmers, and Cederstrand 1957, p.142.

That there must be two kinds of photo events in photosynthesis, one of them specifically associated with chlorophyll *b* (or other accessory pigment).[28]

By 1960, evidence was mounting for not only two pigment systems but also for two photochemical reactions associated with them.

Meanwhile Kok, who had observed a putative reaction center, now termed P700, demonstrated an enhancement type of effect for his system.[29] Far-red light bleached P700 whereas red light restored the pigment. In cyanobacteria, red light oxidized and orange light reduced P700.

The light-activated bleaching of a pigment at around 705 nm had been ascribed by Kok to P700, a pigment present at low concentrations.[30] Later he concluded that the pigment was likely to be a chlorophyll *a* molecule behaving as a single-electron oxidation–reduction system. Subsequently he suggested that it could act as the energy sink (i.e., a pigment to which absorbed energy in the chlorophyll mass would migrate):

> The rate of pigment turnover and some typical correlations with reaction conditions indicate that P700 could play a key role in photosynthesis. The location of its absorption band slightly beyond that of chlorophyll *a* would enable it to function as an efficient light energy collector. If indeed it acted as such a sink, its concentration (about 300–500 times less than the concentration of chlorophyll *a*) agrees well with the size of what has become known as the photosynthetic unit.[31]

The photosynthetic unit was a system of pigment molecules, particularly chlorophylls, that acted as collectors of light energy. This light energy was transferred to the longwave chlorophyll, known as P700, from the collecting molecules that became known as "antenna" molecules. But what was P700? Kok and others concluded it was a chlorophyll molecule. However, in the early 1970s, work with various physical methods, particularly electron-spin resonance and other advanced forms of spectroscopy led to the conclusion that P700 was, in fact, a chlorophyll dimer.[32]

7.5 THE Z-SCHEME AND TWO PHOTOSYSTEMS

In 1960 a diagram of the events associated with the putative two light reactions of photosynthesis was formulated and became a central feature of the understanding of the light reaction. At the outset the diagram was relatively simple but developed in complexity as the understanding of the light reactions and the associated electron-transport chains developed. The diagram, which became known as the Z-scheme, showed the pathways for electron transport in terms of the oxidation–reduction potential of the electron carriers.

[28] Myers and French 1960, p.736.
[29] Kok 1959.
[30] See Kok 1957.
[31] Kok 1961, p.527. The photosynthetic unit referred to is that arising from Emerson and Arnold's 1932 experiments and the work of Gaffron and Wohl.
[32] Norris et al. 1974.

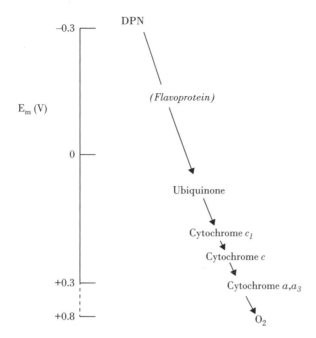

E_m (V)

-0.3 — DPN

(Flavoprotein)

0 —

Ubiquinone

Cytochrome c_1

Cytochrome c

+0.3 — Cytochrome a,a_3

+0.8 — O_2

Figure 7.3 The Electron-Transport Chain of the Mitochondrion Shown in Terms of the Oxidation–Reduction Potentials of the Carriers

The components of the chain from DPN (NAD) at –320 mV through the cytochromes to oxygen at around +800 mV are shown as their midpoint potentials in volts. There was uncertainty about the potential of the flavoprotein.

It was developed by analogy with the mitochondrial electron-transport chain, which is shown plotted against the standard potentials of the electron carriers in Figure 7.3.

The initial formulation of the Z-scheme by Hill and Bendall[33] is shown in Figure 7.4, where experimental results called for cytochrome b_6 to be reduced by light and cytochrome f to be oxidized by light. Such a situation required the input of light energy at two points, demonstrating the requirement for two light reactions in the photosynthetic process. Thus, on thermodynamic grounds two light reactions were confirmed.[34]

The notion of two light reactions in chloroplast photosynthesis was further strengthened during the 1960s. The scheme outlined by Hill and Bendall in 1960 was redrawn in the early 1960s in the form of a "Z," as shown in Figure 7.5, which is more complex than the 1960 version. Later formulations turned the diagram 90°, as shown in Section 7.7.

[33] Hill and Bendall 1960. The Z-scheme was a diagrammatic representation of the sequence of electron carriers in chloroplast electron transport based on their oxidation–reduction potentials. The scheme had the shape of the letter "Z." A version of this scheme from the early 1970s is shown in Figure 7.5.

[34] Hill points out that "the idea of two light reactions was therefore not new to us" (Hill 1965, p.142) as he was aware of the work of French and others leading to the conclusion that there were two light reactions in photosynthesis (see for example, Myers and French 1960).

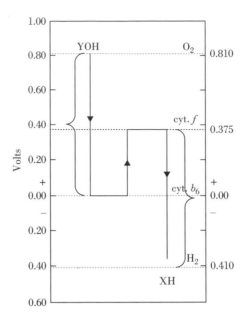

Figure 7.4 The Original Z-Scheme of Hill and Bendall 1960

The scheme shows the oxidation of water at +0.810 V, reducing cytochrome b_6 at approximately 0 V (a photochemical reaction) that reduces cytochrome f at + 0.375 V. Cytochrome f is oxidized by a further photochemical reaction reducing X. Note that X here is not the same as the electronegative electron acceptor of PSI described by Evans et al. in 1975. Reproduced with permission from Hill and Bendall 1960, "Function of the Two Cytochrome Components in Chloroplasts: A Working Hypothesis" (*Nature* 186: 136–137), Nature Publishing Group, fig. 4, p.137.

Commenting on Hill and Bendall's paper, Govindjee and Krogmann noted that the

Z-scheme was a detailed, explicit and testable formulation of the idea that there might be two separate light reactions: it made clear their relation to each other as a connection in series, and identified them with the two pigment systems. The Z-scheme also accounts for the observed minimum quantum requirement of oxygen evolution of eight.[35]

Thus, on theoretical grounds and by examination of the chloroplast oxidation–reduction system, two light reactions began to appear as an essential part of chloroplast photosynthesis. Such an interpretation needed experimental support.

The following year, Louis Duysens and coworkers produced evidence for two separate photochemical reactions by using a red alga, *Porphyridium*. One reaction was involved in the reduction and the other in the oxidation of cytochrome f. The first system was activated by 560-nm light (which activated the accessory pigment, phycoerythrin) and was

[35] Govindjee and Krogmann 2004. Note that the evolution of a molecule of O_2 in the Hill type of reaction involves the transfer of four electrons through the two light reactions, thus requiring eight photochemical events.

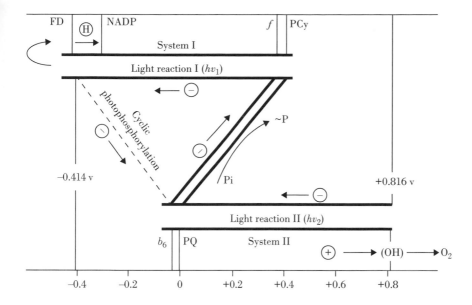

Figure 7.5 The Z-Scheme of Hill 1965

The scheme shows essentially the same process as in Figure 7.5 but in more detail, now drawn as a "Z." This formulation has been used to describe the light reactions but later was drawn as in Figure 7.10. Reproduced from Hill 1965, "The Biochemists' Green Mansion: The Photosynthetic Electron Transport Chain in Plants" (*Essays in Biochemistry* 1: 121–151), Biochemical Society, fig. 4, p.143.

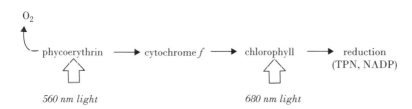

Figure 7.6 Two Light Reactions in Red Algae

A simplified representation of the conclusions from an experiment with a red alga showing two light reactions. Based on Duysens et al. 1961.

associated with O_2 evolution. In the second system, 680-nm light activated chlorophyll (see Fig. 7.6).[36] Such experiments served to link the concept of the Emerson enhancement effect with that of the oxidation–reduction system of the chloroplast. The demonstration by Rajni Govindjee and coworkers that the Hill reaction itself could involve two light reactions further strengthened this understanding.[37]

[36] Duysens, Amesz, and Kamp 1961.
[37] Govindjee, Thomas, and Rabinowitch 1961.

Figure 7.7 Demonstration of Two Light Reactions in Chloroplast Photosynthesis
Reduction of dichlorophenol indophenol (DCPIP) by water involving a first light reaction and oxygen evolution.
Reduction of NADP by reduced DCPIP involving a second light reaction with chlorophyll (Chl). (Based on Losada et al. 1961.)

Additional support for the concept of two light reactions came from Arnon's laboratory, where the oxidation and reduction of the dye dichlorophenol indophenol was used to demonstrate two separate and independent light reactions. The dye was used as both electron acceptor and donor (see Fig. 7.7).

Reduction of the dye was associated with O_2 evolution in a Hill type of reaction. However, the dye could be oxidized in the light and NADP (TPN) reduced, and this process was linked to phosphorylation; here there was no O_2 evolution.[38]

Duysens and his colleagues designated these two photochemical systems as system I and system II, eventually becoming known as photosystems I and II (PSI and PSII). By the early 1960s, system I was characterized by being activated by far-red light and associated with chlorophyll *a*. It was associated with NADP reduction and probably had a reaction center, P700 as described by Kok. Roderick Clayton considered that the primary reactions of P700 and also the bacterial P890 were of the one quantum → one electron-transfer type.[39] System II, about which rather less was known, was activated by shorter wavelengths and was associated with accessory pigments (chlorophyll *b*, and in red algae, phycobilins, etc.) and forms of chlorophyll *a* absorbing light at shorter wavelengths. It was linked to O_2 evolution. Although these interpretations probably represented the majority view, there were alternative short-term proposals, and, in any case, revisions became necessary.

Around 1960, as discussed elsewhere, the use of detergents with mitochondria began to enable the separation of the four complexes of the respiratory chain. Such work seems to have inspired N. Keith Boardman and Jan Anderson who in 1964 managed to separate by centrifugation two fractions from spinach chloroplasts treated with the detergent digitonin. These two fractions appeared to be concentrated in PSI and PSII and also differed in their pigment composition. This partial physical separation of two photosystems gave impetus to the notion that the light reactions were spatially separate and that subchloroplast particles capable of specific activities could be separated

---

[38] Losada, Whatley, and Arnon 1961. Note that Arnon later abandoned two light reactions for NADP reduction.
[39] See Clayton 1963.

out.[40] This approach was further developed successfully in several laboratories in the following years.

The identification of a reaction center, P700, in what became known as PSI, left open the question of whether PSII also involved a comparable reaction center. Horst Witt (1922–2007),[41] who had developed sensitive spectroscopic techniques for photosynthetic study at the Max Planck Institute of Physical Chemistry and from 1962 at the Max-Volmer Institute for Physical Chemistry at the Technical University of Berlin, had earlier also provided evidence for the Z-scheme.[42] In 1969 he observed rapid absorbance changes at 682 nm, which he attributed to a chlorophyll molecule in PSII. Later this became known P680, the reaction center for PSII.[43]

The role of light in photosynthesis could be seen as providing the energy, through two reaction centers, for electron transport from water to NADP, a process that would therefore require two light quanta per electron. This process would be coupled to the synthesis of ATP as seen in the mitochondrial oxidative phosphorylation system. The nature of the electron-transport system was substantially elucidated during the 1960s. The reaction attributed to Robin Hill could then be described as

$$H_2O \rightarrow P680\left[\text{photosystem 2}\right] \rightarrow \text{cytochrome } f \rightarrow P700\left[\text{photosystem 1}\right] \rightarrow NADP.$$
$$\downarrow$$
$$O_2$$

This formulation omits those several electron carriers recognized as intermediates in the photosynthetic electron-transport system in the 1960s (see Section 7.7).

7.6 THE CONTRIBUTION FROM BACTERIAL PHOTOSYNTHESIS

As with the study of oxidative phosphorylation, the study of simpler bacterial systems made noteworthy contributions to understanding photosynthesis. Of particular value was the small size of the photosynthetic unit in purple bacteria compared with that in plant chloroplasts. As discussed earlier, a small number of mostly anaerobic bacteria were found to carry out photosynthesis, and these had been investigated particularly by van Niel.[44] These were the purple sulfur bacteria (*Chromatium*), the purple nonsulfur bacteria (*Rhodopseudomonas, Rhodospirillum, Rhodobacter*), and the green sulfur bacteria (*Chlorobium*). Photosynthesis in these bacteria did not involve O_2 evolution. The introduction of high-resolution electron microscopy in the early 1950s showed that in the purple nonsulfur bacterium, *Rhodospirillum rubrum,* and other species, the cell

---

[40] Boardman and Anderson 1964.
[41] See Junge and Rutherford 2007 for a brief biographical note on Horst Witt.
[42] Witt, Müller, and Rumberg 1961.
[43] Döring et al. 1969. See also Floyd, Chance, and Devault (1971), who used low-temperature studies for detection of this pigment.
[44] Van Niel 1935, 1944; see also Chapter 6.

contained a number of vesicles named chromatophores.[45] These could be isolated and shown to have the bacterial form of chlorophyll, bacteriochlorophyll.[46] The vesicles were subsequently found to have a close association with the cell membrane and in 1954 were shown by Albert Frenkel to carry out photophosphorylation.[47] These experiments demonstrated that chromatophore preparations were capable of photochemical activity and initiated the biochemical study of bacterial photosynthesis in a manner analogous to Hill's development of the use of chloroplasts.

The bacteriochlorophyll absorption spectrum was found to differ from that of plant chlorophyll in that the red absorbance maximum in vivo was shifted into the infrared, in *Chromatium* for example, around 800, 850, and 870 nm. A reaction center for these pigments was first detected by Duysens and mentioned in his 1952 thesis. In purple bacteria he found a bleaching of a small proportion of the bacteriochlorophylls that he attributed to an oxidation, at around 890 nm.[48] This bacteriochlorophyll constituted about 2% of the total and was later named P890.[49] In 1963, with W. J. Vredenberg, he demonstrated from fluorescence studies the transfer of energy to the proposed reaction center. Similar conclusions were reached concurrently by Clayton. Duysens concluded that

> The high efficiency for bleaching of *P*890 indicates that *P*890 may be an intermediate in bacterial photosynthesis. *P*890 is suggested to be a pigment similar to *B*890; it is, however, non- or weakly fluorescent and photochemically active; it may act as the photosynthetic reaction center.

This group also thought that their results could be "quantitatively explained by energy transfer from bacteriochlorophyll to a pigment *P*890 present in small amounts."[50]

Also in the early 1960s, in green photosynthetic bacteria, energy transfer from chlorobium chlorophyll (Chl 660) to chlorophyll 770 was demonstrated in which this pigment also appeared to act as the reaction center although the latter point needed confirmation.[51] In fact, the *Chlorobium* system proved more difficult to work with, and the

---

[45] Schachman, Pardee, and Stanier 1952.

[46] By the 1960s, bacteriochlorophyll was found to occur as a closely related group of pigments in bacteria. Bacteriochlorophyll *a* was found to be widely distributed in photosynthetic bacteria whereas bacteriochlorophyll *b* was found in some purple bacteria. Bacteriochlorophylls *c* and *d* were found in the green bacteria and are referred to here as chlorobium chlorophylls. Further species of bacteriochlorophyll were identified later.

[47] Frenkel 1954, 1956.

[48] Duysens et al. 1956.

[49] The designation of bacterial reaction centers was "P" followed by their absorption maximum. The maximum differed among species, but the designations of P870 and P890 were made for the purple bacterial centers.

[50] Vredenberg and Duysens 1963, p.357.

[51] Sybesma and Olson 1963.

isolation of reaction centers proceeded much more slowly owing in part to the large number of antenna chlorophylls.[52]

If energy was transferred from the chlorophyll mass to the reaction center, what was the nature of the photochemical reaction? Several studies suggested that it was a single-electron transfer. Various laboratories also concluded that a cytochrome c was the likely electron donor to the system. Cytochrome c was known to be a major component of chromatophores, and its oxidation–reduction could readily be followed spectroscopically following a flash of light, giving results that supported the role of this cytochrome as donor.[53] Several candidates for the electron acceptor were proposed in the late 1960s. Early observations of changes in the ubiquinone spectrum suggested that this might be the electron acceptor. However, an iron signal was obtained suggesting that a nonheme Fe could be the acceptor or even a complex of ubiquinone and Fe.

A first and significant step in resolving these problems was the isolation of a reaction center complex from the chromatophores of the purple bacterium *Rhodopseudomonas sphaeroides*[54] achieved by Roderick Clayton (1922–2011) in 1968.[55] This complex obtained by detergent treatment contained the P870, two molecules of an associated pigment, P800, also a bacteriochlorophyll. It also contained ubiquinone, a b-type cytochrome, a c-type cytochrome, and iron and copper atoms. The ability to isolate these complexes opened a new phase in the study of bacterial photosynthesis. Improved preparations were made and the particles further analyzed. They contained three proteins, two of which were essential for photochemical activity.[56] The pigment content for a single reaction center could be refined to four bacteriochlorophyll and two bacteriopheophytin[57] molecules.[58] The P870 pigment, like that in higher plant reaction centers, was composed of a pair of bacteriochlorophyll molecules.[59]

The 1970s were a period of extensive study on the isolated pigment–protein complexes, the reaction centers. These differed slightly from species to species.

As Roderick Clayton later commented,

> The RCs of the non-sulfur purple bacteria *Rhodobacter sphaeroides* and *Rhodopseudomonas viridis* have proven to be especially easy to extract and purify and have been the most rewarding sources of detailed information on chemical patterns, composition, structure, and physical mechanisms.[60]

---

[52] Antenna molecules (chlorophylls and other pigments) had the function of light absorption and passing the energy to other molecules (eventually the reaction center) rather than photochemical activity.

[53] Olson and Chance 1960.

[54] Later this organism became known as *Rhodobacter sphaeroides*.

[55] Reed and Clayton 1968; see also Reed 1969.

[56] Okamura, Steiner, and Feher 1974.

[57] The structure of bacteriopheophytin is similar to that of bacteriochlorophyll but without the central magnesium atom.

[58] Straley et al. 1973.

[59] Norris et al. 1974.

[60] Clayton 2002, p.64. RCs refer to reaction centers.

However, in the purified preparations from the purple nonsulfur bacteria (*Rhodopseudomonas spheroides* or *Rhodospirillum rubrum*), the protein was composed of one each of three subunits (molecular weights of 28,000, 24,000, and 21,000).[61] The principal pigment components were four bacteriochlorophylls, two associated with P870, and two bacteriopheophytin.[62] A carotenoid was also a probable constituent. The nature of the photochemical reaction was also much debated during this period, particularly the nature of the electron acceptor for which iron and a quinone were both possibilities. Feher's group in 1975 concluded that ubiquinone was the primary electron acceptor for the reaction center and that the iron was closely associated but apparently not the acceptor.[63] As previously noted the donor was earlier agreed to be a *c*-type cytochrome although the details of this varied with the organism.

By the mid-1970s, opinion about the nature of the primary electron acceptor from P870 had begun to favor an iron–quinone complex. At the same time picosecond (10^{-12} sec) technology was being developed in photosynthetic studies. This led to the detection in at least two laboratories of an intermediate electron carrier between P870 and the quinone–iron complex, initially known as "I" but soon identified with the bacteriopheophytin.[64] Thus the transfer of electrons from light-excited P870 (P870*) could be shown as

$$P870^* \; BPh \; Fe-Q \rightarrow P870^+ BPh^- Fe-Q \rightarrow P870 \; Bph \; Fe-Q^-.$$

An electron from the cytochrome *c* complex would be transferred to the P870$^+$, returning it to P870. The view of the reactions associated with the reaction center is shown in Figure 7.8. The left-hand scale shows the midpoint potentials. Light activation of the pigment P870 (BChl$_2$) lowers its potential from +450 mV to –550 mV, enabling it to reduce the low-potential bacteriopheophytin. A cycle of electron transfers follows, leading to reduction of the positively charged pigment, P870.

Feher and his colleagues summarized the understanding of the reaction center:

> Thus by the middle of the 1970s, it was established that the primary donor is a bacteriochlorophyll dimer, the primary and secondary acceptors are quinones and a transient intermediate acceptor is a bacteriopheophytin. The secondary donor in *Rb sphaeroides* is an exogenous water soluble cytochrome c_2, whereas in *Rps viridis* it is a multiheme cytochrome which forms an integral part of the reaction center. The reaction center can be viewed as a protein scaffolding that holds the various cofactors (reactants) in their appropriate positions for efficient electron transfer.[65]

Such conclusions required further confirmation, and experimentation continued to clarify the operation of the reaction center. However, the next key step was the full

---

[61] Okamura, Steiner, and Feher 1974.
[62] Reed and Peters 1972.
[63] Okamura, Isaacson, and Feher 1975.
[64] Fajer et al. 1975.
[65] Feher et al. 1989, p.111.

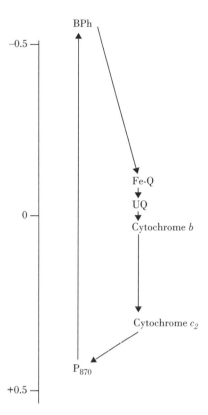

Figure 7.8 Scheme Showing the Reactions Associated With Light Activation of the Reaction Center in the Photosynthetic Bacterium, *Rhodopseudomonas*

When the reaction center, P_{870} of the photosynthetic purple nonsulfur bacterium, *Rhodopseudomonus sphaeroides* is activated by light, it reduces bacteriopheophytin (BPh) and in turn reduces the iron–quinone complex (Fe–UQ). The complex reduces the cytochromes, which in turn reduce the reaction center. E_m is the midpoint potential. Ubiquinone (UQ) is the quinone in this species (Based on Fajer et al. 1975.)

description of the structure of a bacterial reaction center achieved in the mid-1980s (see Chapter 8) and also the isolation and properties of the green bacterial reaction center.

The successful work on the bacterial reaction center gave a strong basis for understanding the fundamental nature of the mechanism of photosynthesis. It was possible because the structure in bacteria was much simpler than that in the chloroplast system. It did provide a strong boost to research on chloroplast bioenergetics.

7.7 THE CHLOROPLAST ELECTRON-TRANSPORT CHAIN

The 1960s was a time when the nature of the chloroplast electron-transport chain took shape. The work of Duysens, Amesz, and Kamp in 1961 had demonstrated that cytochrome *f* (originally discovered by Hill; see Chapter 6) was reduced by one photosystem and oxidized by a second photosystem.[66] In the years that followed it was clearly confirmed that this cytochrome functioned between PSII and PSI. Two b-type cytochromes had also been located in the chloroplast. Cytochrome b_{559} was generally located close to PSII on the reducing side but whether and where it functioned in the main chloroplast electron-transport chain remained a matter for debate to the end of the twentieth century. Cytochrome b_6 also proved a major problem because some results suggested that it was reduced by both photosystems! The difficulties here are reminiscent of those found with cytochrome b in mitochondrial electron transport. As previously noted, Arnon had described a cyclic electron flow around PSI and cytochrome b_6 was seen as possibly part of that cyclic system.

The identification of electron carriers as potential members of the chloroplast electron-transport chain can best be considered in the events surrounding each photosystem. These could be examined in terms of the electron donor to the system and the electron acceptor. The end result of investigations was an oxidation–reduction pathway involving a number of carriers from water as the initial electron donor to NADP as the final electron acceptor.

The reducing side of PSI. In 1952, Davenport, Hill, and Whatley had originally described the "methaemoglobin reducing factor," a protein discovered in their attempt to understand the reducing activities of the Hill reaction. At this stage the protein proved unstable. However, the group isolated the protein in 1960 and described some of its catalytic properties, although not its nature, and showed it was identical with the photosynthetic pyridine nucleotide reductase of San Pietro and Lang.[67] Finally the protein was isolated and purified by Arnon's group, who named it ferredoxin because of its similarity to proteins given that name in nonphotosynthetic bacteria. Ferredoxin was shown to be an iron protein that transferred a single electron from the primary photochemical reaction center of PSI to NADP.[68] The transfer of the electron from the reaction center to NADP required a flavoprotein enzyme (a ferredoxin–NADP reductase) bound to the chloroplast and originally described by Avron and Jagendorf.[69] By the early 1960s the reduction of NADP by P700 could be seen as follows:

$$P700 \: [PS \: I] \rightarrow ferredoxin \rightarrow ferredoxin - NADP \: reductase \rightarrow NADP.[70]$$

Around 1970, questions began to arise on whether ferredoxin was the primary electron acceptor for PSI. A more electronegative electron acceptor was thought to be likely. Richard Malkin and Alan Beardon in 1971 obtained evidence with EPR spectroscopy[71]

---

[66] Duysens, Amesz, and Kamp 1961.
[67] Davenport 1960; Davenport and Hill 1960.
[68] Tagawa and Arnon 1962.
[69] Avron and Jagendorf 1956.
[70] PSI and PSII are used to denote the two photosystems. PSII is concerned with O_2 evolution and PSI initially with pyridine nucleotide reduction and with ferredoxin reduction.
[71] EPR or electron paramagnetic resonance spectroscopy detects unpaired electrons in, for example, organic radicals or transition metal (e.g., iron) ions.

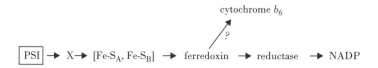

Figure 7.9 Reduction of NADP by PSI

The diagram shows the pathway of electrons from PSI to NADP. Fe-S$_A$ and Fe-S$_B$ are the two iron–sulfur centers reducing ferredoxin and the reductase is the ferredoxin-NADP reductase.

of a bound iron–sulfur protein acting as a primary acceptor and with a lower oxidation–reduction potential than ferredoxin.[72] Concurrently an associated optical change was observed and referred to as P430; it was attributed to an iron–sulfur protein. However, by the mid-1970s, at least two other bound iron–sulfur centers had been detected (centers A and B). Subsequently an even more electronegative component, X, was proposed as the primary electron acceptor and also thought to be an iron–sulfur center.[73] The matter became increasingly complex, leading Richard Malkin to comment in 1982 that

> Although our understanding of the primary charge separation in PSI was greatly advanced by characterization of P700 in the 1960s, the status of the component which accepts the electron released from P700 has been a more controversial matter. In recent years with the application of rapid kinetic spectroscopy and low-temperature EPR techniques, this electron transfer sequence has been understood in much greater detail, and it is now believed that several electron acceptors function in the primary electron acceptor complex of PSI and are involved in producing a stable charge separation.[74]

The details of the mechanisms involved remained an issue for a considerable time.

Arnon had originally described what he termed cyclic photophosphorylation. This was a photophosphorylation characterized by the absence of O$_2$ evolution and a requirement for a cofactor (originally vitamin K or flavin monophosphate). During the studies of ferredoxin it was discovered that this protein was capable of promoting cyclic phosphorylation.[75] Studies suggested that ferredoxin, reduced by PSI, might reduce cytochrome b_6 although there was conflicting evidence. Indeed the understanding of the cyclic pathway of electron transport remained uncertain for a considerable time, and there were questions about whether such a pathway was an artefact. In due course the view that cytochrome b_6 was reduced by PSI in a cyclic process strengthened. Overall, the reduction process by PSI in the mid-1970s appeared to be as shown in Figure 7.9.

Electron donation to PSI. Elucidating the sequence of electron carriers between the two photosystems did not prove an easy task. Among the problems that arose during the

---

[72] Malkin and Bearden 1971.

[73] See for example Evans et al. 1975.

[74] Malkin 1982, p.459.

[75] Arnon 1965.

1960s was Arnon's proposal, eventually rejected, that there were not two but three light reactions in chloroplast photosynthesis, complicating attempts to understand the chloroplast electron-transport chain.

Early work had shown unambiguously that cytochrome f was a donor to PSI but not necessarily the primary donor. Another electron carrier that might be the primary donor was a blue copper protein named plastocyanin. It was first discovered in the alga *Chlorella* by the Japanese worker Sakae Katoh but was soon found in chloroplasts.[76] Its midpoint potential was close to that of cytochrome f, and its location in the electron-transport chain was seen as close to that of the cytochrome.[77] Various studies[78] suggested that plastocyanin was necessary for cytochrome f oxidation by PSI (P700) although contrary results were also obtained, by Arnon's group for example. Progressively, workers concluded that the balance of evidence supported the sequence:

$$\text{cytochrome } f \rightarrow \text{plastocyanin} \rightarrow \text{P700}.$$

Thus plastocyanin became the agreed electron donor to P700.

The reducing side of PSII. As noted when discussing the mitochondrial respiratory chain, the late 1950s was a time when there was substantial interest in quinones. This was briefly discussed earlier, but the discovery of a quinone in leaves, similar to ubiquinone, initiated extensive studies of the role of this compound during the 1960s.[79] It was named plastoquinone, and its importance was suggested by the demonstration that it appeared to be necessary for the Hill reaction. Studies of its oxidation and reduction showed that plastoquinone functioned close to the reducing side of PSII.[80] By the end of the decade it was clear that plastoquinone functioned to transfer electrons between the reducing side of PSII and cytochrome f[81] and that there was a substantial pool of plastoquinone molecules operating in this location.

However, evidence suggested that the plastoquinone pool was not the primary electron acceptor for PSII. The identity of the primary electron acceptor for P680 was a more difficult question. Duysens and his colleague had detected a substance that quenched chlorophyll fluorescence of PSII and referred to it as Q. In the late 1960s, Witt's group detected an absorption change at 320 nm, named X_{320}, which they identified as the primary electron acceptor of PSII and as a semiquinone, Q^-.[82] For several years the matter was debated but in 1977 the issue appeared to be settled by David Knaff and coworkers, who confirmed that the acceptor for P680 in PSII was a special plastoquinone that accepted a single electron, becoming a semiquinone.[83]

---

[76] Katoh 1960.

[77] See Hind 1968.

[78] See for example Avron and Shneyour 1971.

[79] Bishop 1961. The general structure of plastoquinone was known by 1960.

[80] Rumberg et al. 1963.

[81] For a view of the role of plastoquinone as seen at the beginning of the 1970s, see Böhme and Cramer 1972.

[82] Stiehl and Witt 1969.

[83] Knaff et al. 1977.

Electron donation to PSII, photolysis of water. PSII was associated with the photolysis of water and the release of O_2:

$$2H_2O \rightarrow O_2 + 4e + 4H^+.$$

The question of electron donation from water to P680 in PSII nevertheless proved challenging. During the period under consideration, two key features of the process emerged. First, Pierre Joliot used short flashes of light with dark-adapted chloroplasts and measured O_2 evolution. He showed that O_2 was evolved after four photochemical events at the reaction center. Kok proposed that the light reaction caused the accumulation of four positive charges in the O_2-evolving complex, leading to the release of O_2.[84]

Second, there was growing evidence that the process of O_2 release from water involved manganese. Then, in 1970, George M. Cheniae and I. F. Martin showed a clear dependence of O_2 evolution on manganese.[85] Observations have suggested that a manganese protein accumulated the positive charges as a result of PSII photoreactions and that the protein then reacts with water to evolve O_2.[86] The mechanism remained obscure.

The revised Z-scheme. The major events associated with the chloroplast electron-transport chain emerged during the 1960s and 1970s so that a broad framework for O_2 evolution from water and reduction of NADP was now apparent. These are shown in Figure 7.10, which gives the view from the early mid-1970s. This exemplifies the widely used Z-scheme originally proposed by Hill and Bendall, although they drew the diagram somewhat differently (see Fig. 7.5). Many details remained to be resolved, and confirmation for some aspects was still being sought. For example, Arnon had demonstrated a cyclical process around PSI but the details of this pathway remained tentative.

A new view now emerged in photosynthesis, the realization of the close similarity between PSII and the purple bacterial reaction center. In the same way, the slightly later elucidation of the green bacteria, *Chlorobium*, showed a closer link between that reaction center and that of PSI.

7.8 CHLOROPLAST PHOTOPHOSPHORYLATION

Photophosphorylation involving an oxidation–reduction system or electron-transport chain was, by the 1960s, becoming central to the concept of photosynthesis. It is important to realize how much of the appreciation of the mechanisms involved were dependent on mitochondrial studies, which tended to be the guide for those in this field. Indeed, Jagendorf observed in 1959 that the field "had achieved the position of mitochondrial research in 1952.[87] Up until the early 1960s, the chemical hypothesis of Bill Slater that proposed that the energy of respiration was used for the synthesis of a high-energy chemical intermediate that would drive the synthesis of ATP provided the basis for thinking about chloroplast systems. Slater in 1953 had not considered photophosphorylation

[84] Kok, Forbusch, and McGloin 1970, but see also Joliot and Kok 1975 and Joliot 2004.
[85] Cheniae and Martin 1970.
[86] Wydrzynski et al. 1976.
[87] Jagendorf 2002, see p.234.

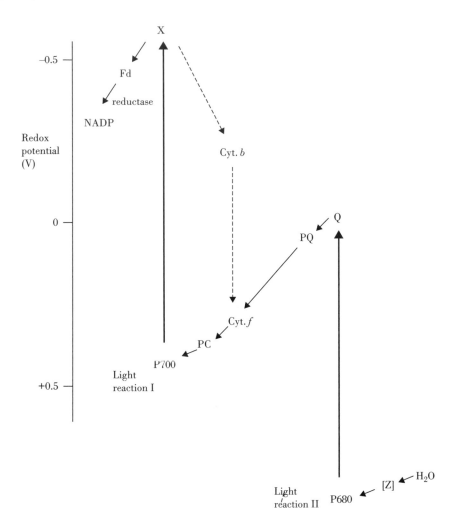

Figure 7.10 The Z-Scheme for Chloroplast Electron Transport as Seen in the Early 1970s
This is an example of this scheme that was widely used to summarize the understanding of the chloroplast electron-transport chain. The left-hand axis gives the midpoint potential of the intermediates. These are Fd: ferredoxin; reductase: the ferredoxin NADP reductase; X: the primary electron acceptor for photosystem I; P700: the reaction center for PSI; PC: plastocyanin (the position of this component was still slightly uncertain); Cyt f: cytochrome f; cyt b_6: cytochrome b_6 (although there was some evidence that this cytochrome was reduced by PSI, there was also a possibility that it was reduced by PSII, and there was significant confusion about its role); PQ: plastoquinone; Q: the primary electron acceptor for PSII; P680: the reaction center for PSII Z: an unidentified intermediate.

(Arnon and Frenkel had yet to publish their evidence for the process). However, the Slater phosphorylation model (known as the chemical theory) was adapted by workers in photosynthesis such as Avron and Jagendorf, who provided a version of the chemical hypothesis for chloroplast phosphorylation:

To account for this phenomenon of coupling, a working hypothesis was suggested based on the widely described schemes for oxidative phosphorylation in mitochondria.[88]

In contrast to Slater's model, Mitchell's mechanism (chemiosmotic hypothesis[89]) was put forward for both oxidative and photosynthetic phosphorylation and after discussion with Daniel Arnon, among others.[90] By the time the chemiosmotic proposal was formulated, it had become clear from the work of Hill, Arnon, and others that the chloroplast system had much in common with the mitochondrial system and that they presented a common problem. As Mitchell commented about his hypothesis,

> It is evident that the basic features of the chemiosmotic coupling concep-
> tion . . . are in accord with much of the circumstantial evidence at present avail-
> able from studies of oxidative and photosynthetic phosphorylation.[91]

However, in the early 1960s, the interpretation of experiments on phosphorylation was generally made in terms of the chemical intermediate of the chemical hypothesis.

This assumption continued until Jagendorf's group noted that the high-energy intermediate of photophosphorylation appeared to be a pH gradient. As he wrote to Mitchell,

> Specifically in the chloroplast phosphorylating system we have discovered the
> existence of a pool of high energy, non-phosphorylated "intermediate;" and
> all of our data so far seem to be consistent with it being a pH gradient, as you
> predicted.[92]

Subsequently, the group showed that a pH gradient could drive ATP synthesis even in the presence of electron-transport inhibitors, such as dichlorophenyldimethylurea (DCMU).[93] These experiments provided the first independent evidence in support of Mitchell's chemiosmotic hypothesis in which a pH gradient normally formed by an electron-transport chain was expected to drive ATP synthesis by an ATP synthase.[94] The experiment was successfully repeated in chloroplasts in other laboratories but, as noted elsewhere, attempts to carry out a similar experiment in mitochondria were unsuccessful for some years, a situation that tended to detract from the value of the chloroplast observation.

[88] Avron and Jagendorf 1959, p.967.

[89] See Chapter 5, Section 5.3.

[90] See letter quoted in Prebble and Weber 2003, p.79.

[91] Mitchell 1961, p.148.

[92] Letter, Jagendorf to Mitchell, May 25, 1964, quoted in Prebble and Weber 2003, p.124.

[93] DCMU, a herbicide, inhibited photosynthesis in the region of PSII.

[94] Jagendorf and Uribe 1966. Jagendorf was one of those who found the chemiosmotic concept as described by Mitchell extremely difficult to understand. Geoffrey Hind, a research student, was valuable in explaining the concept to Jagendorf. See Jagendorf 2002.

Jagendorf sent an advance copy of his manuscript to Mitchell, commenting in the covering letter that with these sorts of data, "the acceptance probability for the chemiosmotic theory was bound to rise. I'm at the 75% level for sure, now."[95] Indeed the chemiosmotic theory seems to have been more readily accepted by those in the photo-synthesis field. Certainly during the 1960s and 1970s other approaches to the mechanism of photophosphorylation were considered, particularly those based on conformational concepts but also Williams's proposals. However, the majority of workers came to find chemiosmotic concepts a valuable way to view photophosphorylation. By 1978, reviewers were able to write that

> There seems to be general agreement that in the chloroplast thylakoid, light in-duced vectorial electron transport causes vectorial proton movement, building up a ΔpH (pH gradient) and/or a $\Delta \psi$ (membrane potential) across or within the membrane. It is also thought that this combination, the PMF (proton motive force) can drive ATP synthesis in vitro, but it is not agreed that it does so in vivo, or if it does, how the process is carried out.[96]

The question of whether laboratory-obtained evidence reflected the in vivo situation bedeviled work on photophosphorylation and also oxidative phosphorylation.

Another area where mitochondrial studies influenced chloroplast research was in coupling factors. Following Racker's work on F_1, Avron isolated from chloroplasts a coupling factor capable of stimulating phosphorylation in depleted chloroplasts, which proved to be unstable.[97] Subsequently, Racker's group, which became interested in photophosphorylation, isolated an ATPase from chloroplasts. This CF_1 (chloroplast factor 1) was shown to strongly stimulate photophosphorylation in subchloroplast particles although the properties of the protein were quite distinct from the mitochon-drial one.[98] Like the mitochondrial enzyme, the chloroplast enzyme could be seen as particles attached to the membrane. Further, they could be removed and recombined with the membrane to restore the original phosphorylation activity.[99] Subsequently ev-idence for a chloroplast CF_0 was obtained, and the ATPase particles could be shown in electron micrographs to attach to membranes containing the presumed CF_0.[100] One marked difference from the mitochondrial enzyme (F_1) was that the purified coupling factor from chloroplasts had no ATPase activity unless activated. Treatment by trypsin and other agents served to induce rapid ATPase activity.

Locating the sites of phosphorylation in the electron-transport chain did not prove easy. The number of ATPs synthesized when two electrons passed from water to NADP was originally estimated by Arnon as less than one. Subsequently higher values were

[95] Letter, Jagendorf to Mitchell, December 17, 1965.
[96] Reeves and Hall 1978.
[97] Avron 1963.
[98] Vambutas and Racker 1965. These workers found a number of differences between the CF_1 of chloroplasts and the F_1 of mitochondria.
[99] Oleszko and Moudrianakis 1974.
[100] Carmeli and Racker 1973.

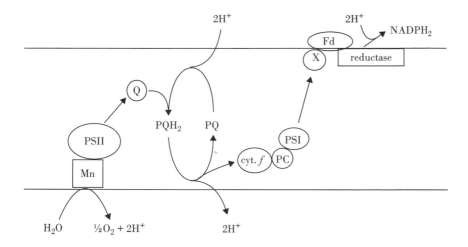

$2H^+$ $2H^+$

Figure 7.11 Noncyclic Photosynthetic Electron Flow From Water to NADP Across the Thylakoid Membrane Showing the Probable Distribution of Components as Understood in the Mid-1970s

The system produced four protons on the inside and took up four protons from the outside for each pair of electrons passing from water to NADP. Fd: ferredoxin; red: Fd-NADP reductase; X: primary electron acceptor in PSI; PC: plastocyanin; Cyt f: cytochrome f; PQ shuttle: the plastoquinone molecules that are reduced by Q and H^+ and oxidized by Cyt f; Q: the primary electron acceptor for P680; Mn: the manganese protein associated with oxygen evolution. (See for example Trebst 1974, fig. 5, p.447.)

obtained in a number of laboratories, leading to the view that the maximum value could be up to two.[101] The location of the sites for ATP synthesis was initially seen by Arnon's group as lying between the two photosystems. Later Avron and Chance showed that there was a site for ATP synthesis between PSII and cytochrome f, a result refined by H. Böhme and W. A. Cramer to between plastoquinone and cytochrome f.[102] Plastoquinone could be seen (in Mitchell's terms) as conveying protons *across* the membrane.

As already noted, many experiments suggested that more than one ATP was synthesized in the pathway from water to NADP. For those, the majority, who accepted Mitchell's chemiosmotic approach, it was necessary to consider the sidedness of the thylakoid[103] membrane because Mitchell's proposals saw transport of protons across the membrane. The reaction that oxidized water, donated electrons to PSII, and released protons was known to be on the inside of the thylakoid membrane whereas the reaction reducing NADP was on the outside.[104] Such a configuration led to the release of protons into the lumen of the thylakoid and uptake from the outside (matrix) contributing to the proton gradient across the membrane (see Fig. 7.11). Thus, if the chemiosmotic

[101] See for an example of higher values, Isawa and Good 1968.

[102] Avron and Chance 1966; Böhme and Cramer 1972.

[103] The structure of the chloroplast was seen as being composed of a number of flattened vesicles or sacs that traversed the chloroplast within its double membrane. These sacs were termed thylakoids. It had become apparent that the light reactions and the electron-transport chain were located in the membranes that formed the thylakoids.

[104] See Junge and Ausländer 1973.

approach was accepted, there were grounds to support the notion that more than one ATP was synthesized in electron transport from water to NADP. However, the lack of a consensus on the role of protons in ATP synthesis left the overall situation uncertain on the number of ATPs synthesized per pair of electrons in noncyclic phosphorylation. There was also debate at this stage about the importance of the cyclic process complicated by the difficulty of measuring the rate of cyclic electron transport. Resolution of these problems required a more detailed understanding of the ATP synthase and the electron-transport chain.

As Robin Hill pointed out, preparations of chloroplasts preceded the availability of comparable mitochondrial preparations by some 10 years. The complications brought about by the two light reactions in photosynthetic electron transport resulted in slower progress than in mitochondria, and questions to which there seemed convincing answers in mitochondria were much less readily resolved in chloroplasts. However, by the late 1970s, the outline mechanism of photosynthesis was understood, but this understanding needed experimental support from the more detailed aspects of the system. There remained the need to investigate the more intricate events in the photosynthetic unit, the reaction centers, electron transport, and photophosphorylation. Progress in this area would depend on the molecular biological approach.

8

The Impact of Protein Technology
1977–1997

In 1967, Bill Slater, taking into account the role of lipid and other hydrophobic areas, pondered on how the mechanism of oxidative phosphorylation (and photophosphorylation) might be elucidated:

> Studies of the role of hydrophobic regions in catalytic proteins might give a clue. I suspect, however, that it will be a long time before we understand the mechanism of oxidative phosphorylation. It is an open bet whether it will be the enzymologist, the membranologist or the protein chemist who will give us the answer.[1]

In 1981, reflecting on this comment, Lars Ernster and Gottfried Schatz wrote

> the answer will come from all three, in a relay, and that the bâton has just passed from the membranologist to the protein chemist.[2]

Many of the advances in the period covered by this chapter were made on the basis of a much more detailed understanding of the nature of the proteins involved in oxidation–reduction and in phosphorylation, particularly the ability to construct three-dimensional models of the various complexes.

8.1 THE FIFTH PERIOD OF INVESTIGATION

This chapter is concerned with the history of both studies of oxidative phosphorylation and studies of photosynthesis. By the 1980s most of the problems in the two fields of research were broadly the same, except for those concerned with the role of light, so that the methods applied were similar. Although the issues raised for oxidative phosphorylation were also those of photosynthesis, the latter field also faced considerable challenges relating to photochemistry. As with earlier periods, major advances tended to be made first in the oxidative phosphorylation field.

---

[1] Slater 1967, p.9.
[2] Ernster and Schatz 1981, p.247s.

Work done in previous years left unresolved a number of fundamental issues derived from earlier approaches to the mechanism of oxidative phosphorylation. Two of these, crucial to the history of bioenergetics, are considered here; they concern the stoichiometric relationships of phosphorylation and proton transport together with the nature of uncoupling between phosphorylation and electron transport. These rather detailed studies aided the comprehension of the mechanisms of bioenergetics, primarily in mitochondria, but their conclusions proved relevant to bioenergetic systems in general.

Perhaps the major question in the late twentieth century concerned the proton. What were the mechanisms of proton movements linked to oxidation–reduction and to phosphorylation? This underlies much of the work described in this chapter. I seek to show the way in which knowledge of proteins provided the missing element in understanding bioenergetic processes, including the mechanism of oxidative phosphorylation and photophosphorylation. Until the late 1970s knowledge had depended primarily on what was known of quinones, flavin groups, iron–sulfur centers, and heme groups, mostly gained from various types of spectroscopy. The advent of protein chemical techniques, especially the ability to isolate membrane proteins and to work successfully with them, transformed the study of the subject, a transformation that continued after the end of the twentieth century.

8.2 THE CHEMIOSMOTIC MECHANISM FOR OXIDATIVE AND PHOTOSYNTHETIC PHOSPHORYLATION

By the end of 1978 there was general, though not universal, agreement that the overall mechanism of oxidative phosphorylation and photophosphorylation could be described in the terms of the chemiosmotic theory: that is, the creation of a proton electrochemical gradient by an electron-transport chain and the utilization of the energy of the gradient for ATP synthesis. It is difficult to identify any experiment at that time that clinched the deal for the proton-based approach, although a large body of experimental evidence had accumulated in the later 1960s and early 1970s.[3] It was more that a consensus appeared to have been achieved partly by the publication of the multiauthor review[4] discussed earlier (although this certainly did not show total agreement on the proposals for the mechanism for phosphorylation) and partly by the award of a Nobel Prize to the author of the chemiosmotic theory. If there was an experiment, then it was probably the Racker–Stoeckenius experiment published as a short note back in 1974.[5]

The acceptance of the chemiosmotic theory represented a pivotal moment in the history of bioenergetics. As Chris Cooper wrote in a review in 2000,

> For those of us who graduated in the late 1980s, the golden age of bioenergetics seemed a thing of the past. Following the universal acceptance of the

---

[3] For a discussion of this issue see Chapter 9.
[4] Boyer et al. 1977.
[5] Racker and Stoeckenius 1974; see Chapter 5, Section 5.7.

chemiosmotic theory, my old departmental chairman even stated that the essential research in this area was complete.[6]

However, it is important to realize that in terms of achieving an understanding of the mechanism of oxidative and photosynthetic phosphorylation, the chemiosmotic theory, although a crucially important achievement, was only part of the answer to the puzzle of phosphorylation. Its achievements were the redirection of metabolic research away from the idea of high-energy intermediates based on classical biochemistry and toward the energetics of ion gradients. Such a step was of signal significance not only for the study of oxidative phosphorylation and photophosphorylation but also for the studies of energy-dependent transport systems (active transport) and for the understanding of the role of biological membranes. However, so far as the mechanism of phosphorylation was concerned, in some ways the new perception left more questions unanswered than it answered. The issues of stoichiometries remained unresolved, although Mitchell and others believed that they had answered the question (even so the answers were not consistent). More important, the mechanisms involved in creating the proton electrochemical gradient were essentially unknown as was the means whereby the gradient drove ATP synthesis.

Mitchell had attempted to answer all these questions. His approach to the mechanism for creating the proton electrochemical gradient was to suppose that the electron-transport chain could be seen as alternating proton and electron carriers—a series of loops. In each loop a proton carrier was reduced on the matrix (inner) side of the membrane involving the uptake of a proton. The carrier was then oxidized on the outside, involving the release of a proton and the reduction of an electron carrier. The electron passed inward across the membrane (see Fig. 8.1).[7] In effect hydrogens ($H^+ + e^-$) moved outward and electrons (e^-) inward. More sophisticated versions of the principle were developed later, particularly the Q cycle. There were significant problems in envisaging the respiratory chain in this way, and most of the schemes failed to carry conviction. They lacked experimental evidence—a situation that undoubtedly made acceptance of Mitchell's overall theory difficult, although a growing appreciation from around 1970 that the two sides of the mitochondrial membrane were very different helped.[8] The solution to these problems would need to come from a detailed knowledge of the structure of the complex proteins.

8.3 THE Q CYCLE

An exception to the preceding criticism was a scheme Mitchell produced in 1975, known as the Q cycle.[9] This achieved both a plausible mechanism for proton transport in the mitochondrial complex III (the oxidation of ubiquinone and reduction of cytochrome c)

---

[6] Cooper 2000, p.455.
[7] Mitchell 1966a.
[8] Racker 1970b.
[9] Mitchell 1975a, 1975b.

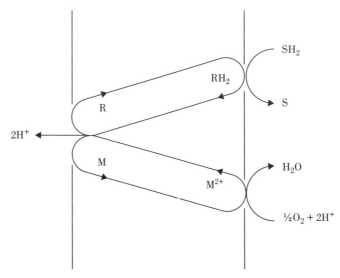

Figure 8.1 Mitchell's Diagram Showing the Principle of Loops of Alternating Hydrogen and Electron Carriers for Transporting Protons Across Membranes

The substrate SH_2 (possibly NADH) is oxidized by the proton carrier R to form RH_2. This is oxidized on the opposite side of the membrane by the electron carrier M, which becomes M^{2+}, and the reaction releases a proton. On the right-hand side M^{2+} reduces oxygen with the uptake of two protons to form water. The overall effect is the transport of two protons across the membrane coupled to the flow of two electrons through the system. Reproduced with permission from Mitchell 1966, *Chemiosmotic Coupling in Oxidative and Photosynthetic Phosphorylation*, Glynn Research Ltd., fig. 2, p.18.

and also explained the puzzling behavior of cytochrome *b*. Its relevance to the bacterial photosynthetic electron-transport chain soon became apparent and to the chloroplast chain a little later. The proposal was not totally novel as Wikström[10] and Jan Berden had already produced a scheme with experimental support covering part of Mitchell's concept (the oxidation of quinone by both cytochrome *c* and *b*).[11] The principle of the Q cycle was that ubiquinone was reduced on the inside of the inner mitochondrial membrane by an electron donor, which also required uptake of two protons from the matrix.[12] On the outside the quinone was oxidized to release two protons. However, in the oxidation, one electron was passed through to the *c* cytochromes and one to cytochrome *b*, which conveyed it back to reduce ubiquinone on the inside (see Fig. 8.2). The backflow

[10] Mårten K. F. Wikström (1945–) is a Finnish bioenergeticist educated at the University of Helsinki who since 1996 has held a professorship in physical biochemistry at the Academy of Finland.

[11] Wikström and Berden 1972.

[12] The modern understanding of membranes is based on the fluid mosaic model of Singer and Nicholson 1972. In this view, which saw the lipid bilayer as essentially fluid, ubiquinone could be seen as being a mobile molecule moving between sites for reduction and sites for oxidation.

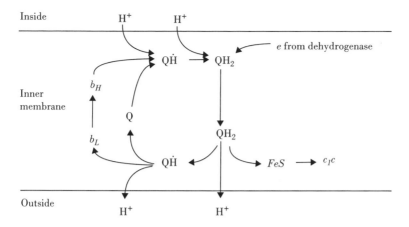

Figure 8.2 The Q-Cycle

Diagrammatic representation of the Q-cycle showing the role of ubiquinone (Q) in the transport of protons (H⁺) across the inner mitochondrial membrane. The reduction of Q takes place on the inside of the membrane by electrons coming from the dehydrogenase complexes or from cytochrome b_H (b_{562}). The quinone (Q, QH_2) is mobile within the membrane lipid, and the reduced form migrates across the membrane and is oxidized initially by the c cytochrome chain (c_1, c) and then by the b cytochrome (b_L, b_{566}). The quinone can then be reduced again on the inside of the membrane. Later versions of the cycle added the iron–sulfur center (*FeS*) that directly oxidizes QH_2 and reduces the c cytochromes. The overall process transfers two protons across the membrane for each electron passing through complex III. Based on Mitchell 1975 and Bowyer and Trumpower 1981.

of electrons accounted for the otherwise puzzling behavior of cytochrome *b* and also increased the proton transport to four protons per two electrons.

The initial acceptance of the cycle was mixed.[13] It certainly attracted much interest at an international conference in Fasano, Italy, in 1975 (where Mitchell used a hand-cranked model) not only among mitochondrial workers but also among those studying photosynthetic bacterial systems. Perhaps the reception of the proposal is summed up by Sergio Papa, who wrote

> I must say that I find your Q cycle extremely ingenious as well as attractive, although too speculative. It is indeed able to explain in a rational way many of the so far unexplained phenomena in the *b-c₁* complex.[14]

Mitchell, who appreciated that there could be a number of variants of the cycle, proposed the mechanism for the mitochondrial system but realized it might have relevance to bacterial and photosynthetic systems.[15] Indeed, some of the earlier evidence

[13] Like the chemiosmotic theory itself, the experimental evidence for the Q cycle was largely lacking when Mitchell proposed it, although it did explain several puzzling experimental results. Consequently some of the detailed elements were soon changed such as, for example, the order in which reduced ubiquinone was oxidized by the b cytochrome and by the c cytochrome chain.

[14] Letter, Papa to Mitchell, October 16, 1975. P. D. Mitchell archive, file G777, University of Cambridge Library.

[15] Mitchell 1976.

supporting the mechanism came from photosynthetic bacterial studies. Confirmation that one of the variants of the Q cycle reflected the mechanism of electron and proton movements in the b–c_1 complex, had to await the understanding of the structure of the complex.

8.4 STOICHIOMETRIC PROBLEMS
IN MITOCHONDRIA

From the time of the earliest experiments on oxidative (aerobic) phosphorylation, stoichiometric[16] measurements were made of the phosphorylation relative to O_2 consumption, giving a P/O ratio (or the equivalent, a P/2e ratio[17]). Such measurements were important in setting the parameters for the mechanisms proposed for oxidative phosphorylation. After the work of Belitzer in 1939, Ochoa's experiments set what became for many years the standard value for the P/O ratio of 3.0. It should be noted that in the then-current theory, the chemical theory required an integer for this ratio. When Mitchell proposed his chemiosmotic theory, he still assumed an integer and accepted that the number of protons translocated by the respiratory chain and by the ATP synthesizing enzyme would need to be consistent with the accepted value of the P/O ratio.

The need to understand the mechanism of proton translocation immediately raised the question of how many protons were translocated when an atom of O_2 was reduced (the H^+/O ratio) for substrates such as NADH and succinate. It proved no easier to measure proton O_2 ratios than it did to measure P/O ratios—possibly more difficult. Similar problems were experienced when the number of protons translocated for the synthesis of a molecule of ATP was measured.

However, during the 1970s, other questions were beginning to arise. By this time it was realized that because ATP is synthesized in the mitochondrial matrix, the transport of ADP and Pi into the mitochondrion and ATP out had to be considered. This transport exchange was found to involve the net transfer of one proton inward (with the phosphate).[18] So the synthesis of ATP would normally require the protons flowing through the synthesizing enzyme and, in addition, those involved in transport of phosphate and adenine nucleotides into and out of the mitochondrion (1 H^+/ATP). Thus the adenine nucleotide exchange, particularly the movement of phosphate, complicated attempts to measure phosphorylation and the movements of protons. In the 1970s Mitchell's value of six protons translocated for each atom of O_2 reduced[19] began to be questioned. In particular, Lehninger's group had obtained values twice those of Mitchell (H^+/O = 12).[20] This

---

[16] The normally fixed numerical relationship between quantities, thus the value of the P/O ratio, the relationship between the quantity of phosphorylation and the quantity of O_2 uptake associated with that phosphorylation.

[17] The P/2e ratio refers to the number of phosphorylations when two electrons pass through a coupling region. If O_2 is the electron acceptor then the P/2e ratio has the same value as the P/O ratio.

[18] See Section 8.10.

[19] Mitchell and Moyle 1967. The figure refers to the oxidation of NADH.

[20] Alexandre, Reynafarje, and Lehninger 1978.

group found that three protons per ATP synthesized were translocated by the synthase it-self. So together with the proton associated with phosphate–adenine nucleotide exchange, ATP synthesis would require four protons per molecule synthesized. Hence Lehninger's measurement of proton movements were still consistent with a P/O ratio of 3.

After proposing the Q cycle in which four protons per pair of electrons were translocated across the membrane in complex III (cytochrome b–c_1 complex), Mitchell concluded that the final complex of the respiratory chain, the cytochrome oxidase, did not have a proton-translocating function.[21] This was reinforced by the lack of an obvious proton-conducting component in the oxidase. In contrast, Mårten Wikström produced evidence that it did.[22] A major dispute occurred on this issue in the late 1970s and early 1980s until in 1985 Mitchell accepted Wikström's experimental results.[23],[24] Mitchell changed his view of proton translocation by the respiratory chain as a whole in 1986, accepting that

> In the absence of significant phosphate uptake, the observed stoichiometry of around 8 . . . appears to be characteristic of NADH oxidation.[25]

Proton stoichiometry could also be measured using partial reactions, such as, for ex-ample, the oxidation of succinate by cytochrome c to measure complex III (with suitable experimental conditions). Another approach was to use isolated complexes of the res-piratory chain incorporated into lipid vesicles and to measure the number of protons translocated per electron (or pair of electrons) passing through the isolated complex. For complex I (the NADH dehydrogenase), the information was not easily gained. In 1985 Youssef Hatefi (1929–) expressed the uncertainty about measurements of proton trans-location in this complex as follows. "The H^+/e^- stoichiometry of proton translocation coupled to electron transfer by complex I has been estimated to be 1, 1.5 or 2."[26] Some 14 years later he made this comment:

> In general, the mechanism of proton translocation by the enzymes of oxidative phosphorylation is unknown. In the case of complex I, the proton stoichiometry of the system is also unclear. However, the available values ($H^+/e \geq 2$) preclude a

[21] Mitchell had shown to his satisfaction that the total number of protons ejected per NADH oxidized was six. Two of these were associated with the first complex of the respiratory chain. If four were linked to the second complex, then there was no need to postulate any proton ejection by the third complex, the cytochrome oxidase. Further, Mitchell was unable to identify any com-ponent in the oxidase that might act in proton translocation.

[22] See Wikström and Krab 1979.

[23] Mitchell et al. 1985. In accepting that the oxidase did pump protons, Mitchell proposed an-other O-loop mechanism, but this proposal did not gain support. A full account of this dispute is given in Prebble and Weber 2003, chapter 11, pp.222–247.

[24] Wikström and Casey 1985.

[25] Mitchell et al. 1986, p.229.

[26] Hatefi 1985, p.1023.

mechanism whereby a single electron transfer step would be coupled to the translocation of one proton.[27]

At the end of the twentieth century, although approximately two protons per electron was the likely stoichiometry of complex I, there remained uncertainty.

In the case of complex III (the ubiquinone–cytochrome c reductase), the situation was better. Reviewing the problem in 1979, Wikström and Krab concluded that there was general agreement on $H^+/e = 2$.[28] In contrast, the situation for the cytochrome oxidase remained unclear, certainly until after the resolution of the dispute among Wikström, Mitchell, and the others previously referred to. In 1979, Wikström had already concluded that the oxidase translocated protons with an H^+/e ratio $= 1$, and this view was confirmed later.[29] However, in the oxidase, consideration needed to be given to the protons required for the reduction of O_2 to water, which were taken up from the inside and utilized close to the center of the membrane, amounting to 0.5 protons per electron. The total proton transfer across the membrane for a *pair* of electrons passing from NADH to O_2 might be 11, close to that found overall by Lehninger's group, previously mentioned. There remained some uncertainty about these figures arising primarily from other ion movements, particularly phosphate, occurring during experiments.

8.5 UNCOUPLING AND UNCOUPLING PROTEINS IN MITOCHONDRIA

One of the early puzzles of oxidative phosphorylation was the phenomenon of uncoupling.[30] Unlike the phosphorylation occurring in glycolysis, for which oxidation was obligatorily coupled to phosphorylation, the link between respiration and phosphorylation appeared fragile, a point noted as early as 1939 by Belitzer and Tsybakova.[31] They found the first chemical compound, arsenate, that was capable of uncoupling, leading to a loss of phosphorylation ability and an increase in respiration. Dinitrophenol began to be used as an uncoupler in experiments from the early 1950s,[32] and later several other compounds were found to uncouple.[33] Explaining uncoupling had been a requirement of plausible theories of oxidative phosphorylation. Mitchell had proposed that such substances uncoupled by conveying protons across the inner mitochondrial membrane, so dissipating the electrochemical proton gradient that would otherwise drive ATP synthesis.[34] Evidence that uncouplers translocated protons across mitochondrial membranes

---

[27] Hatefi 1999, p.28.

[28] Wikström and Krab 1979.

[29] Wikström 1989.

[30] Coupling refers to the link between the electron-transport chain and phosphorylation. It is broken by uncouplers. See Chapter 4, Section 4.6.

[31] See Belitzer and Tsybakova 1939.

[32] See Judah 1951.

[33] Commonly used potent uncouplers included CCCP (carbonylcyanide-*m*-chlorophenylhydrazone) and FCCP (carbonylcyanide-*p*-trifluoromethoxyphenylhydrazone).

[34] See Mitchell 1966a, pp.486–491.

and across artificial lipid membranes was obtained in several laboratories; Skulachev's group[35] in particular showed a clear relationship between uncoupling ability and proton conduction in lipid vesicles.[36] Mitchell's approach also provided a ready explanation for the uncoupling activity of ions such as calcium that can be accumulated in mitochondria at the expense of the proton electrochemical gradient.

Although such organic substances were not natural uncouplers and were mostly not found in cells, artificial uncouplers were valuable experimental tools. During the 1970s, natural uncouplers began to be found. Thus unesterified fatty acids were found to have uncoupling properties. Viewing uncouplers as substances that were capable of conducting protons across membranes appeared to explain their action. However, experiments began to suggest that the understanding of the mode of action of classical uncouplers might be slightly more complex, involving a role for proteins as well.[37]

In the 1960s and 1970s, a natural type of uncoupling was found in brown fat associated with nonshivering heat production, particularly in hibernating and newborn animals. The mitochondria in this tissue appeared to be largely uncoupled in their native state and function by using the energy available from respiration to provide heat rather than ATP synthesis. It soon became clear that the uncoupling was due to proton migration across the membrane that could be substantially reduced (increased coupling) by treatment with purine nucleotides, ATP and GTP (guanosine triphosphate). The nucleotides interacted with a site on the outside of the inner mitochondrial membrane, and this was used to identify proteins to which they bound. David Nicholls's group identified two proteins that reacted with the nucleotides with molecular weights of 30,000 and 32,000.[38] They were able to demonstrate that the smaller one was concerned with nucleotide transport across the membrane, leaving the larger one as the putative proton transporter.[39] In view of its role in the thermogenic action of brown fat, this uncoupling protein acquired the name thermogenin. In 1985, Martin Klingenberg (1928–) and his associates at the University of Munich isolated this protein and reconstituted it into phospholipid vesicles whereby they were able to demonstrate its ability to transport protons.[40] The protein was shown to have six helices and to have sequences that relate it to other mitochondrial-transport proteins. Further work using mutant proteins (those with altered amino acids) has begun to elucidate the way in which the protein functions.[41]

In earlier studies the transport of chloride ions also appeared to be associated with the uncoupling protein. This has led to a different, though disputed, view of the uncoupling protein for which it has been suggested that its real role is to convey anions across the

[35] Vladimir Petrovich Skulachev (1935–) of the Moscow State University and member of the Russian Academy of Sciences since 1975 is probably the leading Russian bioenergeticist of the twentieth century.

[36] Skulachev, Sharaf, and Liberman 1967.

[37] Skulachev 1999.

[38] For a history of work on the uncoupler protein in brown-fat mitochondria, see Nicholls and Rial 1999.

[39] See Nicholls 1979.

[40] Klingenberg and Winkler 1985.

[41] See Bouillaud et al. 1994.

mitochondrial membrane. These anions would include the unprotonated fatty acids that would transfer across the membrane without a proton but return as electroneutral protonated form.[42]

The discovery of an uncoupling protein (designated uncoupling protein 1, UCP1) in brown-fat mitochondria raised the question of whether similar proteins might function in other tissues. Studies found two further proteins, UCP2 and UCP3, which appear similar to UCP1 and occur in many mammalian tissue mitochondria.

The history of the study of uncoupling in the second half of the twentieth century has many of the qualities of bioenergetics studies of the period. The early phase of work concerned the effect of small molecules (such as dinitrophenol) on the process of oxidative phosphorylation in mitochondria and submitochondrial particles. With improved techniques uncoupling was seen as the transfer of protons across the membrane. Further work became possible when the relevant proteins (the uncoupling proteins) could be isolated and tested in vesicles. The availability of the protein's structure then enables studies directed to the mechanism of its function. Thus work had progressed from the small-molecule effects to the properties of membranes to the study of the structure, function, and mechanism of action of the proteins themselves. That said, further work may well substantially change our understanding of this problem.

8.6 BACTERIORHODOPSIN, A BIOENERGETIC PROTEIN

Before looking at the influence of protein structure on the growth of ideas on proton translocation, we examine the role of a bacterial protein capable of acting as a proton pump on the development of this subject. The membrane protein, bacteriorhodopsin, found only in a relatively obscure group of bacteria, proved of great interest to bioenergeticists.[43] Starting in 1967, Walther Stoeckenius and his colleagues drew attention to a purple pigment synthesized by a bacterium, *Halobacterium halobium*, which is capable of survival in extremely high-salt solutions in which the solubility of O_2 is low.[44] The organism is able to use light energy for growth although it also has a respiratory system. The cell membrane was shown to contain a purple pigment similar to the eye pigment, rhodopsin. The pigment forms patches in the membrane that can be isolated as relatively pure protein, bacteriorhodopsin. The similarity to rhodopsin enabled early rapid progress in the study of this protein. The chromophore was identified as retinal (vitamin A aldehyde), which appeared to be linked to a lysine residue in the protein.[45]

In the light bacteriorhodopsin was able to drive ATP synthesis. When placed in lipid vesicles, the protein could be shown to pump protons when activated by light energy, a process that takes place independently of any electron-transport chain. Bacteriorhodopsin

---

[42] Garlid et al. 1996.
[43] For a recent account of the bacteriorhodopsin story, see Grote 2013.
[44] For an early review, Henderson 1977. Note that this pigment was briefly discussed in Chapter 5, Section 5.7.
[45] Oesterhelt and Stoeckenius 1971.

itself is capable of converting light energy into an electrochemical proton gradient that can be used by the membrane-bound ATP synthase to form ATP.[46]

The importance of this pigment to the experimental study of bioenergetics was early appreciated:

> The purple membrane is a very simple system that hopefully will provide the answers to a number of basic biophysical questions. It forms a crystalline array containing only one protein of about 26,000 MW, from which detailed structural information about the folding of a membrane protein and the arrangement of lipid should be obtained. It contains the chromophore retinal, found elsewhere only in the visual pigments, so its light reactions have added significance in addition to having the practical benefit of occurring in a closed photochemical cycle. It is an ion pump requiring no substrates other than water and light, which nevertheless may function in principle in a general way common to other pumps involved in both ionic and molecular transport.[47]

The relevance of the energetics of this system to supporting Mitchell's chemiosmotic theory has already been noted earlier in this chapter in relation to the Racker–Stoeckenius experiment of 1974.

It was found by Stoeckenius and others that, following brief activation of bacteriorhodopsin by light, a series of spectral changes could be observed, indicating different forms of the protein. A cycle of events was identified, starting with an activated pigment passing through various forms back to the initial ground state. A metastable intermediate known as M was identified, whose formation was associated with the loss of a proton followed by uptake when M was converted back to the ground state. The cycle was first formulated in the early 1970s[48] and was often revised and expanded but was a valuable tool in trying to understand the mechanism for more than two decades.

The understanding of the role of bacteriorhodopsin developed along two lines, one being concerned with its role in membrane transport and the general biochemistry of *Halobacterium*[49] and one in the mechanism of action of the protein. The first line revealed the importance of the sodium pump[50] (sodium/proton exchange) in *Halobacterium* together with other membrane-bound pump systems dependent on the formation of the proton gradient. The more important line of investigation approached the mechanism of proton pumping through an understanding of the structure of the protein and the interaction of retinal within it.

A significant development concerned the role of retinal. At the end of the 1970s, a controversy arose over whether during the photocycle retinal underwent a change from the all-trans to the 13-cis form. The matter was settled in 1981, and it was agreed that the

---

[46] Oesterhelt and Stoeckenius 1973.

[47] Henderson 1977, p.107.

[48] See Lozier, Bogomolni, and Stoeckenius 1975.

[49] Some of these issues are discussed in Lanyi 1978.

[50] Note that the ecological niche exploited by this organism is pools above the high tide mark where high salt concentrations are found.

result of light activation involved a trans→cis isomerization of the retinal to form the M intermediate and a cis→trans change when returning to the original state.[51] The all-trans form of the molecule was rod shaped whereas the 13-cis form had a bend close to the reactive aldehyde end; the change would therefore result in a significant movement of the aldehyde group at the center of the protein where it was thought that the retinal was located.

Although spectral studies provided important information on how bacteriorho-dopsin worked as a light-driven proton pump, it was obvious that progress toward a real understanding of the mechanism of the system would depend on understanding the protein. The protein forms a two-dimensional lattice in the purple membrane, and this was analyzed by electron-diffraction studies, which showed that it consisted of seven transmembrane helices.[52] The amino acid sequence of the protein was determined in 1981, both by protein sequencing and gene sequencing, showing forty-eight amino acid residues that could be modeled in the seven helices.[53] A model of the protein structure showing the position of the amino acid residues was finally achieved by Henderson and his colleagues in 1990 (see Fig. 8.3) using refined high-resolution electron microscopy at low temperatures.[54] The model showed the β-ionone ring end of the retinene located in a pocket with the reactive end projecting into a proton-conducting channel. The reactive aldehyde group would physically move relative to those amino acid residues in its environment as a consequence of the trans→cis→trans isomerization of the photocycle of reactions. This physical movement of the retinene would initiate the proton movements.

Much further work on this protein followed Henderson's 1990 structure. Site-specific mutations were used to reveal the involvement of some amino acids in the mechanism.[55] However, even at the end of the century, bacteriorhodopsin had not given up all its secrets:

> Much information has been compiled over the past 28 years; no doubt this mol-ecule is now the best characterized active ion translocator. However, its structure is not solved to atomic resolution, and the structural changes responsible for the accessibility change of the active center for protons are not known. Further, the time-resolved migration of protons along the conduction pathways in the struc-ture has eluded a complete description so far. For this knowledge of the complete structural dynamics of the molecule is required.[56]

The value of this protein to a general understanding of bioenergetics is substantial, but like so many themes discussed in this chapter, the story remains incomplete.

The project of understanding the mechanism of bacteriorhodopsin activity has not proceeded as rapidly as was initially expected. When the purple pigment was discovered

[51] Mowery and Stoeckenius 1981.
[52] Henderson and Unwin 1975.
[53] Khorana at al. 1979; Dunn et al. 1981.
[54] Henderson et al. 1990.
[55] Krebs and Khorana 1993.
[56] Haupts, Tittor, and Oesterhelt 1999, pp.391–392.

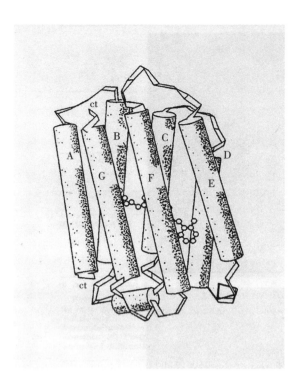

Figure 8.3 Model of Bacteriorhodopsin

The protein consists of a single polypeptide chain in the form of seven transmembrane helices (A–G) shown as rods. The cytoplasm is at the top. The retinene is shown behind helices F and G. Reproduced with permission from Henderson, Baldwin, Ceska, Beckmann, and Downing 1990, "Model for the Structure of Bacteriorhodopsin Based on High Resolution Electron Cryo-Microscopy" (*Journal of Molecular Biology* 213. 899–929), Elsevier, fig. 14, p.918.

as a biological phenomenon, very rapid progress was made in defining the role of the pigment in the membrane. By virtue of the ease of preparation of the purple membrane containing the two-dimensional mosaic of the protein, it was only some 8 years before the general structure of bacteriorhodopsin in the membrane could be determined by electron microscopy. At the same time a first photoreaction sequence was formulated. However, it took a further 15 years to gain a knowledge of the special arrangement of the amino acid residues. The central problem of bacteriorhodopsin research was to understand the mechanisms involved in proton pumping driven by light energy. Such a task depended primarily on a detailed knowledge of protein structure and of the photochemical reaction. Neither was easy to gain adequately, but at the end of the century atomic structures of the protein were obtained at increasingly higher resolutions. This led to the hope that the long-sought description of the mechanism of the protein was within grasp.[57]

[57] Luecke et al. 1999; Lanyi 1998.

8.7 UNDERSTANDING THE RESPIRATORY CHAIN

By the late 1970s, it had become clear to most workers in the field that the way forward in understanding oxidative phosphorylation would require an understanding of the proteins themselves. Slater's group had commented in relation to complex III:

> In order to gain a better insight into the mechanism of electron transfer and energy conservation in QH_2:cytochrome c oxidoreductase [complex III], a detailed knowledge of the polypeptide composition is essential.[58]

Indeed, the first step in this process was to determine the molecular size and number of subunits in each of the respiratory complexes and the assignment of a role to each if possible. This work became possible around 1970 following the development of improved methods for studying membrane proteins, particularly using polyacrylamide gel electrophoresis with sodium dodecyl sulfate, as exemplified by work with erythrocyte membranes.[59]

However, progress in this field was slow, and in the late 1970s when Slater made the previous comments, his work had failed to detect all the subunits in the complex and there was uncertainty about the identity of cytochrome b. During the 1980s, the subunit composition of the complexes of the respiratory chain were satisfactorily resolved with the exception of the large complex I. In the years that followed, the amino acid sequence of the subunits was determined and the effect of alterations of specific amino acids on protein behavior observed, throwing some light on the role of proteins. Ultimately models of the structure of the complexes themselves were prepared.

Nevertheless, the essential knowledge of proteins still had to be combined with other studies, including various now-advanced forms of spectroscopy. There were also new approaches to the study of the mechanism of oxidative phosphorylation, which became important in the 1980s. The comparative biochemistry of respiratory chains was one such approach. This was particularly so in bacteria, where a variety of respiratory chains, often involving different oxidases, were found. The respiratory system in *Paracoccus denitrificans* proved to be quite similar to that in the mammalian mitochondrion. However, the bacterial complexes tended to possess many fewer subunits while carrying out the same basic process of oxidative phosphorylation as the mammalian enzyme. The cytochrome oxidase of *Paracoccus* was found to contain only four subunits but carried out the same role of oxidizing cytochrome c and reduction of O_2 coupled to the pumping of protons across the membrane as the mammalian enzyme with thirteen subunits. Some bacteria were found to possess even fewer subunits than *Paracoccus*. The bacterial enzymes often provided a much-simplified model for studies of the mechanism of the complexes.

Another element to have an impact on the study of the respiratory complexes (and particularly the ATP synthase) was the genetic one. This was strongly stimulated in the 1960s by the realization that mitochondria possessed their own DNA. In 1965 Nass and

---

[58] Marres and Slater 1977, p.532.
[59] Rosenberg and Guidotti 1968.

coworkers concluded that "DNA is an integral part of most, probably all mitochondria."[60] The DNA was found to be circular and shown to be of a prokaryotic type. The mitochondrion was shown to have the full apparatus for protein synthesis (tRNA, ribosomes, RNA polymerase, etc.) during the 1970s. Genetic mapping of the DNA became possible toward the end of the 1970s. This showed that a small number of proteins associated with oxidative phosphorylation were encoded in the mitochondrial DNA. For example, in the cytochrome oxidase, it was the large subunits that were coded in the mitochondrial DNA and the smaller ones by the nuclear DNA.[61] A key development in the study of the molecular biology of the mitochondrion was the sequencing of the mitochondrial genome, achieved in 1981.[62]

8.8 ELUCIDATING THE MECHANISM OF RESPIRATORY CHAIN COMPLEXES

8.8.1 NADH Dehydrogenase

Essentially it was the isolated complexes that became the basis of studies of the mechanism of electron transfer and proton pumping. Of the three main energy-conserving complexes of the respiratory chain (complex II, succinate dehydrogenase, was not energy conserving), the first, the NADH dehydrogenase, proved much the most difficult. Despite many efforts during the period under consideration, little progress was made owing primarily to its complexity and large size. It possessed a number of iron–sulfur centers, ubiquinone as well as a flavin, and unequivocal evidence for the sequence of electron carriers was not forthcoming. Attempts to identify the subunits was equally problematic; the complex was found to have around sixteen subunits in the late 1970s but at the turn of the century the number was around forty-six. Little was known for certain about proton pumping except that four protons per pair of electrons passing through the complex were transported across the membrane.

8.8.2 Cytochrome Oxidase (Complex IV)

In early work this enzyme, the terminal member of the respiratory chain, was found to reduce O_2 to two molecules of water using four electrons drawn from cytochrome c. As noted earlier, the number of protons transported by the complex was debated, particularly in the 1980s. The enzyme has a long history of intense experimentation from Warburg's study of his *Atmungsferment* onward.[63] The application of the techniques of protein chemistry did much to clarify significant aspects of the mechanism. Initially analysis of the subunits produced somewhat variable results[64] but in the 1980s it was shown that the mammalian enzyme possessed thirteen subunits,[65] although even then three of

---

[60] Nass, Nass, and Afzelius 1965, p.537.

[61] Mason and Schatz 1973.

[62] See Anderson et al. 1981 and associated papers.

[63] See Chapter 3.

[64] Malmström 1979 lists seven subunits for the mammalian enzyme

[65] See table 1 in Musser, Stowell, and Chan 1995.

these were initially seen as questionable. Reviewing the situation in 1983, Capaldi and colleagues commented

> The subunit structure of eukaryotic cytochrome c oxidases is a point of continuing debate. Two factors have led to the present confusion. First the enzyme contains several polypeptides of similar molecular weight and these are difficult to separate for identification. Second conclusions about what are bona fide subunits and what are impurities of the preparation depend on how cytochrome c oxidase is defined. [66]

Amino acid sequences of the main subunits were prepared at the end of the 1970s and the sequences of the large subunits (I, II, and III) were confirmed from the genetic information obtaining from sequencing the DNA of the mitochondrial genome.

When it became possible to prepare two-dimensional crystals suitable for electron microscopy, a three-dimensional model of the enzyme was produced.[67]

The role of the subunits was also a matter for debate. The bacterium *Paracoccus* was found to have a fully functional oxidase enzyme with only subunits I and II. (Also, an unusual preparation of the mammalian enzyme was found to be functional but had only subunits I and II). Such results suggested that these subunits must be the location for the two hemes (a and a_3) and the two copper atoms detected in the complex (Cu_A and Cu_B).[68] Other arguments led to the view that the hemes and Cu_B were in subunit I and Cu_A was in subunit II, a view that was accepted in the 1990s. A high-affinity binding site for cytochrome c (which reduced cytochrome oxidase) was found on subunit II. Such information gave a broad overview of the operation of the enzyme. However, understanding the catalytic events in the oxidase proved a major challenge. Initially the reduction of heme a by cytochrome c was agreed; it was in any case Keilin's view. Whether Cu_A was involved in the transfer of the electron from cytochrome c to heme a was debated during the 1980s, but in the 1990s its involvement was largely settled. It was soon realized that the mechanism would involve the binding of O_2 to the active site (Cu_B–a_3) and its reduction,[69] but understanding how this occurred without producing destructive free-radical O_2 species proved difficult (the sequence of electron carriers can be seen in Fig. 8.4). By use of mainly spectroscopic methods, a cycle of events at the active site of the enzyme was proposed.[70] The nature of the mechanism of the active site was discussed and clarified during the 1980s and 1990s as experimental information accumulated and was refined.[71]

After the major dispute between Mårten Wikström and Peter Mitchell on whether the oxidase pumped protons, there was agreement that the enzyme did pump protons.[72] How

---

[66] Capaldi, Malatesta, and Darley-Usmar 1983, p.138.

[67] Ibid.

[68] Ludwig and Schatz 1980; Saraste et al. 1980. The view was influential, but the bacterial enzyme was later shown to have three and then four subunits.

[69] Greenwood, Wilson, and Brunori 1974. Note that the heme–iron–copper center was discovered by Helmut Beinert's group (van Gelder and Beinert 1969).

[70] See for example Clore et al. 1980.

[71] See Babcock 1999.

[72] I have discussed this dispute in Prebble and Weber 2003, chapter 11.

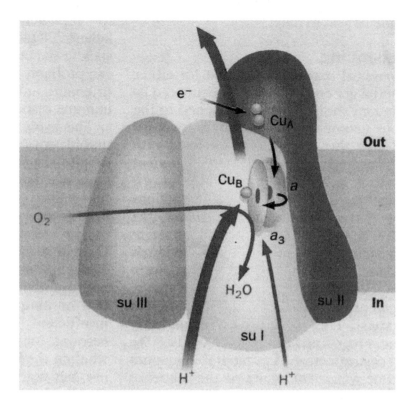

Figure 8.4 Cytochrome Oxidase

The model shows the three principal subunits of the *Paracoccus* enzyme, all of which traverse the membrane. Cu_A on subunit II receives electrons from cytochrome c (not shown) and passes them through haem a to the haem a_3-Cu_B pair where oxygen is reduced to water. The two proton channels are shown as heavy arrow and light arrow. The figure demonstrates the role of structure in contributing to the understanding of the enzyme mechanism. Reproduced with permission from a Correction (*Nature* 378: 235), Nature Publishing Group, 1995. (Published diagram provided by H. Michel.)

this process linked with the redox reactions associated with the heme a-Cu_A pair proved to be very debatable. Having reviewed a number of the more detailed proposals, Musser and colleagues came to this conclusion:

> The plethora of proposed chemical models attempting to explain the proton pumping reactions catalysed by the CcO complex, especially the number of recent models, makes it clear that the problem is far from solved . . . . It is clear that proton pumping is effected by conformational changes induced by oxidation/reduction of the various redox centres in the CcO complex.[73]

Conformational changes had been observed in the enzyme since the 1970s and have often been linked with proton pumping.[74]

[73] Musser, Stowell, and Chan 1995, p.60. The shorthand "CcO" refers to the cytochrome oxidase.
[74] See Wikström 1977.

A major breakthrough in understanding the cytochrome oxidase came in 1995 when a high-resolution structure of the enzyme from *Paracoccus denitrificans* was described by Iwata and coworkers.[75] This bacterial enzyme possessed just four subunits as opposed to thirteen in the mammalian enzyme. The function of the transmembrane subunit III and the small subunit IV were not known. Subunit II possessed the Cu_A center composed of two copper atoms and subunit I possessed the binuclear center, heme a–Cu_B. (The relationships of the three large subunits [but not subunit IV] are shown in Fig. 8.4) Of significant interest was the identification of two putative proton pathways through the complex, one associated with the uptake of protons for water formation and the other seen as the proton-pumping pathway. It is this latter pathway that has been seen as problematic.

The following year, the high-resolution crystal structure of the mammalian enzyme (having a molecular size of 200,000) was published, confirming the thirteen subunits.[76] The sites of the metal centers remain similar to those of the bacterial enzyme but the proton pathways seem to follow different routes. These structures confirm much previous work on the enzyme as well as offer new perspectives of its mechanism. The isolation of such a large membrane protein complex marked a major biochemical achievement. It still left many unanswered questions, including how protons were translocated. Inevitably, further structural studies were required.

8.8.3 Ubiquinone–Cytochrome c Reductase (Complex III)

The complex oxidizes ubiquinone and reduces cytochrome c while pumping four protons for every two electrons passing through the system. The earlier work showed that it contained, in addition to ubiquinone, two b cytochromes, cytochrome c_1 and an iron–sulfur center (the Rieske iron–sulfur center). In the mid-1970s, Peter Mitchell proposed a scheme, described as the Q cycle (see Fig. 3.2 and Section 3.3), showing how the protons might be transported across the membrane and also accounting for the anomalous behavior of the b cytochromes. The scheme was relatively complex, and evidence in support of it accumulated during the period under discussion so that Bernard Trumpower could write this in 1981:

> The proton-motive Q cycle accounts for all of the currently available information relating to how the iron-sulfur protein functions in the cytochrome b-c_1 complex.[77]

However, much later Hatefi expressed some reservations about whether the Q cycle could yet be regarded as totally supported experimentally.[78] Nevertheless by the end of the century the cycle looked relatively secure despite the many detailed questions it had raised.

---

[75] Iwata et al. 1995.
[76] Tsukihara et al. 1996.
[77] Trumpower 1981, p.152.
[78] Hatefi 1999.

The isolation of complex III was first achieved in the early 1960s but, as with the other complexes, improved methodology produced purer preparations. Work in the late 1970s produced complexes with about eight subunits.[79] Better procedures showed that the mammalian complex, a dimer, actually contained a total of eleven subunits.[80] The bacterial enzyme was simpler, possessing only three subunits in *Paracoccus denitrificans*: the *b* cytochrome, the c_1 cytochrome, and the iron–sulfur protein. These are similar to three of the subunits in the mammalian enzyme, and because both enzymes catalyze the same basic process, the other eight subunits of the mammalian enzyme were presumed to have functions not essential to the basic process of energy conservation. The mechanism of the complex, broadly predicted by Mitchell's Q cycle, was progressively elucidated, particularly by the use of a group of inhibitors that blocked activity at three sites in the cycle.

The subunit identified as cytochrome *b* was the only subunit in the complex coded by a mitochondrial gene. On the basis of the mitochondrial genome sequence studies, its amino acid sequence was determined—it had a molecular size of 42,540. It was assumed that this polypeptide possessed the two hemes (b_{562}, b_{566}) but proving the point was not a trivial task, and there remained the possibility that there were two separate cytochromes. A detailed study of the primary sequence of the protein from various sources led to the prediction that it consisted of nine helices spanning the membrane. Later predictions (based on better hydrophobicity of the sequences) favored eight transmembrane helices and one additional helix. The two hemes were liganded between two of the helices.[81]

The structural predictions of the 1980s and early 1990s were generally confirmed, including the eight transmembrane helices of the *b* cytochrome, when the crystals of the protein were prepared suitably for x-ray analysis. A high-resolution model of the enzyme dimer showed the spatial relationship of the hemes and the iron–sulfur center together with cytochrome c_1.[82] The inhibitors, much used in the elucidation of the functioning of the b–c_1 complex, were also located in the model, which gave a further basis for the understanding the mechanism. Of particular interest was evidence of mobility of the Rieske iron–sulfur protein that was seen as relevant to understanding the process of quinone oxidation by cytochrome *b* and the Rieske protein, each accepting just one electron.[83]

The 1980s were a period during which the methodology applied earlier to the preparation of complexes and separation of subunits gave satisfactory and convincing evidence about the subunits. This work was greatly aided by the use of simpler bacterial enzymes that were still able to carry out the basic bioenergetic functions of the enzymes. The protein structures that emerged during the 1990s provided a strong basis for understanding the mechanism of energy conservation in the respiratory chain. Although these models did much to confirm earlier work, some of which was speculative, and to clarify the nature of what was occurring in the complexes, there still remained substantial further steps before the molecular mechanisms of oxidative phosphorylation could be fully understood. Advanced forms of spectroscopy were greatly aiding these developments.

[79] See von Jagow and Sebald 1980, table 1, p.294.
[80] See González-Halphen, Lindorfer, and Capaldi 1988.
[81] Widger et al. 1984.
[82] Xia et al. 1997.
[83] Yu et al. 1998.

When taking an overview of the work on the respiratory chain from Keilin to the end of the century, Peter Rich (1951–),[84] formerly director of Mitchell's Research Institute at Glynn, commented that with all the information gathered, his discussion highlighted

> The fact that numerous questions remained, particularly in relation to catalytic-cycle intermediates, protonation processes and questions of dynamic aspects of individual amino acids and, in some cases, whole protein domains that might be key to the mechanism.[85]

He felt that Fourier-transform infrared spectroscopy, which had been used to probe the role of individual cofactors and amino acids, was reaching a level technically capable of addressing many of the problems.

8.9 PHOTOSYNTHETIC COMPLEXES

Many of the major advances in understanding photosynthesis in the 1980s and 1990s were associated with the ability to isolate and analyze the protein complexes of the system. The skills developed particularly in the early 1980s for isolating, separating, and studying membrane proteins, leading eventually to a knowledge of their three-dimensional structure, were applied to photosynthetic systems. Work already successfully carried out with the respiratory electron-transport chain inspired this approach to the chloroplast electron-transport chain and also to the similar systems in photosynthetic bacteria. The development of these structural studies was crucial to the enhancement of understanding of the mechanism of photosynthesis. Earlier spectroscopic techniques were developed further during this period, but placing this knowledge within the new structures became the key to appreciating mechanism.

8.9.1 The $b_6 f$ Complex

Of significant interest was the realization that the chloroplast possessed a complex that appeared to have strong similarities with complex III of the respiratory system. This $b_6 f$ complex oxidized the quinone (plastoquinone) and reduced a c-type cytochrome, cytochrome f, and proved to be analogous to the mitochondrial system oxidizing ubiquinone and reducing cytochrome c_1. Similar complexes were also found in photosynthetic bacteria (*Rhodopseudomonas*) where $b–c_1$ complexes that reduced cytochrome c_2 were isolated in the early 1980s.[86] The plant chloroplast system was isolated at about the same time, and the main protein subunits, four in number, were separated at the beginning of the 1980s from spinach chloroplasts in Günter Hauska's laboratory.[87]

---

[84] Rich worked at Cambridge University before joining Mitchell at Glynn and later founded the Glynn Laboratory at University College, London.

[85] Rich 2003.

[86] See for example Gabellini et al. 1982.

[87] Hurt and Hauska 1982; Hauska et al. 1983.

Comparison of the chloroplast complex with that from mitochondria showed that the cytochrome b subunits were similar, the plant protein possessing two hemes as with the animal one, although not so readily distinguished spectrally. However, there were two subunits in chloroplasts that together corresponded to the animal cytochrome b.[88] Of the other two main subunits, one was the Rieske iron–sulfur protein and the other cytochrome f. The latter protein, which plays a role analogous to cytochrome c_1 of complex III, does not seem to share homology with the animal cytochrome.[89] The complex does nevertheless appear to carry out the same catalytic functions as its mammalian counterpart. Probably nowhere else is the influence of mitochondrial research on photosynthetic studies as clear and strong as in the b–c_1 complex.

8.9.2 The Purple Bacterium Reaction Center

Of great significance in elucidating the mechanism of photosynthesis was the determination of the three-dimensional structure of the reaction center of a purple photosynthetic bacterium. Ever since Emerson and Arnold formulated the idea of a photosynthetic unit in 1932,[90] there had been the implication of a reaction center. The idea developed during the 1950s, particularly with the work of Louis Duysens, who had considered the possibility of a reaction center in photosynthetic bacteria.[91] In the late 1960s Roderick Clayton and his colleagues isolated the first reaction center complex that retained photoactivity from membranes in *Rhodopseudomonas sphaeroides* using detergents.[92] In the early 1970s, George Feher was able to make improved preparations[93] that showed the center to possess three subunits.

The three-dimensional structure of the reaction center was not determined until the early 1980s because of the difficulty of obtaining ordered crystals of membrane-bound proteins. In 1982, Hartmut Michel (1948–) succeeded in preparing highly ordered crystals of the reaction center from a related organism, *Rhodopseudomonas viridis*. This enabled Johann Deisenhofer (1943–) and Robert Huber (1937–) to determine the three-dimensional structure of the complex.[94]

The *R. viridis* reaction center consisted of two central subunits (named L and M), each of which had five membrane spanning helices; it possessed the special pair of bacteriochlorophyll b molecules, which is the functional center of the complex where oxidation–reduction is initiated (see Fig. 8.5). To the central subunits is bound the c-type cytochrome (with four hemes) on one side and another protein, H, on the other side, which has a single transmembrane helix. The central part of the complex has a number of small molecules attached to it in addition to the bacteriochlorophylls.[95]

---

[88] Widger et al. 1984.
[89] Prince and George 1995.
[90] See Chapter 6.
[91] Duysens et al. 1956.
[92] See Clayton and Yau 1972.
[93] Feher 1971.
[94] See Deisenhofer et al. 1985
[95] These include two further bacteriochlorophyll b's, two bacteriopheophytin b's, carotenoid, quinones, and a nonheme iron group.

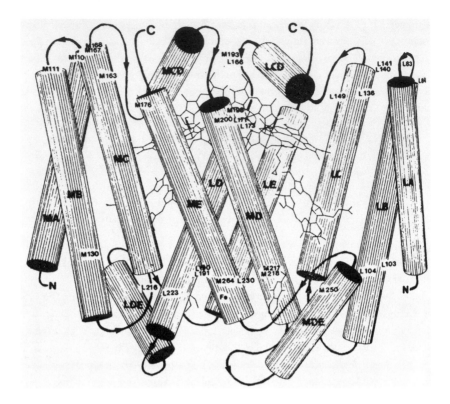

Figure 8.5 A Representation of the Core of the Photosynthetic Reaction Center of *Rhodopseudomonas viridis*

The diagram shows five transmembrane α helices of the two subunits, L and M, labeled LA to LE and MA to ME (represented as columns). In addition, there are two connecting helices in each subunit that do not traverse the membranes labeled LCD, LCE, MCD, and MCE. The special pair of bacteriochlorophyll molecules is shown at the interface of L and M (between the D and E α helices). Reprinted with permission from Michel and Deisenhofer 1988, "Relevance of the Photosynthetic Reaction Center From Purple Bacteria to the Structure of Photosystem II" (*Biochemistry* 27:1–7), American Chemical Society.

Deisenhofer, Huber, and Michel were awarded the Nobel Prize for Chemistry in 1988.[96] In introducing the work for the presentation, Bo Malmström described the determination of the structure of the complex as "a giant leap in our understanding of fundamental reactions in photosynthesis, the most important chemical reaction in the biosphere of our earth." As George Feher and others have commented, "The reaction center can be viewed as a protein scaffolding that holds the various cofactors (reactants) in their appropriate positions for efficient electron transfer."[97] Understanding the structure provided the basis for understanding the mechanism whereby light energy is converted into chemical energy by means of appropriate electron transfers.

[96] The citation was "For the determination of the three-dimensional structure of a photosynthetic reaction center."

[97] Feher et al. 1989.

However, the work on the reaction center complex was a double achievement. Not only did it open up new possibilities for the understanding of photosynthesis, it was also the first membrane-bound protein to be crystallized and a high-resolution three-dimensional structure obtained.[98]

8.9.3 Chloroplast Photosystems and Reaction Centers

The pigment protein complexes associated with the photosystems were relatively small and the structure of the light-harvesting *a/b* protein was determined in the 1990s. Indeed, the first significant protein associated with the photochemistry of the chloroplast to have its structure determined at high resolution was the light-harvesting chlorophyll *a/b* protein complex. This protein, associated with PSII, was already known in the 1980s to possess 232 amino acids and was the most abundant protein in chloroplast membranes. In 1994, its detailed structure (at 3.4-Å resolution) was determined as shown in Figure 8.6. The protein possessed three transmembrane helices with twelve chlorophylls and two carotenoids associated with it. The determination of its structure marked the first step in elucidating the complex membrane-bound structure of the chloroplast photosystems. This development gave a valuable boost to the understanding of the mechanisms of harvesting solar energy by these complexes.[99]

The photosynthetic reaction centers of plant chloroplasts are much bigger than their bacterial counterparts and consequently more difficult to handle. However, not very long after the commencement of studies on separated respiratory complexes, the fractionation of digitonin-treated spinach chloroplasts was described, giving a partial separation of PSI and PSII activities.[100] During the late 1960s and 1970s, progressive improvements of the separation techniques enabled analysis of the pigment and protein contents of the two photosystems.

In considering PSII, Hankamer, Barber, and Boekema described the importance of understanding the mechanism of this complex:

> Photosystem II . . . uses light energy to catalyze a series of electron transfer reactions resulting in the splitting of water into molecular oxygen, protons and electrons. These reactions occur on an enormous scale and are responsible for the production of atmospheric oxygen and indirectly for almost all the biomass on the planet. Despite its importance, the catalytic properties of PSII have never been reproduced artificially. Understanding PSII's unique chemistry is important and could have implications for agriculture as PSII is a main site of damage during environmental stress.[101]

---

[98] Note that, in 1980, the first crystallizations of membrane proteins, bacteriorhodopsin and porin, were achieved although they did not provide the basis for high-resolution x-ray crystallography.

[99] Kühlbrandt, Wang, and Fujiyoshi 1994.

[100] Boardman and Anderson 1964.

[101] Hankamer, Barber, and Boekema 1997, p.642.

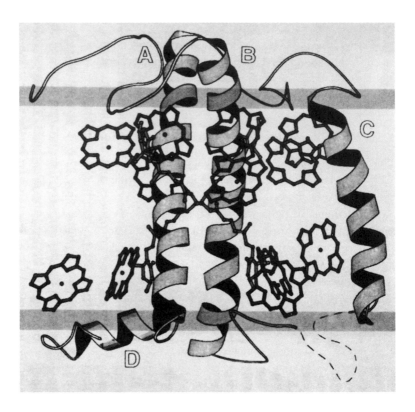

Figure 8.6 Molecular Model of the Light-Harvesting Chlorophyll a/b Protein

The model shows the polypeptide chain with three helices that cross the membrane (A, B, and C) with a fourth helix (D) at the membrane surface. The position of twelve chlorophyll molecules and two carotenoids are also shown. Reproduced with permission from Kühlbrandt, Wang, and Fujiyoshi 1994, "Atomic Model of Pant-Light Harvesting Complex by Electron Crystallography" (*Nature* 367: 614–621), Nature Publishing Group, fig. 3a, p.618.

Thus the importance of establishing the structure and mechanism of PSII is a clear goal for bioenergetics. However, the complexity of the system remained a major challenge for workers in the field.

By the mid-1990s, more than twenty subunits were known to be involved in the PSII complex. However, the isolation of a reaction center for PSII comparable to that previously isolated from purple bacteria was seen as an achievable goal in the 1980s. Initially some six polypeptide subunits were shown to be involved, two large chlorophyll-carrying polypeptides, polypeptides D-1 and D-2, and two polypeptides of cytochrome b_{559} (no longer seen as part of the electron-transport chain). There was a debate about whether D-1 or one of the chlorophyll-carrying polypeptides was associated with the photochemical reaction. This was settled when a complex with photochemical activity consisting of only the D polypeptides and the cytochrome was isolated.[102] The view that the D-1 and D-2 polypeptides were central to the PSII reaction center was strongly supported by the

---

[102] Nanba and Satoh 1987.

similarity between the structures of these proteins and those of the L and M subunits of purple bacteria. Thus the purple bacterial reaction center has been seen as relevant to understanding the processes in the PSII reaction center.[103] The full structure of the PSII complex remained a project for a later time. A proper three-dimensional structure for the complex at 3.8 Å emerged at the beginning of the twenty-first century.[104]

The study of PSI followed a course not dissimilar to that of PSII. Early preparations of the PSI reaction center capable of the basic photochemical reactions were found to possess six subunits,[105] although these results were seen as questionable at the time. Mapping of the chloroplast genome led to the identification of a number of genes associated with the photosystems. In 1985, the first of these genes, that for reaction center proteins in PSI, was sequenced.[106] Later, crystals of the proteins were prepared, leading to an extensive study, lasting more than a decade, to determine the structure of the complex by using material from a cyanobacterium.[107] The conclusions from this work confirmed earlier ideas on the mechanism of the reaction center.

This period showed the importance of understanding molecular structure as a means of understanding mechanism in photosynthesis. The size and intricate nature of the photosystems in O_2-evolving photosynthesis presented a major challenge that has been addressed partly on the basis of the bacterial reaction centers and partly on the techniques developed for mitochondrial complexes. In most cases earlier investigations at lower resolutions provided a partial understanding. It is these structural studies that were essential to the proper development of an understanding of the mechanisms operating in the two photosystems.

8.10 ADENINE NUCLEOTIDE TRANSPORT

In the 1960s, the importance of carrier systems operating across the inner mitochondrial membrane began to be appreciated. Ions such as calcium were of particular interest.[108] Mitchell's view that transport systems were a core part of the mechanism of oxidative phosphorylation underlined the importance of translocation of ions across the inner mitochondrial membrane. The chemiosmotic theory implied a close link between phosphorylation and transport, although such an approach was of limited influence at the time. A number of systems responsible for conveying substances into and out of the mitochondrion, including phosphate, were detected.[109] The history of one of these that exemplifies many of the issues involved, the adenine nucleotide translocator, is considered here.

---

[103] Michel and Deisenhofer 1988.

[104] Zouni et al. 2001. This complex was from a cyanobacterium; cyanobacteria carry-out O_2-evolving photosynthesis similar to that in plant chloroplasts.

[105] Bengis and Belson 1977.

[106] Fish, Kück, and Bogorad 1985.

[107] Jordan et al. 2001.

[108] The question of transport into and out of the mitochondria was raised in Chapter 5.

[109] Chappell 1968. The major thrust of this research was in animal mitochondria but transport systems in plant mitochondria and in chloroplasts were soon being described. The early developments in plant studies are described in Prebble 1988.

The importance of structural studies seems to have been clear at an early stage (around 1970) to its key investigator, Martin Klingenberg (1928–). He recalled that he

> became convinced that only the molecular approach, i.e. defining carrier sites and eventually isolating the carrier protein, would advance our understanding of the mechanism.[110]

Klingenberg's group detected a system for the exchange of adenine nucleotides across the mitochondrial membranes in 1965.[111] The transport was inhibited by atractyloside,[112] an inhibitor of oxidative phosphorylation previously thought to inhibit the ATP synthase enzyme. Atractyloside proved a powerful tool in the study of the system. It soon became clear that in the energized (coupled) mitochondrion, the transport system favored the entry of ADP and the exit of ATP, a phenomenon attributed to the membrane potential influencing the exchange of ionized adenine nucleotides.[113] The question of whether the adenine nucleotide exchange totally or only partially responded to the membrane potential was argued during most of the 1970s. In 1978 LaNoue together with Klingenberg produced evidence to confirm that the exchange was fully electrogenic, involving charged adenine nucleotides responding to the membrane potential (i.e., protons were not transported as a part of the exchange).[114] The effect of the inner mitochondrial membrane potential (resulting from respiratory activity) was to increase the internal concentration of ADP. This led to a decreased energy requirement for ATP synthesis by the synthase enzyme. A comparable argument applied to the intramitochondrial phosphate concentration, which was increased internally by the proton gradient. Such a situation meant that about a quarter of the energy required for ATP synthesis was provided by the protonmotive force across the membrane (proton gradient and membrane potential) rather than through the ATP synthase. Such a situation exemplified the intimate relationship between membrane transport and the process of oxidative phosphorylation, a principle originally derived from the chemiosmotic theory.

A second important inhibitor for the adenine nucleotide exchange with quite different properties was found subsequently, bongkrekate, a branched unsaturated fatty acid.[115] Whereas the atractylosides inhibited the carrier system competitively with adenine nucleotides, bongkrekate inhibition was promoted by ADP. It was concluded that the atractylosides inhibited the system externally and bongkrekate internally. This led to the conclusion that the carrier protein had two conformations, one when opening inward and one when opening outward.

---

[110] Klingenberg 1986, p.364.
[111] Pfaff, Klingenberg, and Heldt 1965. Two other groups could also be described as discovering this system.
[112] Atractylosides were found in *Atractylis gummifera*, a Mediterranean thistle, and in other plants.
[113] Pfaff and Klingenberg 1968.
[114] LaNoue, Mizani, and Klingenberg 1978.
[115] Henderson and Lardy 1970.

The binding to the carrier of these inhibitors labeled with radioactive isotopes was exploited in the isolation of the carrier protein itself. Both the carboxyatractyloside form and the bongkrekate-binding forms were isolated in 1978.[116] Further studies showed that the protein was in fact a dimer, a property common to carriers isolated in this period. The molecular weight of the monomer was 30,000.[117] The isolated protein was incorporated into phospholipid vesicles and shown to convey adenine nucleotides across the membrane.[118] All this gave a good view of the protein's overall operation but failed to reveal its detailed mechanism. The need for a structural study was evident.

The isolated protein was purified and sequenced.[119] From the sequence it was possible to predict the structure, which was primarily six transmembrane helices, a structure common to other mitochondrial-transport proteins.[120] However, as Palmieri and van Ommen noted, the clue to the mechanism of transport systems was seen as being in the structure of the carriers:

> The research area that is still awaiting its breakthrough is the structural biochemistry of the mitochondrial carrier proteins, closely related to the understanding of the transport mechanism. Being integral membrane proteins, until now it has been impossible to obtain either crystals or a sufficiently high protein density in artificial membranes to allow for electron microscopic analysis of its structure (two dimensional crystallization), as has been successfully applied for a small number of other integral membrane proteins.[121]

The structure was eventually determined.[122] Work on other carriers confirmed the view that understanding mitochondrial metabolism was integrally involved in appreciating the transmembrane-transport systems. More important, it had now been shown that the complex mechanisms of mitochondrial bioenergetics were a dynamic property of mitochondrial membranes, particularly the inner membrane. The proton motive force was not only concerned with the synthesis of adenine nucleotides but also with the transport of substances across the inner mitochondrial membrane.

8.11 THE ATP SYNTHASE

The characteristics of this period of the history of bioenergetics is perhaps best exemplified by the successful elucidation of the mechanism of the ATP synthase. Some basic consideration of the early history of the enzyme has already been given.[123]

---

[116] Klingenberg, Riccio, and Aquila 1978; Aquila et al. 1978. Note there were earlier, though less satisfactory, isolations of the protein in the mid-1970s.
[117] Hackenberg and Klingenberg 1980.
[118] Krämer and Klingenberg 1977.
[119] Aquila et al. 1982.
[120] Walker and Runswick 1993.
[121] Palmieri and van Ommen 1999.
[122] Pebay-Peyroula et al. 2003.
[123] See Chapter 5.

There were several lines of investigation that led to an understanding of this remarkable enzyme. As already noted, Paul Boyer, influenced by the strong burst of interest in protein conformation that occurred in the early 1960s, had proposed a conformational theory to explain the overall mechanism of oxidative phosphorylation. Although this proved unsustainable, he developed his ideas on protein conformation as a means of understanding aspects of the ATPase (ATP synthase). His experiments on the effect of uncouplers on the ATPase reaction led him to the conclusion that the provision of energy for ATP synthesis was associated with a conformational change in the enzyme protein, leading to release of already-formed ATP.[124] Such a view was unexpected as it was assumed that the energy would be required for the synthesis of ATP itself. Acquiring evidence for this revolutionary approach took some time. Further experiments led Boyer to the view that there was probably more than one catalytic site on the ATPase for ATP synthesis and that two sites worked cooperatively.[125]

This approach left questions about the role of protons in the ATPase reaction without any hint of an answer. Mitchell retained the view that the proton must operate at the active site of the enzyme whereas Boyer progressively accommodated a role for protons (and the chemiosmotic theory) as the cause of the conformational change in his proposals for the mechanism of the ATPase:

> Involvement of electrical potential or proton gradients in transmission of energy from oxidations to the phosphorylation complex is regarded as an attractive possibility.[126]

Although allowing for the possibility that the proton motive force could induce conformational changes in the enzyme that would lead to ATP synthesis, he continued to argue for his original conformational theory of oxidative phosphorylation:

> Also attractive at this stage is the possibility of direct conformational transfer of energy from electron transport to ATP synthesis in oxidative phosphorylation and photophosphorylation.[127]

Even in his 1977 review, Boyer continued to press his conformational theory as a *possible* explanation for the mechanism of energy transfer from electron transport to phosphorylation.[128] However, beyond this point Boyer concentrated on the conformational processes in the ATPase itself.

The structural understanding of the enzyme was initiated by the electron microscopy of Fernández-Morán but strongly supported by the dissociation and reconstitution

---

[124] Boyer et al. 1973. These ideas were introduced in Chapter 5, Section 5.8.

[125] Boyer et al. 1977; Kayalar, Rosing, and Boyer 1977.

[126] Boyer 1975b, p.5.

[127] Boyer 1975b. See p.6.

[128] For a discussion of the debate between Boyer and Mitchell on the role of protons and their relationship to ATP synthesis, see Prebble 2013.

experiments of Racker's group in the 1960s.[129] Essentially the enzyme was seen as a "knob" containing the catalytic site for ATP synthesis and hydrolysis, F_1, attached to a protein component that was integral with the membrane, F_o. In 1972, Alan Senior and Harvey Penefsky purified the mitochondrial F_1 complex, and the latter showed that it contained five different subunits.[130] The molecular weight of each subunit was determined. The best estimate of the molecular weight of F_1 was about 360,000 but there was uncertainty about the numbers of each subunit present in the F_1 complex.[131]

The 1970s also saw the beginnings of the application of genetic techniques to the study of the ATP synthase. As with other bioenergetic complexes, the bacterial system proved simpler and initially easier to work with than the mammalian one. In *Escherichia coli*, a number of uncoupled mutants unable to synthesize ATP by oxidative phosphorylation (*unc* mutants) were isolated that mapped in a region designated the *unc* operon, which proved to code for the ATPase. This was eventually shown to be composed of nine genes, five coding for F_1 proteins, three for F_o and one additional one postulated to be involved in assembly of the complex.[132] The sequencing of these bacterial genes was achieved in the early 1980s. From these gene DNA sequences, amino acid sequences were derived. This then gave a precise molecular size for each subunit. They also showed that the α and β subunits were similar and must have evolved from a single ancestor.[133] The complete sequence of the beef enzyme was achieved slightly later, in 1985.[134]

Further application of genetic techniques included the substitution of one amino acid by another, which proved valuable in elucidating protein-based mechanisms in a variety of ways. For example, several studies have identified an amino acid (aspartic acid) in the *c* protein of F_o as essential for proton conductance by this part of the ATPase. Changing the amino acid to glycine or asparagine rendered F_o impermeable to protons so the aspartic acid might be assumed to be part of the mechanism of proton conduction.

Although the types of subunit in F_1 were relatively easily identified, the number of each subunit in the complex was more difficult. Even in 1983 Amzel and Pedersen felt that the subunit composition was still uncertain:

It is fair to say that the subunit stoichiometry of the ATP-synthase complex is still unknown. In the case of F_1, subunit stoichiometries of the type $\alpha_2\beta_2\gamma_2\delta_2\varepsilon_2$ and $\alpha_3\beta_3\gamma\delta\varepsilon$ have been proposed.[135]

Both formulations gave molecular weights for the total complex sufficiently close to the experimentally determined value of 360,000 to be possible. Several lines of evidence

[129] Chapter 5, Sections 5.5 and 5.6.
[130] Brooks and Senior 1972; Knowles and Penefsky 1972. Although these authors referred to their subunits numerically as 1–5, in due course the five subunits were named in order of decreasing size: α, β, γ, δ, ε.
[131] See Senior 1973.
[132] See Downie, Gibson, and Cox 1979.
[133] Gay and Walker 1981a, 1981b; Saraste et al. 1981.
[134] Walker et al. 1985.
[135] Amzel and Pedersen 1983, p.808.

suggested that the catalytic site for ATP synthesis was on the β subunit. Nucleotides such as ATP bound to both α and β subunits but those binding to the former were tightly bound whereas to the latter, binding was more readily reversible. Although, as previously noted, there was a lack of agreement on the stoichiometry of the F_1 subunits, Boyer was influenced particularly by the work of Kagawa's group in Japan, who had concluded that there were three β subunits.[136] This contributed to his thinking about the ATPase mechanism and a growing realization that the enzyme had three rather than two active sites. It should be noted that although three nucleotide binding sites were demonstrable, there was no agreement that they were all catalytic; proposals included the possibility of one catalytic and two regulatory sites[137] and, even in the early 1990s, a system with only two catalytic sites.[138] However, toward the end of the 1980s, general agreement on a subunit stoichiometry for the *E. coli* enzyme of $α_3β_3γδε$ was reached whereas the stoichiometry of the F_o subunits was ab_2c_{10-12}. The enzyme from beef heart was more complex, Walker's group reporting a total of thirteen different subunits in 1987.[139]

The outcome of Boyer's attempts to rationalize the data on the enzyme was the binding change mechanism.[140] Each of the three catalytic sites would be at a different stage in catalysis. The change in conformation of the β subunit (which would affect the catalytic site) would be induced by the rotation of core protein(s) of which the γ, δ, and ε subunits were the candidates.[141] The model was progressively developed during the 1980s. The role of protons was seen as driving the rotation of the F_1 core subunit (γ) through processes taking place in the membrane-bound base of the enzyme, F_o. In *E. coli* three subunits had been identified in F_o, named *a, b,* and *c*. Attempts to understand the events in the membrane part of the enzyme resulted in various models for the proton mechanism within F_o being proposed, although these remained largely speculative.[142]

The binding change mechanism proposed by Boyer was supported by a significant amount of evidence, mainly of an enzymological nature.[143] It made predictions about the synthesis of ATP dependent on structural changes in the conformation of the enzyme proteins and implied rotation of some of the proteins, primarily the γ subunit, which was seen as being in the center of the ring of alternating α and β subunits. Structural evidence to support these proposals became an absolute requirement.

During the 1980s some limited information was gathered from electron microscopy and x-ray crystallography of relatively low resolution. But understanding the structure of the enzyme would be dependent on x-ray diffraction studies of good enzyme crystals at high resolution. The laboratory of John Walker in Cambridge sought to grow suitable crystals of the bovine F_1 enzyme, which contains 2983 amino acids! Although

[136] Kagawa et al. 1976.
[137] Wang, Joshi, and Wu 1986.
[138] See the review by Berden, Hartog, and Edel 1991.
[139] Walker, Runswick, and Poulter 1987.
[140] Gresser, Myers, and Boyer 1982.
[141] Mitchell also proposed a rotary mechanism that delivered protons to each of the catalytic sites; Mitchell 1985.
[142] See for example Cox et al. 1986.
[143] See Boyer 1993.

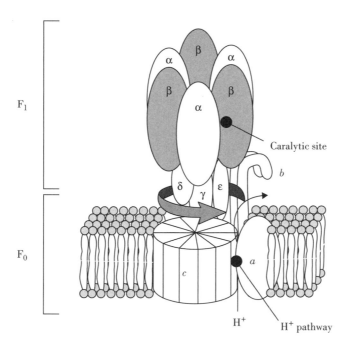

F₁

Caralytic site

b

δ ε γ

F₀

c

a

H⁺   H⁺ pathway

Figure 8.7 The Bacterial ATPase as Seen at the End of the 1990s by Futai and Omote

The upper part (F₁) is composed of a ring of 3α and 3β subunits, the catalytic sites for ATP synthesis being on the β subunits. A γ subunit runs from the lower ring of c subunits through the middle of the αβ ring. The "stalk" has the δε subunits with the γ subunit in the middle. The αβ ring is also attached to the base by the b subunits linked to the a subunit.

The enzyme was presumed to function by the rotation of the c subunits linked to γ. This is driven by the passage of protons through the membrane at the junction between c and a. The rotation of the γ subunit brings about conformational changes in the β subunit leading to ATP synthesis. Reproduced with permission from Futai and Omote 1999, *Mutational Analysis of ATP Synthase. An Approach to Catalysis and Energy Coupling* (in *Frontiers of Cellular Bioenergetics*, ed. Papa et al.), Springer, fig 1, p 400

at the beginning these crystals diffracted x-rays "rather poorly," suitable crystals were obtained in 1990 after 7 years of work. By using this material, high-resolution structures were obtained. This confirmed earlier views of an alternating ring of α and β subunits around the central γ subunit (see Fig. 8.7). There were three different conformations of the β subunits consistent with the predictions of the binding change mechanism. Intriguingly, the γ subunit is rigidly curved, and this region appears to impose the appropriate conformations on the β subunits in accord with the binding change mechanism.[144] It was the proposal and enzymological evidence for the binding change mechanism and its confirmation by elucidation of the structure of F₁ that won the Nobel Prize for Boyer and Walker in 1997.[145]

[144] Abrahams et al. 1994.
[145] The Nobel Prize in Chemistry for 1997 was shared, half to Jens Skou "for the first discovery of an ion-transporting enzyme, Na⁺, K⁺-ATPase" and the other half to Paul Boyer and John Walker "for their elucidation of the enzymatic mechanism underlying the synthesis of adenosine triphosphate (ATP)."

Although the highly significant achievements of Boyer and Walker strongly supported the idea of rotation as proposed by Boyer and also Mitchell, it did not show that rotation of the core of the enzyme actually occurred. Attempts in more than one laboratory showed that partial rotation of the γ subunit occurred but demonstration of full rotation was achieved by a group at the Tokyo Institute of Technology led by Masasuke Yoshida.[146] These workers succeeded in fixing part of the enzyme ($\alpha_3\beta_3\gamma$) to a glass surface and attaching a large fluorescent molecule to the γ subunit. The large molecule was just big enough to be seen under a fluorescence microscope. It was seen to rotate when ATP was applied, demonstrating the rotation of the γ subunit to which it was attached.[147] Thus the mechanism for ATP synthesis in cellular phosphorylation turns out to involve what is nature's smallest rotary motor.

There remained the major question of how the proton motive force drove the rotation of the core of the enzyme. This was presumed to occur in F_0 and remained a challenge for the twenty-first century. However, in 1996, Wolfgang Junge and colleagues put forward a valuable proposal in which the c subunits in F_0 reacted with protons and rotated against subunit a. This rotation of the c ring was seen as rotating the γ subunit

Although the developments described here concerned mitochondria and bacteria, evidence accumulated concurrently to show that the chloroplast enzyme mechanism was very similar. Here again the work on chloroplasts followed the path set by mitochondrial studies.

The mechanism of this enzyme, which Paul Boyer referred to as "a truly remarkable molecular machine," marks the pinnacle of bioenergetics research up to the end of the twentieth century. In particular, the story of its elucidation gives a clear demonstration of the path of bioenergetics research, particularly from the period of trying to understand the overall mechanism of oxidative phosphorylation to the period of the application of molecular biological techniques to bioenergetic problems. As Alan Senior pointed out about research on the enzyme at the end of the 1980s:

> The study of ATP synthesis by oxidative phosphorylation started approximately 50 years ago. It has been a major field of biology since its inception, attracting some of the most industrious and innovative researchers. In the last decade there has been a revitalization, largely due to the impact of genetic and molecular biology techniques, which have yielded huge amounts of primary structural information, and to the impact of direct enzymological techniques, which have revealed intrinsic properties of catalytic sites and their cooperativity.[148]

---

[146] Noji et al. 1997.

[147] About one revolution per second was observed, the slow rate being due to the load of the large molecule and the absence of the δ and ε subunits; the rate in vivo is estimated at around fifty times faster.

[148] Senior 1988, p.219.

Although at the end of the century much remained to be understood about the mechanism of maintaining a proton motive force by the respiratory chain, a broad molecular understanding of cell bioenergetics had been achieved. A developing field, the relationship of this knowledge to the understanding and treatment of various diseases, was now beginning to attract attention.

9

The Search for Mechanism

This chapter seeks to examine some of the themes that underlie the development of bioenergetics history. The field has been created by multiple mergers of lines of research. The two major fields of oxidative and photosynthetic biochemistry influenced each other after the Second World War and ultimately merged. But apart from this there are other significant mergers or what have been seen by some as interfield excursions. During these events, significant ideas and techniques were introduced, and these drove the research forward. Ultimately it was the challenge of the biochemistry of membranes that had to be met to achieve an in-depth appreciation of mechanism. But to get this far, the field had to negotiate a major crisis, and perhaps this aspect marks out the history more than any other.

9.1 PHOTOSYNTHETIC AND OXIDATIVE BIOCHEMISTRY—THE RELATIONSHIP

There were two almost-independent pathways that led to the creation of an identifiable field of bioenergetics in the second half of the twentieth century. One concerned the process that came to be known as photosynthesis, and the other concerned respiration, which became oxidative phosphorylation. Indeed, it was only in the second half of the twentieth century that developments in these two fields began to merge.

In the formative period of the eighteenth century, both fields coincided in the classical experiments of Joseph Priestley, in which respiration and photosynthesis appeared to work in opposition. A sprig of mint was shown to reverse the negative effects of respiration (or combustion). Respiration and photosynthesis were seen as very different processes, and understanding of the two diverged—after all, one was associated with animals and man whereas the other was a property of green plants. For more than the next 100 years progress in understanding the mechanism of photosynthesis was limited, being concerned with an appreciation of light as energy, the role of chlorophyll, and the division of the process into light and dark reactions. Much thought was given to possible light reactions involving chlorophyll and CO_2. Such lines of investigation had little to do with studies of basic metabolism (fermentation, glycolysis, etc.) and respiration.

Possibly the first influence of studies of respiration and oxidation on photosynthesis studies that can be detected came from Thunberg, who thought that an oxidation–reduction might be involved in photosynthesis. The view persisted, but more significant was the work of Wood and Werkman, who showed that assimilation of CO_2 could occur as a straightforward metabolic event. This at least raised the possibility that there were alternatives to the chlorophyll–CO_2 photochemical reaction favored in the nineteenth and early twentieth centuries. Further undermining of the traditional view of photosynthesis came from the work of van Niel on photosynthetic bacteria, which suggested that the photochemical event could indeed be an oxidation–reduction. However, it was the discovery of the Hill reaction that finally destroyed the idea of the chlorophyll–CO_2 reaction for all but a very few die-hard supporters such as Warburg.

There is little doubt that both fields received significant stimulation from the ability to isolate the relevant particle, the chloroplast and the mitochondrion, both in the 1930s although the biochemical significance of the mitochondrion did not become clear until after the Second World War. The influence of the isolation of the particle on research proceeded in two phases. First it provided a valuable means of making an active preparation, and second it led to a growing realization that understanding the biochemistry could not be achieved without understanding the fundamental role of the particle, essentially its membranes. However, the isolation of mitochondria and chloroplasts proceeded independently, apparently without influencing each other.

The demonstration of a chloroplast oxidation–reduction did suggest some similarity between respiratory studies and photosynthesis. However, the discovery of chloroplast cytochromes by Hill in the 1950s, Arnon's ideas about photophosphorylation, the proposal of a photosynthetic electron-transport chain in the late 1950s, together with the Z-scheme of 1960 all suggested a similarity between chloroplast and mitochondrion. Mitochondrial studies were now a clear influence on research in photosynthesis. The 1950s was a time when workers in photosynthesis realized that the chloroplast had an electron-transport chain that appeared to behave like the one already described in outline for the mitochondrion. There became available to those working on photosynthesis a wealth of knowledge, including theories of mechanism about electron transport and its link to phosphorylation.

Essentially a new field had been created: bioenergetics. The proposals put forward in the 1960s for the mechanism of phosphorylation, unlike Slater's 1953 chemical theory, were proposed for oxidative *and* photosynthetic phosphorylation. It was Mitchell's theory that had the greatest impact on those in the photosynthetic field at least in part because of the discoveries in Andre Jagendorf's laboratory. These workers discovered that a pH gradient appeared to be the intermediate between electron transport and phosphorylation, but more particularly they had shown that a pH gradient would lead to ATP synthesis in the dark in their chloroplast preparations.

This discovery was a very positive influence of the photosynthesis field on the mitochondrial world and the first independent evidence for the chemiosmotic approach. But its influence was muted by the inability to carry out comparable experiments on mitochondria—at least for some time.

The new field faced major methodological problems in dealing with membranes. However, the methods devised for one part of bioenergetics were also applicable to the other. In general, methods were developed initially in work on mitochondria. That having been said, the first membrane complex to be isolated, crystallized, and its structure determined to high resolution was a bacterial photosynthetic reaction center that demonstrated what could be achieved. A common feature of both studies in oxidative phosphorylation and photosynthesis was that the bacterial equivalents were normally simpler, smaller, and therefore easier to work with and to comprehend. Studies with bacteria had a substantial influence on the progress of both chloroplast and mitochondrial research from the time of van Niel in the 1930s onward.

9.2 THE COURSE OF BIOENERGETICS HISTORY

When David Keilin reflected on the development of research on respiration, he took a positive view of the events leading up to the formulation of the respiratory chain and its further investigation. He described it as follows:

> In looking back at the development of our knowledge of respiration we can see that it did not follow a straight or logical course. On the contrary, it can be compared to the flow of several convergent streams winding in their course.[1]

Converging streams have been a crucial part of bioenergetics history particularly in the years after the Second World War. This period also provided the right environment for the birth of the new field of bioenergetics. As Fruton points out in respect to the United States, there was

> Public expectation that the victorious and wealthy nation that produced the atomic bomb could also produce cures for cancer and other hitherto incurable diseases. Sizeable financial aid for research and both predoctoral and postdoctoral education became available.[2]

Such developments were mirrored to a lesser extent elsewhere, in the UK for example.[3] Coupled with this new public understanding of the importance of science was the release of a number of able scientists from war duties who now sought to develop their reputations in civil science.

Various streams of research including the demonstration that the mitochondrion was the organelle of oxidative phosphorylation and of the major cell oxidations brought cell biology and bioenergetics together to face a common set of problems. This coincided with a merging of work on respiration (such as that of Keilin) and that on aerobic phosphorylation (such as that of Kalckar, Belitzer, and Ochoa) to begin to solve the questions posed by oxidative phosphorylation. These convergences in the late 1940s were followed by another significant convergence in the 1960s, the merging of ideas from cell biology on "active transport" with those of oxidative phosphorylation. At this point it had become clear that the understanding of either field was dependent on understanding the other.

The interrelation of cell biology and biochemistry over this period has been explored in detail by William Bechtel and his colleagues:

---

[1] Keilin 1966, p.3.
[2] Fruton 1999, p.96.
[3] Postwar funding of research in the UK is discussed by de Chadarevian 2002, see pp.33ff.

Interfield excursions between biochemistry and cell biology played a crucial role during the period 1940–1965. Scientists in these fields jointly achieved an interdisciplinary account in which each respiratory mechanism was situated in the appropriate cell structure, and the component reactions and their organization had been specified to varying degrees of adequacy.[4]

Interfield theories that bridge two branches of science such as cell biology and biochemistry were explored by Darden and Maull.[5] The mutual development of the understanding of the subcellular particle, the mitochondrion, was seen as a part of both fields after around 1950. This certainly stimulated research as noted by George Palade who had been central to the work on the mitochondrion at the Rockefeller Institute in the late 1940s and 1950s:

> Historically the opening of the "submicroscopic" realm of biological organization by the electron microscope followed closely after the development of cytochemical techniques, especially of techniques for the separation of cell components by differential centrifugation. The ample information obtained in each of these two fields has stimulated research in the other, with the result that a number of cell components have acquired a new biochemical and physiological significance.[6]

Here the essential point is that this influence of mitochondrial research converging with studies of the respiratory chain and aerobic phosphorylation generated the emerging field of mitochondrial bioenergetics in the 1950s. Another element in this confluence was the understanding of phosphorylation derived from a key reaction in glycolysis, the glyceraldehyde phosphate dehydrogenase reaction, and also the ATP synthesizing reaction in the citric acid cycle.[7] It was an analysis of this oxidative reaction associated with ATP synthesis that formed the basis for understanding the overall process of oxidative phosphorylation.

The creation of the field that became known as bioenergetics has elements in common with the emergence of molecular biology over a similar time frame. Although the formation of molecular biology was a larger enterprise, it is possible to trace its origins in disparate research activities such as the phage group, the work on bacterial transformation by Griffiths and Avery, the study of the one gene–one enzyme hypothesis, and others.[8] In fact, the antecedents of molecular biology were much wider. As Judson comments,

> Molecular biology arose in the synthesis of particular lines from five distinct disciplines. The synthesis began in the mid-thirties; the merger of genetics and

---

[4] Bechtel and Abrahamsen 2007, p.2. A detailed examination of the development of cell biology, particularly in relation to cell bioenergetics, is given in Bechtel 2006.

[5] Darden and Maull 1977.

[6] Palade 1956, pp.186–187.

[7] This reaction, the succinate thiokinase reaction, is linked to ATP synthesis in bacteria but to guanosine triphosphate synthesis in mammalian mitochondria.

[8] Morange 1998, in exploring the "Roots of the New Science," Molecular Biology considers a number of factors including those I have mentioned but also the influence of the Rockefeller Foundation and further the influence of physics and physicists. See part 1.

x-ray crystallography, with Watson and Crick and the structure of DNA was a triumph of that synthesis approaching its maturity. Each of the disciplines in the synthesis was developing vigorously even as their coming together began.[9]

These are some of the positive elements in the history of bioenergetics of a type referred to by Keilin. However, science also has apparent dead ends that have held up the progress of the field, although as Hill pointed out for one of them, the nineteenth-century approach to the mechanism of photosynthesis, they normally generate a great deal of creative experimentation.

In the nineteenth century there were two major diversions into paths that ultimately led nowhere. The conviction that dominated the photosynthetic field was a belief that the central process involved a light reaction with chlorophyll and CO_2. This was coupled with a view about formaldehyde as an early product of carbon assimilation. The view lasted until the middle of the twentieth century, being advocated by the great German biochemist Otto Warburg.

Meanwhile studies in respiration were impeded, perhaps less significantly, by another misunderstanding, this time of the nature of oxidation. In the nineteenth century it seemed obvious that the process of oxidation would involve the addition of O_2 to organic molecules. Such a view lasted until at least the first third of the twentieth century, although a new approach to biological oxidation was provided by Thunberg and Wieland early in the century. They showed that biological oxidation involved the removal of hydrogen rather than the addition of O_2.

Other lines of research that had negative outcomes included several theories of oxidative phosphorylation but particularly the prediction that there would be a high-energy chemical substance formed by respiration and used for the synthesis of ATP. Much of the mitochondrial work in the 1950s was involved in investigating in one way or another this chemical theory, but it also gave a full basic description of the properties of oxidative phosphorylation. There were also a number of proposals of Peter Mitchell that could also be seen as dead ends. All of these theoretical ideas necessitated experimental testing, which not only brought evidence to bear on the main issue, but also gave valuable information about the systems investigated.

9.3 DOES METHODOLOGY DRIVE BIOENERGETICS?

A summary of many of the significant techniques employed in the history of bioenergetics is shown in Table 9.1. This indicates the very wide range of methods applied in the field, although spectroscopy in various forms has been a major feature of progress in both respiratory studies and photosynthesis.

In this account of the evolution of bioenergetics research, five phases have been identified. Each phase has been initiated by something approaching a revolution in

---

[9] Judson 1996, p.581. Judson's five disciplines from which particular lines emerged also included physical chemistry, microbiology, and biochemistry.

thinking and methodology. Previous patterns of work had begun to fail in their ability to lead to fresh understanding, resulting in a need for a novel approach. However, more important, new methodology or new perception has created the fresh approach. This description of progress is similar to but not the same as that associated with Kuhn's concept of paradigms and paradigm change. Freeman Dyson has drawn attention to the way progress in physics can be viewed. He contrasts Kuhn's approach to the history of science based on revolutions driven by ideas with that of Peter Galison in which major advances are seen as dependent on tools. Dyson feels "we need both of them to give us a complete picture."[10] The same seems to be true in the history of bioenergetics in which in the 1960s theories of phosphorylation were the driving force, but later the tools of protein chemistry and molecular biology created the new approach.

A major early influence on this field of study was the chemical revolution, which can certainly be regarded as a paradigm change. Much of the work in the nineteenth century derived from the work of Lavoisier, which included rethinking the nature of respiration and combustion. As Perrin commented when considering the chemical revolution:

> Lavoisier had a larger goal than a new combustion theory. His vision embraced a new system of chemistry in which atmospheric air or some component of air entered by traceable chemical paths into operations of the animal vegetable and mineral kingdom.[11]

Lavoisier's views dominated much thinking in the nineteenth century and also affected those of Ingen-Housz who had a not-dissimilar influence on nineteenth-century thinking on photosynthesis.

Perhaps the single event in this story that had the greatest influence was the experiments carried out by the Buchner brothers. There were two aspects to this discovery, which can be seen in Nobel Laureate Otto Loewi's[12] recollection of this event when he was 24 years old:

> I vividly remember the tremendous sensation that this discovery elicited, far beyond the circle of biologists and chemists, for instance also in the field of the philosophers, for it once and for all contradicted the thesis of the great Pasteur, that a chemical process as complex as fermentation would only be possible in the living cell.[13]

---

[10] Dyson 2012, p.1426. The importance of tools in driving scientific research has been developed by Peter Galison in a book, *Image and Logic*.

[11] Perrin 1990, p.270.

[12] Otto Loewe (1873–1961) shared the Nobel Prize in Physiology in 1936 with Henry Dale for their discoveries relating to chemical transmission of nerve impulses.

[13] Translated and quoted in Friedmann 1997, p.113. This recollection was given to the Fourth International Congress of Biochemistry in Vienna in 1958.

Table 9.1 Some of the Major Methods Driving Forward the Development of Cell Bioenergetics

Approximate Date	Method	Application	Principal Investigator(s)
17th century	Vivisection experiments on respiration	Physiology of respiration	Oxford physiologists: Boyle, Hooke, Mayow, Lower,
18th century	Study of gases	Basic chemistry of respiration and photosynthesis	Hales, Priestley, Lavoisier, Black
Late 19th century	Simple spectroscopy	Properties of animal pigments, role of chlorophyll	Stokes, MacMunn, Engelmann (Warburg)
1897 onward	The use of cell-free systems	Demonstration of basic metabolic processes	Buchner, Harden, etc.
ca. 1900	Chromatography	Initially separation of plant pigments but the principles influenced many methods later	Tswett
1908 onward	Phosphate chemistry	Identification of phosphate compounds in metabolism including ATP	Harden, Meyerhof, the Eggletons, Fiske & Subbarow
ca. 1910 onward	Warburg apparatus	Respiration Cell oxidations Photosynthesis	Warburg, Krebs
ca. 1918	Thunberg Tube	Cell oxidations	Thunberg
1925 onward	Microspectroscope	Respiratory chain	Keilin
1932 onward	Introduction of flashing-light techniques	Light reaction (photosynthesis)	Emerson
1937	Use of chloroplast preparations	Photosynthesis	Hill
1940 onward	Introduction of radioisotopes	CO_2 fixation in photosynthesis. ATPase reaction	Ruben, Benson, Cohn, Calvin, Boyer
ca. 1940	Monochromatic light	Analysis of role of pigments (photosynthesis)	Dutton, Haxo, Duysens, and others
Mid-1940s	Paper chromatography	Separation of the products of CO_2 assimilation.	Consden et al.

Table 9.1 (Continued)

Approximate Date	Method	Application	Principal Investigator(s)
ca. 1948 onward	Use of mitochondrial preparations. Use of cooled centrifuges	Respiration, oxidative phosphorylation	Lehninger, Slater, Chance ,etc.
1950s	Electron microscopy	Structure of mitochondria & chloroplasts	Palade, Sjostrand, Park, Weier
ca. 1950s	O_2 electrodes replaced Warburg apparatus	Oxidations	Chance and others
ca. 1950s	Introduction of commercial spectrophotometers	Analysis of electron-transport chains	Chance, also Slater and others
ca. 1960s onward	Use of detergents for disaggregation of membranes	Separation of complexes in the respiratory chain, chloroplast light reactions	Green's group, Boardman & Anderson, and others
1960s onward	Application of advanced spectroscopic techniques	Analysis of events in photosynthesis, etc.	Witt, Kok, and others
1960s–1970s	Advanced electron microscopy of membranes; freeze-fracturing techniques	Structure of mitochondrial and chloroplast membranes	Fernández Morán, Park, Bullivant, Packer, and others.
1960s onward	Introduction of SDS-electrophoretic separation of subunits	Analysis of mitochondrial and chloroplast electron transport chain subunits	Green, Racker, and others
1960s onward	Electron paramagnetic resonance spectroscopy	Iron–sulfur proteins in mitochondria and chloroplasts	Beinert, Malkin, and others
ca. 1975 onward	Introduction of picosecond technology	Detection of transients in photosynthesis	Fajer, Feher, and others.
1985s onward	Preparation of crystals of membrane-bound proteins	Bacterial, mitochondrial, and chloroplast proteins	Marcel

(continued)

Table 9.1 (Continued)

Approximate Date	Method	Application	Principal Investigator(s)
1985 onward	X-ray crystallography of membrane-bound proteins	Bacterial, mitochondrial, and chloroplast proteins	Deisenhofer and Huber.
ca. 1990	High-resolution electron microscopy of two-dimensional crystals	Structure of bacteriorhodopsin	Henderson

In fact, Buchner's work was important both in terms of ideas and technique. The demonstration that metabolic processes could be studied outside cells put paid to the ideas of vitalism and opened the door to thinking about metabolic biochemistry and enzymology. It also provided an experimental technique that, for bioenergetics, led to the appreciation of the importance of phosphate in metabolism to the discovery of the adenosine phosphates, ADP and ATP, and to the discovery of aerobic phosphorylation (oxidative phosphorylation).

For bioenergetics itself, the technique with arguably the greatest influence over almost 100 years was spectroscopy. The mathematician George Stokes had successfully applied a simple form of spectroscopy to blood pigments. He also initiated a study of fluorescence that proved so important in understanding the photochemical events of photosynthesis. Hoppe-Seyler was another who carried out early studies with spectroscopic methods. Charles MacMunn made a significant contribution with the application of the technique to biological systems in his book *Spectrum Analysis Applied to Biology and Medicine*. Significantly his histohaematins and myohaematins were the first accounts of respiratory pigments now known as cytochromes.

Optical spectroscopy more than any other technique established the nature of the respiratory chain. In the early twentieth century the same principles as applied by MacMunn were applied by Keilin but with improved equipment. Following the Second World War, the spectroscopic equipment was immensely improved with the introduction of various forms of spectroscopy. In particular, Britton Chance, having worked on the electronics of radar during the war, was then able to apply these skills to the development of various forms of high-quality spectrophotometry. Other forms of spectroscopy followed, especially electron-spin resonance spectroscopy, which in the hands of Helmut Beinert was used to demonstrate the importance of iron–sulfur proteins in the respiratory chain. The technique also became an essential tool in elucidating photosynthetic systems.

The great advantage of the spectroscopic methods was that they were essentially noninvasive. They could be applied to respiratory systems without significant chemical manipulation. In a period when the standard biochemical methodology was unable to handle membrane systems effectively, they enabled substantial progress to be made.

Although in this discussion of bioenergetics history a number of techniques have been identified as influencing the course of research, it was spectroscopy and cell-free systems that enabled most of the discoveries that created bioenergetics. They were, however, strongly supported by electron microscopy and the development and availability of the refrigerated centrifuge after the Second World War, which made possible the visualization and preparation of active mitochondria preparations. The same techniques stimulated photosynthetic studies, particularly the spectrophotometric ones. Chloroplasts could initially be prepared without a centrifuge, but the centrifuge soon became a part of methodology in that field. Of significance specifically in photosynthesis was the use of the heavy isotope of O_2 and particularly radioisotopes of carbon, which made possible analysis of carbon assimilation.

The methodology that began to be used in the 1960s and came to fruition in the 1980s and 1990s was of course the techniques of handling and analyzing membrane proteins. The development dominated research toward the end of the twentieth century. There can be little doubt that both ideas and new experimental techniques dictated the progress in bioenergetics.

9.4 THE SIGNIFICANCE OF THE MEMBRANE

The great successes of biochemistry during the first half of the twentieth century, glycolysis, urea cycle, citric acid (Krebs) cycle, and much work on enzymes and proteins, had concerned soluble systems. Those who pursued the unraveling of oxidative phosphorylation still retained this traditional approach to biochemical problems. In the 1950s there were numerous attempts to solubilize the oxidative phosphorylation system, and those few apparently successful attempts could not be sustained—in fact "soluble" oxidative phosphorylation generally turned out to be a property of tiny membrane vesicles. The approach to oxidative phosphorylation in the 1950s assumed that the role of the membrane was solely to provide scaffolding to which the system was attached. From around 1960, that situation changed, broadening the view of bioenergetics and stimulating new approaches. Membranes and their properties, including their role in selective transport, began to be taken into account. The review by Robertson in 1960 (referred to in Chapters 4 and 5) highlighted the important link he observed between transport, particularly "active transport," and oxidative phosphorylation. Although the arguments were based primarily on the uptake of ions by plant cells and secretion of hydrochloric acid into the stomach, both of which he associated with respiration, he concluded that the link with oxidative phosphorylation occurs in all living cells.

In the mitochondrion itself, interest in the ability of the particle to retain ions such as potassium had developed during the 1950s. In the early 1960s, the study of the ability of mammalian mitochondria to accumulate large amounts of calcium showed this system to be an integral part of oxidative phosphorylation. Taking into account the role of the respiratory chain, the phosphorylation system, and now membrane transport, Lehninger suggested

> That transformation of respiratory energy is trimodal. This conception suggests that the mechanism of oxidative phosphorylation should reveal something of the mechanism of transport and mechanochemical activities[14] and, conversely,

[14] This refers to the ability of mitochondria to change shape (seen as a mechanical activity of the membrane) and to swell and contract, taking up and releasing water. Such activities also appeared to be linked to oxidative phosphorylation.

that study of the membrane changes and active transport should shed light on oxidative phosphorylation.[15]

Thus the membrane as a location for transport systems was now a serious subject for investigation within the overall study of mitochondrial bioenergetics.

The shift in approach to the membrane is also seen in the hypotheses being used by bioenergeticists to understand events in mitochondria. The 1950s were dominated by a single approach, the chemical theory (in various forms): Oxidative phosphorylation was to be understood in terms of explanations based on the soluble glycolytic glyceraldehyde phosphate phosphorylation as proposed initially by Slater. In the 1960s, Mitchell's chemiosmotic theory was directly based on transmembrane reactions, and in due course the protons were seen to be translocated across the membrane. Williams's theory was based on events that required the anhydrous interior of the membrane. Green subsequently considered the changes in structure of the mitochondrial membranes to be the clue to understanding oxidative phosphorylation. Boyer's theory saw energy transferred between proteins in an integrated membrane structure.

Thus understanding oxidative phosphorylation would also involve understanding active membrane transport. The membrane now had an integral role in the thinking of bioenergeticists. This was perhaps the most crucial contribution to bioenergetics from cell biology, whether seen as a confluence or as interfield excursion. It opened up new research possibilities and challenged workers ever more powerfully to resolve the fundamental mechanism of oxidative phosphorylation. It initiated the fourth phase of research in the oxidative phosphorylation field.

9.5 MEMBRANE PROTEINS

The research that stemmed from the convergences in the 1940s and 1950s and from the cell biological contributions led to major advances in understanding the process of oxidative phosphorylation. However, successes were limited by the inability to examine the component parts of the system, the protein complexes, and individual proteins, which was the normal procedure in much biochemical research. The use of various types of spectroscopy had generated much knowledge of both the mitochondrial and the chloroplast electron-transport chains. Nevertheless, beginning at the very end of the 1950s and early 1960s, an ability to separate crude respiratory complexes, although poor at first, was developed particularly in David Green's laboratory in Wisconsin. The quality of the isolated complexes progressively improved over the ensuing years. In due course this led to the analysis of the protein components of the complexes. The introduction of the detergent sodium dodecyl sulfate (SDS) with polyacrylamide gels made possible analysis of membrane proteins, initially with limited results, but in due course the methodology

---

[15] Lehninger and Wadkins 1962, p.72.

was much improved. At about the same time in Efraim Racker's laboratory, the isolation of the F_1 complex was achieved, although this was not an integral membrane protein but rather an attachment to the membrane; nevertheless, its function was performed in association with proteins integral to the membrane.

Bioenergeticists were now fully aware of their need to gain an understanding of the membrane-bound proteins. The principle that Wikström and Krab had articulated required the skill to separate and examine them:

> One of the golden rules of biochemical research states that few things can be proved unless you can take them apart and put them together again with preservation of function.[16]

However, not until the 1980s did bioenergeticists fully acquire the techniques to separate adequately the proteins of the membrane-bound bioenergetic systems to reveal their structure. It was this acquisition of techniques to analyze membrane proteins that stimulated a new phase in the study of bioenergetics. Although the early methodology came from studies of membrane proteins of blood cells, bioenergeticists developed these methodologies to meet their own needs. Perhaps the earliest major success was the determination of the three-dimensional structure of the reaction center in the bacterium *Rhodopseudomonas viridis*. For those working in the field of photosynthesis, the core problem was to understand the photosynthetic reaction center, and this was their unique challenge along with the structure of pigment complexes that revealed the spatial relationships of the individual pigment molecules. Other structural achievements followed, although good model structures were not easily obtained, and in some cases success did not come until the twenty-first century. Nevertheless, detailed structures aided but did not necessarily solve the problem of mechanism. In cytochrome oxidase, for example, the use of mutant proteins became an essential tool.[17]

The new approach opened the way for separating proteins, for sequencing them either directly or from analysis of their genes, and in due course their crystallization for x-ray analysis. The methods associated with molecular biology provided fresh stimulation for research, leading to a much deeper understanding of the mechanisms of oxidative phosphorylation and also photophosphorylation. Such developments are epitomized by the analysis of the ATP synthase and the representation of its mechanism as a molecular motor.

9.6 THE CONCEPT OF MECHANISM

The foregoing has explored the way in which the search for mechanism was pursued historically in bioenergetics. However, there are some issues in this history that impinge on the nature of mechanism itself as it is explored by biological scientists, and these need to be noted.

[16] Wikström and Krab 1979, p.206.
[17] See Rich 1999 for example.

The story told here is fundamentally a search for an understanding of respiration and photosynthesis, an explanation of its biological significance, especially in relation to cellular energetic questions. Essentially it is seeking a mechanism. Biologists and particularly those working in the field of bioenergetics have generally seen their labor as concerned with understanding a mechanism. Indeed, in recent years there has been increasing interest in the concept of mechanism among those working in the history and philosophy of science. Lindley Darden comments that

> We are not alone in thinking that the concept of "mechanism" is central to an adequate philosophical understanding of the biological sciences[18] although this view has to be modified by noting that this approach is not appropriate for such fields as taxonomy.

Bechtel in his investigation of the history of cell biology notes that

> recognizing that the goal of many scientific inquiries is to describe the mechanism responsible for the phenomenon of interest provides a different perspective on many aspects of scientific inquiry.[19]

He provides a helpful definition of his view of mechanism:

> A mechanism is a structure performing a function in virtue of its component parts, component operations, and their organization. The orchestrated functioning of the mechanism is responsible for one or more phenomena.[20]

The quest for mechanisms in the field of bioenergetics and its antecedents has shifted substantially over the four centuries discussed here. For Mayow, the seventeenth-century Oxford physiologist, it concerned a portion of the air that was necessary for life. With Lavoisier the mechanism of respiration was concerned with slow combustion. As noted earlier, his understanding of the two processes, combustion and respiration, that fixed air were developed together. His conclusions provided the basis for much of the debate in the nineteenth century until they were challenged, particularly by Claude Bernard. Toward the end of the nineteenth century, interest in the cell began to shift attention to cellular mechanisms. With the work of David Keilin, Otto Warburg, and others, the mechanism of respiration was now seen at a very different level and concerned with enzymes. Concurrently and independently questions about the mechanism of aerobic phosphorylation began to arise. As early as 1939 Belitzer and Tsybakova noted that

> The problem of the mechanism of the "respiratory" phosphorylation is bound intimately to the problem of the mechanism of cellular respiration.[21]

[18] Darden 2006, p.15.
[19] Bechtel 2006, p.5.
[20] Ibid., p.26.
[21] Belitzer and Tsybakova 1939, p.225.

However, it was not until the 1950s that this relationship was taken seriously enough to stimulate effective experimentation. By then the key mechanistic question was the relationship of respiration to phosphorylation, seen in terms of a high-energy chemical intermediate (the chemical theory). For the 1960s and part of the 1970s, the issue was a choice between theories and therefore mechanisms of oxidative phosphorylation. Finally, in the 1980s and 1990s, with the electrochemical proton gradient identified as the link between the phosphorylation process and the respiratory process, the question now concerned how protons interacted with the membrane. In essence this was the relationship of protons to the proteins of the respiratory chain and the phosphorylating enzyme.

The level of the mechanism, in the sense of detail, changed over time from Mayow to the late twentieth century. It was of course influenced by what was happening elsewhere in biochemistry and in biology more widely. But that change in level was associated with conceptually different questions not necessarily implicit in the previous period. As Francis Crick pointed out,

> What is found in biology is *mechanisms,* mechanisms built with chemical components and that are often modified by other, later, mechanisms added to the earlier ones.[22]

Thus the levels seen from a historical perspective are not quite the same as those described by Cramer and Darden for example, who provide a general description:

> Biological mechanisms typically span multiple levels. Scientists working at higher levels work on ecosystems, populations and the behaviors of organisms within their environments. Others study mechanisms within organisms , . . . and ultimately mechanisms with smaller entities such as macromolecules, small molecules and ions. Different fields of biology are often (to a first approximation) associated with different levels.[23]

Mostly their discussion concerns a nonhistorical consideration. Here the level of the mechanism changes from very general physiology through combustion (no doubt influenced by the development of the steam engine at the beginning of the nineteenth century) to that of the cell as a complex chemical machine to the enzymological approach to membrane-based mechanisms. Biological mechanisms develop from the very broadly physiological to the more precise physiological to the enzymological to those of protein structure. In addition, some mechanisms are eclipsed by history, for example, photochemical reaction of the chlorophyll–CO_2 complex and the chemical hypotheses for phosphorylation.

The question of structure is closely linked with mechanism. It emerges in mechanism discussions in terms of localization of mechanisms for example. For Lavoisier and many of those in the first half of the nineteenth century, this was the question of which tissue

---

[22] Crick 1988, p.138.
[23] Craver and Darden 2013, p.21.

housed the respiratory combustion process. Was it the lungs, the blood, or, as it later emerged, a property of cells in general? Beyond that point the structural issue associated with the mechanism of respiration and phosphorylation was somewhat eclipsed by the biochemists' enthusiasm for homogenates. Some of the rationale for this approach and its justification is given by David Green:

> One may ask with good reason what is the point of imitating the cell with mixtures of the components in test tubes. Is it egotism and vanity on the part of the biochemist or a flair for chemical engineering? The study of mechanism, perforce, must be extremely limited in dealing with intact tissues. The variation of conditions which is essential to studies of mechanism must lie within the confines of those tolerated by living material. The biochemist has therefore to resort to the disorganization of the cell in order to puzzle out mechanisms of reaction. The major discoveries of the mechanisms which cells utilize for their reactions have practically all been made by the analyses of the behavior of cell extracts and of enzyme systems.[24]

Similar concerns arise in photosynthesis. An overall view, no doubt satisfying to those at the beginning of the nineteenth century, provided a working explanation that was due to the work of Ingen-Housz, Senebier, and de Saussure. Attempts to understand this further met with only limited success until Hill began his work with chloroplasts and a more detailed mechanism began to emerge. Here again the structural element began to play a significant role.

The structureless approach was indeed very successful, generating the glycolytic pathway, the urea cycle, and the citric acid cycle. However, it had major limitations.

As Peter Mitchell wrote in 1991,

> Fifty years ago, the bag-of-enzymes view of cell metabolism was prevalent, and the chemical actions of metabolism were generally looked upon purely as processes of primary chemical transformation. However, the elucidation of the biochemistry of the transfer of inorganic ions, metabolites and energy could not make substantial progress without an appreciation of the relationship between chemical transformation and chemical transport.[25]

Indeed Mitchell had struggled against that view, believing that an understanding of cellular systems necessitated an appreciation of the spatial directionality of biological processes that were in fact vectorial; such an approach underlined the importance of cellular membranes. Structure became important when the mitochondrion was found to be the location of oxidative phosphorylation itself at the end of the 1940s. Nevertheless, the full significance of the mitochondrion was not realized until the 1960s, particularly as a result of the work of Mitchell with his chemiosmotic hypothesis. It was really only at this

---

[24] Green 1939, p.185.
[25] Mitchell 1991, pp.297–298.

point, the 1960s, that the fundamental importance of the relationship of cellular structure to mechanism was fully appreciated. This point has also recently been considered by Matlin, who, in a historical study of oxidative phosphorylation in the 1950s and 1960s, felt that structure was important in the design and interpretation of experiments.[26] However, he found considerable variation in the use of structure. Such issues were a part of the developing revolution and consequent crisis that engulfed bioenergetics at this time.

9.7 REVOLUTION, HYPOTHESIS, AND CRISIS

To the extent that any scientific research can be said to reach a state of crisis, there was an acute crisis in the field of bioenergetics in the mid-1970s that had its roots in the 1960s. The underlying issue was one of mechanism that permeated much of the work in this period. It was not so much the search for a mechanism but rather making a choice among several mechanistic proposals, each one with its own advocates. The resulting tensions influenced the approach of workers and created an aspect of mechanism search not seen so obviously elsewhere. The history of bioenergetics cannot be considered without noting this aspect that not only influenced the science but particularly affected the scientists and their personal relationships.

Revolutions occurred during the history of bioenergetics; indeed the first clear-cut description of the process as slow combustion was the product of what is often described as the chemical revolution. The demonstration that cell-free extracts could carry out processes such as fermentation associated with living cells constituted a major revolution, giving birth to metabolic biochemistry. However, it was the slow collapse of the chemical theory and the development of the chemiosmotic hypothesis that provided a conceptual revolution and constituted the crisis period in the history of bioenergetics. Here I comment on the way in which the crisis developed and was resolved in terms of Kuhn's view of scientific revolutions. Although I do not propose to offer any critique of Kuhn's views in light of the bioenergetics story, Kuhn's approach to science history provides a useful basis for examining the events associated with the crisis in the field during the 1960s and 1970s.

The idea of a crisis in the development of scientific ideas was described by Thomas Kuhn (1922–1996) from an analysis of scientific revolutions almost exclusively in the physical sciences. Kuhn wrote,

> Let me first try to clarify what I mean by an "abnormal situation" or what I am elsewhere calling a "crisis state". I have already indicated that it is a response by some part of the scientific community to its awareness of an anomaly in the ordinary concordant relationship between theory and experiment.[27]

Note that not all anomalies lead to a crisis, and it is the reaction of the scientific community to the crisis that is significant for Kuhn. In the case of bioenergetics, the crisis was

[26] Matlin 2016.
[27] Kuhn 1977, p.202.

occasioned by three issues. First, the failure of the community to have an agreed theory to relate to experiment; second, the perception that other scientists in the discipline (biochemistry) regarded the field as in chaos; and third, a feeling among those in the United States at least that this was leading to difficulties in research funding.

The 1950s had been a time when "normal science" as described by Thomas Kuhn had prevailed. Kuhn had defined normal science as "research firmly based upon one or more past scientific achievements."[28] The achievements are those discussed earlier, particularly the work of Belitzer and Ochoa on oxidative phosphorylation and the understanding of oxidation and phosphorylation in glycolysis as developed by Needham and Pillai and by Warburg. The idea of normal science was closely associated by Kuhn with the notion of "paradigm," a term most readily understood in terms of shared methods, modes of explanation, and theories. The term embraces the idea of the community that shares this paradigm.[29] Essentially during the 1950s there was one central theory of oxidative phosphorylation that commanded the allegiance of workers in the field. This (with its variations) was known as the chemical theory and was proposed by Bill Slater in 1953 and was very frequently quoted in the literature. This was an update on an earlier proposal by Fritz Lipmann, which became necessary when it was realized that phosphate was not necessary for respiration and therefore the phosphorylation process itself had to be independent of oxidation–reduction. Paul Boyer remarked to the author that any graduate student could have made Slater's proposal at the time but, although this is probably an exaggeration, the value of Slater's paper lay more in its being a clear statement of what many were thinking rather than in its sheer originality.

Slater's proposals were based on the biochemical understanding of known phosphorylation systems (substrate-level phosphorylation), particularly that associated with phosphoglyceraldehyde oxidation in the glycolytic pathway but also noted other substrate-level phosphorylations such as that associated with succinyl coenzyme A in the citric acid cycle phosphorylation. The proposal therefore had an enviable pedigree and stood as the basis for interpreting experiments in the 1950s and for much of the 1960s. It predicted a *high-energy* chemical intermediate whose chemical identity was unclear but some of whose predicted properties gave a basis for seeking and identifying it. Although numerous substances were proposed as the high-energy intermediate, none survived rigorous testing. Nevertheless, there was a clear objective for investigations, and work continued to amass information both on the process of oxidative phosphorylation and on the nature of the suspected chemical intermediate. This period indeed could be classified as "normal science" in the Kuhnian tradition. It was this stable situation that collapsed in the 1960s and led to the crisis.

By 1960, the hitherto solid position occupied by Slater's chemical theory (and its various modifications) was beginning to look slightly less secure. The failure to find the predicted intermediate was beginning to raise the question, does it exist? In conversation with leaders in the field at a conference in Stockholm in 1960, Peter Mitchell established

---

[28] Kuhn 1970, p.10.
[29] It should be noted that it has been claimed that Kuhn used the term paradigm in more than twenty different ways in his seminal book, *The Structure of Scientific Revolutions*.

that the time was now thought to be ripe for consideration of an alternative approach to oxidative phosphorylation, leading him to propose his chemiosmotic theory in 1961.[30] Many other theories also emerged during the ensuing years. But the solid basis of 1950s research, the chemical theory, began to crumble very slowly, at first almost imperceptibly. However, by 1974 Slater concluded that the chemical theory was no longer viable on the grounds that the predicted high-energy intermediates "do not exist."[31] It is very questionable whether any experiments can be identified that could be seen as discrediting the chemical theory. If one is to identify the anomalies (in Kuhn's terms) they must lie in the following. The theory had two key weaknesses: It predicted a "high-energy" intermediate that could not be found despite extensive experimentation in many laboratories, and the proposal did not explain why experimental demonstrations of oxidative phosphorylation always involved a membranous system. There were several claims to have found either a high-energy intermediate or a soluble membrane-free system, but these were not confirmed. In addition to these limitations, the theory did not provide a very convincing explanation for the mechanism of action of uncouplers such as dinitrophenol.

From the early 1960s to the mid-1970s there was no single paradigm for oxidative phosphorylation. There was a tendency for individual laboratories to align themselves with one particular view, leading to antagonisms in the bioenergetics community. This whole period generated often heated debates, particularly in the United States, and this became known as the period of the Ox-phos wars.[32] Gilbert and Mulkay draw attention to a comment from a scientist they interviewed that illustrates the strains in the field:

> Particularly in the 60s, the oxidative phosphorylation field had the reputation that if you went to a Federation meeting, all the meetings were crowded because everyone went along because they knew there would be a damned good fight there . . . . I think it basically relates to the fact that progress was so slow . . . the field wasn't moving at all.[33]

During this period of uncertainty, in addition to the chemical theory whose importance was slowly receding, there was the chemiosmotic theory that was gaining adherents particularly from the younger generation, and the conformational theory that was particularly attractive to those who lost faith in the chemical theory (some were able to see the conformational theory as a modification of the chemical theory). Further, there were several rather ephemeral theories based on mitochondrial morphology. This last group lost credibility when the osmotic effects of ion movements in and out of the mitochondrial matrix were properly appreciated and provided an explanation for the observed changes in mitochondrial structure. All of the theories after the chemical theory involved the

---

[30] The Stockholm discussion is briefly recorded in a letter from Mitchell to the editor of *Nature* accompanying his initial paper on the chemiosmotic theory. It is quoted in more detail in Prebble and Weber 2003, see pp.79–80.

[31] Slater 1974, p.1153. It should perhaps be noted that a proposal based on a high-energy chemical intermediate appeared briefly in the late 1970s associated with lipoic acid (see Griffiths 1976.)

[32] See Prebble 2002.

[33] Gilbert and Mulkay 1984, p.36.

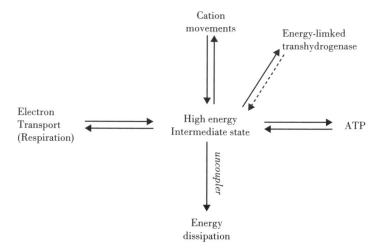

Figure 9.1 Summary of the Energy Relationships Associated With Oxidative Phosphorylation Around 1970

The diagram shows the role of the high-energy intermediate state (high-energy chemical intermediate, proton gradient, or conformational state) in relation to cation transport, ATP synthesis, or hydrolysis and electron transport. The role of an uncoupler is seen to be the dissipation of the energy in the system. The transhydrogenase enzyme of the mitochondrion was also found to be energy linked and is included for completeness. The diagram is typical of attempts to summarize oxidative phosphorylation at the time. See for example Slater et al. 1970.

membrane in one way or another, normally the inner mitochondrial membrane. Parallel proposals appeared for the chloroplast membranes and photophosphorylation.

Such unity in the field as could be found is seen in the summary schemes such as is shown in Figure 9.1, developed by several workers. Slater for example produced one such scheme to introduce the subject at a conference on Electron Transport and Energy Conservation in 1970. This outlined the contemporary view of the relationship among respiration, ATP synthesis, ion transport, and the transhydrogenase,[34] all linked to what here is called energy pressure but elsewhere is referred to as a high-energy state—a physical or chemical intermediate state. The candidates for this state considered by the authors were the formation of A~C (chemical theory), the translocation of protons and consequent membrane potential (chemiosmotic theory), a conformational change in a protein (conformational theory), or "something else."[35] Did schemes such as that in Figure 9.1 provide the basis for the paradigm for the 1960s and 1970s? Probably not, it was too vague.

In due course, at least the American workers began to feel the field was in crisis, although this was less evident among the Europeans. The American view was expressed by Efraim Racker in a letter to the leaders in the field,[36] arguing that the bioenergetics

[34] The transhydrogenase enzyme, a single-membrane protein, was found to be linked to the energy-conservation system of the respiratory chain.

[35] Slater et al. 1970, see p.3.

[36] Letter, Racker to Boyer, Chance, Ernster, Green, King, Lardy, Lehninger, Mitchell, Sanadi, and Slater; dated March 11, 1974. Quoted in Section 5.11. See also Prebble 2002, p.211.

community was seen as excessively engaging in confusing controversies and that funding of research was suffering as a consequence. As a solution to the problem he suggested a joint statement by the leaders of the field.[37]

Such a statement required a measure of agreement and cooperation, which was not readily achieved. In fact, such a proposal did not find favor with several bioenergeticists, particularly Mitchell, whose participation was now essential. An acceptable compromise was found eventually. However, as we will see, there was a consequence of this activity that was probably unintended.

9.8 RESOLVING THE CRISIS: ACCEPTING THE CHEMIOSMOTIC THEORY

The route out of this crisis is perhaps unusual. The new paradigm, the theory eventually agreed by the bioenergetics community to best fit their observations, was Mitchell's chemiosmotic theory. It sought to move the basis for conceptual thinking from organic chemical compounds as the link between respiration and phosphorylation to a physical ion gradient, an electrochemical proton gradient. Several leading workers in the field found the shift difficult to appreciate initially, wishing to see the proton as the intermediate in oxidative phosphorylation, not the proton gradient with its accompanying electric potential.[38] However, the shift was fundamental not just for oxidative phosphorylation but for the understanding of many aspects of membrane transport. Bioenergetics now embraced transport as well as ATP synthesis.

Many people found the chemiosmotic theory very difficult to comprehend.[39] The theory itself was not straightforward. The initial formulation in 1961 had to undergo substantial revision in 1966. Clear independent experimental evidence for the new hypothesis did not really begin to emerge until around 1966 and then not in mitochondria but only in chloroplasts. Further, what was meant by the theory was itself not without problems. Was it the formation of the proton electrochemical gradient by respiration that then drove ATP synthesis? Or did it also include the numerous schemes for detailed aspects of the mechanism of respiratory generation of the gradient proposed by Mitchell?

---

[37] The notion of some sort of common statement was not new. Some 18 months earlier Lars Ernster, the leading Stockholm bioenergeticists, had proposed something similar, but this had not been pursued.

[38] In a letter early in 1968, Mitchell commented on how in "much of the recent discussion about the mechanism of coupling in oxidative and photosynthetic phosphorylation (e.g. papers by Lehninger's group, Chance's group, Vernon's group, Pressman and Harris, and by Slater) has been based on a misconception about the identity of ΔpH [proton gradient] and Δp [proton motive force]." (Part of this letter is reproduced in Prebble and Weber 2003, p.138.)

[39] Rutherford (Bob) Robertson on visiting London in 1967 found that "Peter Mitchell's elegant chemiosmotic hypothesis was so little understood and, by some regarded with suspicion." A book by Vladimir Skulachev found that a book published by Robertson in 1968 "was not worth translating [into Russian] because it accepted the Mitchell hypothesis which most of the world's leading biochemists did not!" Robertson 1992, pp.16–17.

Did the theory include Mitchell's proposal for the mechanisms of the ATP synthase? Such questions hampered resolution of the problem.

Ordinarily one would expect the acceptance of the theory to depend on a critical experiment or group of experiments. Yet it is difficult to identify such an experiment in the mid to late 1970s associated with final resolution of the crisis. So what else might be involved? The events that need to be considered are the Racker–Stoeckenius experiment of 1974 (see Chapter 5, Section 5.7), the combined review, the genesis of which has already been examined, and finally, possibly, the award of a Nobel Prize to Mitchell.

The publication of the Racker–Stoeckenius experiment was in 1974, the year Slater admitted that the chemical theory was no longer viable but long before the arguments on mechanism began to die down. The Racker and Stoeckenius paper published in the *Journal of Biological Chemistry* was only a brief two-page note. The brevity needs to be set against the series of published experiments by both Racker's group and Stoeckenius's group of which this experiment could be seen as the culmination. Understanding the role of this experiment in resolving the bioenergetics crisis is perhaps clarified somewhat by a comment of Mitchell when asked by the author about its influence on the acceptance of the chemiosmotic theory. Mitchell felt it had had little impact on those working in the field but was positively influential on those outside the bioenergetics field. Although finding the critical experiment may seem difficult, it should be remembered that there was a large accumulation of experimental information by the mid-1970s, much of it supporting the chemiosmotic theory but also, in the eyes of some, capable of alternative interpretation.

A second event, previously noted, is the publication of the collection of reviews by most of the senior leaders in the field as a single composite review in the *Annual Reviews of Biochemistry* in 1977—each author making his own statement, by Boyer, Chance, Mitchell, Racker, and Slater.[40] Although many have seen this as the point when accord was achieved, a careful reading of the papers suggests that this is not quite so, although the tone is much more conciliatory. Some authors, particularly Boyer and Slater, still wished to consider the conformational hypothesis as a plausible alternative. Slater also considered a compromise theory in which the ATP synthase was driven by conformational changes effected by the respiratory system but that ion movements were the result of a proton gradient set up by the respiratory system. He did, however, concede that "conformational change and proton gradient are two sides of the same coin." It is possible to see this review as providing the basis on which the Nobel committee recommended Mitchell for the prize.

In view of these uncertainties, a third event, the award of the Nobel Prize in Chemistry to Mitchell in 1978, needs to be considered. In view of the complexity of the matter, the citation for Mitchell's prize "read rather vaguely" as noted by the chairman of the recommending Chemistry Committee.[41] At least to some extent, the award of the prize (inevitably criticized by some as being premature and by others that it should have been shared) finally seems to have resolved the issue.

[40] Boyer et al. 1977.
[41] Malmström 2000; see Chapter 5, Section 5.11.

Hence it is difficult to identify a point when the chemiosmotic theory emerged as the accepted paradigm for bioenergetics. In terms of experiments, although work continued to appear in support of the chemiosmotic theory after 1974, it is difficult to identify a significant experimental event after that point but at that date the matter was not settled. Three years later the review seems to have had substantial influence but unanimity is still lacking. A year after that the Nobel Prize seems to have confirmed the position. However, it does appear that the process of resolving the crisis may have been social rather than experimental, that is, the question of resolving the arguments about the chemiosmotic theory was ultimately a matter of persuasion. It was a question of persuading the workers in the field that the large body of experimental evidence, the Racker–Stoeckenius experiment, the arguments set out in the review did add up to adequate evidence for the validity of the theory. In the view of the Nobel committee the answer was yes, they did.

The excitement that bioenergetics generated in searching for its central mechanism died down at the end of the 1970s. The big question seemed to have been solved, and the necessary exploration of the detailed mechanism of photophosphorylation and oxidative phosphorylation attracted much less attention. Mitochondrial biochemistry no longer occupied a central place on the biochemical stage. Indeed, it was the study of photosynthesis that produced the first outstanding result in the analysis of membrane proteins, the structure of a bacterial reaction center. But few could have foreseen the beauty that emerged in the model for the ATP synthase rotating machine at the end of the century.

References

Abrahams, Jan Pieter, Andrew G. W. Leslie, René Lutter, and John E. Walker. 1994. "Structure at 2.8 Å Resolution of F_1-ATPase From Bovine Heart Mitochondria." *Nature* 370: 621–628.

Adler, Erich, Hans von Euler, and G. Günther. 1939. "Diaphorase I and II." *Nature* 143: 641–642.

Alexandre, Adolfo, Baltazar Reynafarje, and Albert L. Lehninger. 1978. "Stoichiometry of Vectorial H^+ Movements Coupled to Electron Transport and to ATP Synthesis in Mitochondria." *Proceedings of the National Academy of Science USA* 75: 5296–5300.

Allchin, Douglas. 1997. "A Twentieth-Century Phlogiston: Constructing Error and Differentiating Domains." *Perspectives on Science* 5: 81–127.

——— 2002. "To Err and Win a Nobel Prize: Paul Boyer, ATP Synthase and the Emergence of Bioenergetics." *Journal of the History of Biology* 35: 149–172.

Allen, Mary B., Frederick R. Whatley, and Daniel I. Arnon. 1958. "Photosynthesis by Isolated Chloroplasts. VI. Rates of Conversion of Light Into Chemical Energy in Photosynthetic Phosphorylation." *Biochimica et Biophysica Acta* 27: 16–23.

Amesz, Jan. 1973. "The Function of Plastoquinone in Photosynthetic Electron Transport." *Biochimica et Biophysica Acta* 301: 35–51.

Amzel, L. Mario and Peter L. Pedersen. 1983. "Proton ATPases: Structure and Mechanism." *Annual Review of Biochemistry* 52: 801–824.

Anderson, S., A. T. Bankier, B. G. Barrell, M. H. L. de Bruijn, A. R. Coulson, J. Drouin, I. C. Eperon, D. P. Nierlich, B. A. Roe, Frederick Sanger, P. H. Schreier, A. J. H. Smith, R. Staden, and I. G. Young. 1981. "Sequence and Organization of the Human Mitochondrial Genome." *Nature* 290: 457–465.

Andreoli, Thomas E., K.-W. Lam, and D. Rao Sanadi. 1965. "Studies on Oxidative Phosphorylation. X. A Coupling Enzyme Which Activates Reversed Electron Transfer." *Journal of Biological Chemistry* 240: 2644–2653.

Annau, E., Ilona Banga, B. Gozsy, St. Huszak, K. Laki, Ferenc B. Straub, and Albert Szent-Györgyi. 1935. "Über die Bedeutung der Fumarsäure für die Tierische Gewebsatmung: Einleitung, Übersicht, Methoden." *Hoppe-Seyler's Zeitschrift für Physiologische Chemie* 236: 1–68.

Aquila, Heinrich, Wolfgang Eiermann, Wilfried Babel, and Martin Klingenberg. 1978. "Isolation of the ADP/ATP Translocator From Beef Heart Mitochondria as the Bongkrekate-Protein Complex." *European Journal of Biochemistry* 85: 549–560.

Aquila, Heinrich, D. Misra, M. Eulitz, and Martin Klingenberg. 1982. "Complete Amino Acid Sequence of the ADP/ATP Carrier From Beef Heart Mitochondria." *Hoppe-Seyler's Zeitschrift für Physiologische Chemie* 363: 345–349.

Arnon, Daniel I. 1959. "Conversion of Light Energy Into Chemical Energy in Photosynthesis." *Nature* 184: 10–21.

——— 1961. "Changing Concepts of Photosynthesis." *Bulletin of the Torrey Botanical Club* 68: 215–259.

——— 1965. "Ferredoxin and Photosynthesis." *Science* 149: 1460–1469.

Arnon, Daniel I., Mary B. Allen, and F. Robert Whatley. 1954. "Photosynthesis by Isolated Chloroplasts." *Nature* 174: 394–396.

Arnon, Daniel I., F. Robert Whatley, and Mary B. Allen. 1959. "Photosynthesis by Isolated Chloroplasts. VIII. Photosynthetic Phosphorylation and the Generation of Assimilatory Power." *Biochimica et Biophysica Acta* 32: 47–57.

Avron, Mordhay. 1963. "A Coupling Factor in Photophosphorylation." *Biochimica et Biophysica Acta* 77: 699–702.

Avron, Mordhay and Britton Chance. 1966. "Relation of Phosphorylation to Electron Transport in Isolated Chloroplasts." *Brookhaven Symposia in Biology* 19: 149–160.

Avron, Mordhay and Andre T. Jagendorf. 1956. "A TPNH Diaphorase From Chloroplasts." *Archives of Biochemistry and Biophysics* 65: 475–490.

――― 1959. "Evidence Concerning the Mechanism of Adenosine Triphosphate Formation by Spinach Chloroplasts." *Journal of Biological Chemistry* 234: 967–972.

Avron, Mordhay, David W. Krogmann, and André T. Jagendorf. 1958. "The Relation of Photosynthetic Phosphorylation to the Hill Reaction." *Biochimica et Biophysica Acta* 30: 144–153.

Avron, Mordhay and Amir Shneyour. 1971. "On the Site of Action of Plastocyanin in Isolated Chloroplasts." *Biochimica et Biophysica Acta* 226: 498–500.

Azzone, Giovanni F. and Stefano Massari. 1973. "Active Transport and Binding in Mitochondria." *Biochimica et Biophysica Acta* 301: 195–226.

Babcock, Gerald T. 1999. "How Oxygen Is Activated and Reduced in Respiration." *Proceedings of the National Academy of Sciences USA* 96: 12971–12973.

Baeyer, Adolf von. 1870. Ueber die Wasserentziehung und ihre Bedeutung für das Pflanzenleben die Gärung. *Berichte der deutschen chemischen Gesellschaft* 3: 63–75.

Ball, Eric G. 1939. "The Role of Flavoproteins in Biological Oxidations." *Cold Spring Harbor Symposium* 7: 100–110.

――― 1975. "The Development of Our Current Concepts of Biological Oxidations." In *Proceedings of the Conference of the Historical Development of Bioenergetics*, John T. Edsall (ed.), pp. 92–103. Boston: American Academy of Arts and Sciences.

Banks, Barbara E. C. and Charles A. Vernon 1970. "Reassessment of the Role of ATP in vivo." *Journal of Theoretical Biology* 29: 301–326.

Barker, Horace A. and Robert E. Hungate. 1990. "Cornelius Bernardus van Niel. 1897–1985." In *Biographical Memoirs*, pp. 388–402. Washington, DC: National Academy of Sciences.

Barron, Eleazar S. G. and George A. Harrop Jr. 1928. "Studies of Blood Cell Metabolism II. The Effect of Methylene Blue and Other Dyes Upon the Glycolysis and Lactic Acid Formation of Mammalian and Avian Erythrocytes." *Journal of Biological Chemistry* 79: 65–87.

Bartley, Walter and Robert E. Davies. 1954. "Active Transport of Ions by Sub-Cellular Particles." *Biochemical Journal* 57: 37–49.

Bartley, Walter, Robert E. Davies, and Hans Krebs. 1954. "Active Transport in Animal Tissues and Subcellular Particles." *Proceedings of the Royal Society of London B* 142: 187–196.

Bassham, James A., Andrew A. Benson, Lorel D. Kay, Anne Z. Harris, A. T. Wilson, and Melvin Calvin. 1954. "The Path of Carbon in Photosynthesis. XXI. The Cyclic Regeneration of Carbon Dioxide Acceptor." *Journal of the American Chemical Society* 76: 1760–1770.

Bassham, James A. and Melvin Calvin. 1957. *The Path of Carbon in Photosynthesis.* Englewood Cliffs, NJ: Prentice-Hall.

Bay, J. Christian. 1931. "Jean Senebier, 1742–1808." *Plant Physiology* 6: 188–193.

Bechtel, William. 2006. *Discovering Cell Mechanisms. The Creation of Modern Cell Biology.* Cambridge: Cambridge University Press.

Bechtel, William and Adele Abrahamsen. 2007. "In Search of Mitochondrial Mechanisms: Interfield Excursions Between Cell Biology and Biochemistry." *Journal of the History of Biology* 40: 1–33.

Beinert, Helmut, Graham Palmer Terenzio Cremona, and Thomas P. Singer. 1965. "Kinetic Studies on Reduced Diphosphopyridine Nucleotide Dehydrogenase by Electron Paramagnetic Resonance Spectroscopy." *Journal of Biological Chemistry* 240: 475–480.

Belitzer, Vladimir. 1939. "La régulation de la respiration musculaire par les transformations du phosphagène." *Enzymologia* 6: 1–8.

Belitzer, Vladimir and E. T. Tsybakova. 1939. "The Mechanism of Phosphorylation Associated With Respiration." *Biokhimiya* 4: 516–534. Reproduced in English translation in Kalckar 1969, pp. 211–227.

Bendall, Derek S. 1994. "Robert Hill." Biographical Memoirs of Fellows of the Royal Society 40: 141–170.

Bengis, C. and Nathan Nelson. 1977. "Subunit Structure of Chloroplast Photosystem I Reaction Center." *Journal of Biological Chemistry* 252: 4564–4569.

Bensley, Robert R. and Normand L. Hoerr. 1934. "Studies on Cell Structure by the Freezing Drying Method. V. The Chemical Basis of the Organization of the Cell; VI. The Preparation and Properties of Mitochondria." *Anatomical Record* 60: 251–266, 449–455.

Benson, Andrew A. 2002. "Following the Path of Carbon in Photosynthesis: A Personal Story." *Photosynthesis Research* 73: 29–49.

Benson, Andrew A., James A. Bassham, Melvin Calvin, T. C. Goodale, V. A. Haas, and W. Stepka. 1950. "The Path of Carbon in Photosynthesis. V. Paper Chromatography and Radioautography of the Products." *Journal of the American Chemical Society* 72: 1710–1718.

Berden, J. A., A. F. Hartog, and C. M. Edel. 1991. "Hydrolysis of ATP by F_1 Can Be Described Only on the Basis of a Dual-Site Mechanism." *Biochimica et Biophysica Acta* 1057: 151–156.

Bernard, Claude. 1878. *Lectures on the Phenomena of Life Common to Plants and Animals.* Paris: Librairie J.-B Baillière et fils. Trans. H. E. Hoff, R. Guillemin, and L. Guillemin, 1974. Springfield, IL: C.C. Thomas.

Bishop, Norman I. 1961. "The Possible Rôle of Plastoquinone (Q_{-254}) in the Electron Transport System of Photosynthesis." In *Symposium on Quinones in Electron Transport*, G. E. W. Wolstenholme and Cecilia M. O'Connor (eds.), pp. 385–404. London: Churchill.

Black, Clanton C. and C. Barry Osmond. 2003. "Crassulacean Acid Metabolism Photosynthesis: 'Working the Night Shift.'" *Photosynthesis Research* 76: 329–341.

Black, Joseph. 1803. *Lectures on the Elements of Chemistry.* Ed. John Robision. Edinburgh: Mundell.

Blackman, Frederick F. 1905. "Optima and Limiting Factors." *Annals of Botany* 19: 281–295.

Boardman, N. Keith and Jan M. Anderson. 1964. "Isolation From Spinach Chloroplasts of Particles Containing Different Proportions of Chlorophyll *a* and Chlorophyll *b* and Their Possible Role in the Light Reactions of Photosynthesis." *Nature* 203: 166–167.

Böhme, H. and W. A. Cramer. 1972. "Localization of a Site of Energy Coupling Between Plastoquinone and Cytochrome *f* in the Electron-Transport Chain of Spinach Chloroplasts." *Biochemistry* 11: 1155–1160.

Boltzmann, Ludwig. 1886. "Der zweite Hauptsatz der mechanischen Wärmetheorie." In L. Boltzmann, *Populäre Schriften.* Reprint (1979) of 1905 edition, pp. 26–46. Leipzig: J. A. Barth.

Borgström, Bengt, Herschel C. Sudduth, and Albert L. Lehninger. 1955. "Phosphorylation Coupled to Reduction of Cytochrome *c* by β-Hydroxybutyrate." *Journal of Biological Chemistry* 215: 571–577.

Bowyer, John R. and Bernard L. Trumpower. 1981. "Pathways of Electron Transfer in the Cytochrome b-c$_1$ Complexes of Mitochondria and Photosynthetic Bacteria." In *Chemiosmotic Proton Circuits in Biological Membranes*, Vladimir P. Skulachev and Peter C. Hinkle (eds.), pp. 105–122. Reading, MA: Addison-Wesley.

Boyer, Paul D. 1965. "Carboxyl Activation as a Possible Common Reaction in Substrate-Level and Oxidative Phosphorylation and in Muscle Contraction." In *Oxidases and Related Redox Systems*, Tsoo E. King, H. S. Mason, and M. Morrison (eds.), Vol. 2, pp. 994–1017. New York: Wiley.

——— 1975a. "Energy Transduction and Proton Translocation by Adenosine Triphosphatases." *FEBS Letters* 50: 91–94.

——— 1975b. "A Model for Conformational Coupling of Membrane Potential and Proton Translocation to ATP Synthesis and to Active Transport." *FEBS Letters* 58: 1–6.

——— 1981. "An Autobiographical Sketch Related to My Efforts to Understand Oxidative Phosphorylation." In *Of Oxygen, Fuels and Living Matter Part 1*, Giorgio Semenza (ed.), pp. 229–264. Chichester, UK: Wiley.

——— 1993. "The Binding Change Mechanism for ATP Synthases—Some Probabilities and Possibilities." *Biochimica et Biophysica Acta* 1140: 215–250.

——— 1998. "Paul D. Boyer." In *Les Prix Nobel 1977*, pp. 177–185. Stockholm: The Nobel Foundation.

Boyer, Paul D., Britton Chance, Lars Ernster, Peter Mitchell, Efraim Racker, and Edward C. Slater. 1977. "Oxidative Phosphorylation and Photophosphorylation." *Annual Review of Biochemistry* 46: 955–1026.

Boyer, Paul D., Richard L. Cross, and William Momsen. 1973. "A New Concept for Energy Coupling in Oxidative Phosphorylation Based on a Molecular Explanation of the Oxygen Exchange Reactions." *Proceedings of the National Academy of Sciences USA* 70: 2837–2839.

Bouillaud, Frédéric, I. Arechaga, P. X. Petit, S. Raimbault, C. Levi-Meyrueis, L. Casteilla, M. Laurent, Eduardo Rial, and D. Ricquier. 1994. "A Sequence Related to a DNA Recognition Element Is Essential for the Inhibition by Nucleotides of Proton Transport Through the Mitochondrial Uncoupling Protein." *EMBO Journal* 13: 1990–1997.

Brooks, J. C. and Alan E. Senior. 1972. "Methods for Purification of Each Subunit of the Mitochondrial Oligomycin-Insensitive Adenosine Triphosphatase." *Biochemistry* 11: 4675–4678.

Brown, Jeanette S. and C. Stacy French. 1959. "Absorption Spectra and Relative Photostability of the Different Forms of Chlorophyll in *Chlorella*." *Plant Physiology* 34: 305–309.

Buchner, Eduard. 1897. "Alkoholische Gährung ohne Hefezellen" *Berichte der Deutschen Chemischen Gessellschaft* 30: 117–124. Reprinted and translated in A. Cornish-Bowden (ed.), *New Beer in an Old Bottle*, pp. 25–31. Valencia: Universitat de València.

Caneva, Kenneth. 1993. *Robert Mayer and the Conservation of Energy*. Princeton, NJ: Princeton University Press.

Capaldi, Roderick A., F. Malatesta, and V. M. Darley-Usmar. 1983. "Structure of Cytochrome c Oxidase." *Biochimica et Biophysica Acta* 726: 135–148.

Carmeli, C. and Efraim Racker. 1973. "Partial Resolution of the Enzymes Catalyzing Photophosphorylation. XIV. Reconstitution of Chlorophyll-Deficient Vesicles Catalyzing Phosphate-Adenosine Triphosphate Exchange." *Journal of Biological Chemistry* 248: 8281–8287.

Cerecedo, Leopold R. 1933. "The Chemistry and Metabolism of the Nucleic Acids, Purines and Pyrimidines." *Annual Review of Biochemistry* 2: 109–128.

Chance, Britton. 1958. "The Kinetics and Inhibition of Cytochrome Components of the Succinic Oxidase System: III. Cytochrome b." *Journal of Biological Chemistry* 233: 1223–1229.

——— 1961. "Energy-Linked Cytochrome Oxidation in Mitochondria." *Nature* 189: 719–725.

——— 1991. "Optical Method." *Annual Review of Biophysics and Biophysical Chemistry* 20: 1–28.

Chance, Britton and B. Hess. 1959. "Metabolic Control Mechanisms I. Electron Transfer in the Mammalian Cell." *Journal of Biological Chemistry* 234: 2404–2412.

Chance, Britton and Gunnar Hollunger. 1960. "Energy-Linked Reduction of Mitochondrial Pyridine Nucleotide." *Nature* 185: 666–672.

Chance, Britton and Leena Mela. 1966. "Proton Movements in Mitochondrial Membranes." *Nature* 212: 372–376.

Chance, Britton, Donald F. Parsons, and G. Ronald Williams. 1964. "Cytochrome Content of Mitochondria Stripped of Inner Membrane Structure." *Science* 143: 136–139.

Chance, Britton and G. Ronald Williams. 1955. "Respiratory Enzymes in Oxidative Phosphorylation. IV. The Respiratory Chain." *Journal of Biological Chemistry* 217: 429–438.

——— 1956. "The Respiratory Chain and Oxidative Phosphorylation." *Advances in Enzymology* 17: 65–134.

Chance, Britton, David F. Wilson, P. Leslie Dutton, and Maria Erecińska. 1970. "Energy-Coupling Mechanisms in Mitochondria: Kinetic, Spectroscopic and Thermodynamic Properties of an Energy-Transducing Form of Cytochrome b." *Proceedings of the National Academy of Sciences* 66: 1175–1182.

Chappell, J. Brian. 1968. "Systems Used for the Transport of Substrates Into Mitochondria." *British Medical Bulletin* 24: 150–157.

Cheniae, G. M. and I. F. Martin. 1970. "Sites of Function of Manganese Within Photosystem II. Roles in O_2 Evolution and System II." *Biochimica et Biophysica Acta* 197: 219–239.

Claude, Albert. 1949. "Studies on Cell Morphology, Chemical Constitution, and Distribution of Biochemical Functions." *Harvey Lectures* 43 (1947–1948): 121–164.

Clayton, Roderick K. 1963. "Photosynthesis: Primary Physical and Chemical Processes." *Annual Review of Plant Physiology* 14: 159–180.

——— 2002. "Research on Photosynthetic Reaction Centers From 1932 to 1937." *Photosynthesis Research* 73: 63–71.

Clayton, Roderick K. and Hon Fai Yau. 1972. "Photochemical Electron Transport in Photosynthetic Reaction Centers From *Rhodopseudomonas spheroides*. I. Kinetics of the Oxidation and Reduction of P-870 as affected by External Factors." *Biophysical Journal* 12: 867–881.

Clore, G. Marius, Lars-Erik Andréasson, Bo Karlsson, Roland Aasa, and Bo G. Malmström. 1980. "Characterization of the Intermediates in the Reaction of Mixed-Valence-State Soluble Cytochrome Oxidase With Oxygen At Low Temperatures by Optical and Electron-Paramagnetic-Resonance Spectroscopy." *Biochemical Journal* 185: 155–167.

Cohn, Mildred. 1953. "A Study of Oxidative Phosphorylation With O^{18}-Labelled Inorganic Phosphate." *Journal of Biological Chemistry* 201: 735–750.

Colowick, Sidney P., Mary S. Welch, and Carl F. Cori. 1940. "Phosphorylation of Glucose in Kidney Extract." *Journal of Biological Chemistry* 133: 359–373.

Consden, R., A. H. Gordon, and A. J. P. Martin. 1944. "Qualitative Analysis of Proteins: A Partition Chromatographic Method Using Paper." *Biochemical Journal* 38: 224–232.

Cooper, Chris. 2000. "The Two Faces of Bioenergetics." *Trends in Biochemical Sciences* 25: 455–456.

Copenhaver, J. H. Jr and Henry A. Lardy. 1952. "Oxidative Phosphorylations: Pathways and Yield in Mitochondrial Preparations." *Journal of Biological Chemistry* 195: 225–238.

Cox, Graeme B., A. L. Fimmel, Frank Gibson, and L. Hatch. 1986. "The Mechanism of ATP Synthase: A Reassessment of the Functions of the *b* and *a* Subunits." *Biochimica et Biophysica Acta* 849: 62–69.

Crane, Frederick L., Youssef Hatefi, Robert L. Lester, and Carl Widmer. 1957. "Isolation of a Quinone From Beef Heart Mitochondria." *Biochimica et Biophysica Acta*, 25: 220–221.

Craver, Carl F. and Lindley Darden. 2013. *In Search of Mechanisms. Discoveries Across the Life Sciences*. Chicago: University of Chicago Press.

Crick, Francis. 1988. *What Mad Pursuit*. New York: Basic Books.

Culotta, Charles A. 1972. "Respiration and the Lavoisier Tradition: Theory and Modification." *Transactions of the American Philosophical Society* 62: 3–41.

Dalziel, Keith. 1983. "Axel Hugo Theodor Theorell. 6 July 1903–15 August 1982." *Biographical Memoirs of Fellows of the Royal Society* 29: 385–621.

Darden, Lindley. 2006. *Reasoning in Biological Discoveries*. Cambridge: Cambridge University Press.

Darden, Lindley and Nancy Maull. 1977. "Interfield Theories." *Philosophy of Science* 43, 44–64.

Davenport, H. E. 1960. "A Protein From Leaves Catalysing the Reduction of Metmyoglobin and Triphosphopyridine Nucleotide by Illuminated Chloroplasts." *Biochemical Journal* 77: 471–477.

Davenport, H. E. and Robert Hill. 1952. "The Preparation and Some Properties of Cytochrome *f*." *Proceedings of the Royal Society B* 139: 327–345.

———— 1960. "A Protein From Leaves Catalysing the Reduction of Haem-Protein Compounds in Illuminated Chloroplasts." *Biochemical Journal* 74: 493–501.

Davenport, H. E., Robert Hill, and Frederick R. Whatley. 1952. "A Natural Factor Catalysing Reduction of Methaemoglobin by Isolated Chloroplasts." *Proceedings of the Royal Society B* 139: 346–358.

Davies, Robert E. and Hans Krebs. 1952. "The Biochemical Aspects of the Transport of Ions by Nervous Tissue." *Biochemical Society Symposium* 8: 77–92.

Davis, K. A. and Youssef Hatefi. 1971. "Succinate Dehydrogenase. I. Purification, Molecular Properties and Substructure." *Biochemistry* 10: 2509–2516.

de Chadarevian, Soraya. 2002. *Designs for Life. Molecular Biology After World War II*. Cambridge: Cambridge University Press.

De Duve, Christian. 2013. "The Other Revolution in the Life Sciences." *Science* 339: 1148.

Deisenhofer, Johann, Otto Epp, Kunio Miki, Robert Huber, and Hartmut Michel. 1985. "Structure of the Protein Subunits in the Photosynthetic Reaction Centre of *Rhodopseudomonas viridis* at 3Å Resolution." *Nature* 318: 618–624.

Dewan, John G. and David E. Green. 1938. "Coenzyme Factor. A New Oxidation Catalyst." *Biochemical Journal* 32: 626–639.

Dhar, N. R. 1935. "Chemistry of Photosynthesis." *Cold Spring Harbor Symposia on Quantitative Biology* III: 151–164.

Döring, Gunter, Gernot Renger, J. Vater, and Horst T. Witt. 1969. "Properties of the Photoactive Chlorophyll-a_{II} in Photosynthesis." *Zeitschrift für Naturforschung* 24b: 1139–1143.

Downie, J. Allan, Frank Gibson, and Graeme B. Cox. 1979. "Membrane Adenosine Triphosphatases of Prokaryotic Cells." *Annual Review of Biochemistry* 48: 103–131.

Dunn, Robert, John McCoy, Mehmet Simsek, Alokes Majumdar, Simon H. Chang, Uttam L. Raj Bhandary, and H. Gobind Khorana. 1981. "The Bacteriorhodopsin Gene." *Proceedings of the National Academy of Sciences USA* 78: 6744–6748.

Dutton, Herbert J. Winston M. Manning, and Benjamin M. Duggar. 1943. "Chlorophyll Fluorescence and Energy Transfer in the Diatom *Nitzschia closterium*." *Journal of Physical Chemistry* 47: 308–313.

Duysens, Louis N. M. 1951. "Transfer of Light Energy Within the Pigment Systems Present in Photosynthesizing Cells." *Nature* 168: 548–550.

Duysens, Louis N. M., Jan Amesz, and B. M. Kamp. 1961. "Two Photochemical Systems in Photosynthesis." *Nature* 190: 510–511.

Duysens, Louis N. M., W. J. Huiskamp, J. J. Vos, and J. M. van der Hart. 1956. "Reversible Changes in Bacteriochlorophyll in Purple Bacteria Upon Illumination." *Biochimica et Biophysica Acta* 19: 188–190.

Dyson, Freeman J. 2012. "Is Science Mostly Driven by Ideas or by Tools?" *Science* 338: 1426–1427.

Edsall, John T. 1973. "Introduction." In *The American Academy of Arts and Sciences, Conference on the Historical Development of Bioenergetics*, J. T. Edsall (ed.), pp. vi–xi. Boston: American Academy of Arts and Sciences.

Eggleton, Philip and Grace P. Eggleton. 1927a. "The Physiological Significance of Phosphagen." *Journal of Physiology* 63: 155–161.

——— 1927b. "The Inorganic Phosphate and a Labile Form of Organic Phosphate in the Gastrocnemius of the Frog." *Biochemical Journal* 21: 190–195.

Einbeck, Hans. 1914. "Ueber das Vorkommen der Fumarsäure in frischen Fleische." *Zeitschrift für physiologische Chemie* 90: 301–308.

Embden, Gustav, E. Griesbach, and E. Schmitz. 1914. "Über Milchsäurebildung und Phosphorsäurebildung im Muskelpressaft." *Zeitschrift für physiologische Chemie* 93: 1–45.

Emerson, Robert and William Arnold. 1932a. "A Separation of the Reactions in Photosynthesis by Means of Intermittent Light." *Journal of General Physiology* 15: 391–420.

——— 1932b. "The Photochemical Reaction in Photosynthesis." *Journal of General Physiology* 16: 191–205.

Emerson, Robert, Ruth Chalmers, and Carl Cederstrand. 1957. "Some Factors Influencing the Long Wave-Limit of Photosynthesis." *Proceedings of the National Academy of Sciences USA* 43: 133–143.

Emerson, Robert and Charlton M. Lewis. 1943. "The Dependence of the Quantum Yield of *Chlorella* Photosynthesis on Wave Length of Light." *American Journal of Botany* 30: 165–178.

Engelhardt, Vladimir. 1930. "Ortho- and Pyrophosphat im aeroben und anaeroben Stoffwechsel der Blutzellen." *Biochemische Zeitschrift* 227: 16–38.

——— 1932. "Die Beziehungen zwischen Atmung und Pyrophosphatumsatz in Vogelerythrocyten." *Biochemische Zeitschrift* 251: 243–268.

——— 1975. "On the Dual Role of Respiration." In *Conference on the Historical Development of Bioenergetics*, J. T. Edsall (ed.), pp. 61–74. Boston: American Academy of Arts and Sciences.

Engelhardt, Vladimir and Aleksandr E. Braunstein. 1928. "Über die Beziehungen zwischen der Phosphorsäure und der Glykolyse im Blut." *Biochemische Zeitschrift* 201: 48–65.

Engelhardt, Vladimir and M. N. Ljubimova, M.N. 1930. "Glycolyse und Phosphorsäureumsatz in den Blutzellen verschiedener Tiere." *Biochemische Zeitschrift* 227: 6–15.

——— 1939. "Myosin and Adenosine Triphosphatase." *Nature* 144: 668–669.

Engelmann, Theodor W. 1881. "A New Method for the Investigation of Oxygen Liberation by Plant and Animal Organisms." *Botanische Zeitung* 39: 441–448.

―――― 1882. "On the Production of Oxygen by Plant Cells in a Microspectrum." *Botanische Zeitung* 40: 419–426.

Ernster, Lars and Gottfried Schatz. 1981. "Mitochondria: A Historical Review." *Journal of Cell Biology* 91: 227s–255s.

Euler, Hans von and Karl Myrbäck. 1923. "Gärungs-co-Enzym (Co-Zymase) der Hefe. I. *Zeitschrift für physiologische Chemie* 131: 179–203.

Evans, Michael C. W., Bob B. Buchanan, and Daniel I. Arnon. 1966. "A New Ferredoxin-Dependent Carbon Reduction Cycle in a Photosynthetic Bacterium." *Proceedings of the National Academy of Sciences USA* 55: 928–934.

Evans, Michael C. W., C. K. Sihra, James R. Bolton, and Richard Cammack. 1975. "Primary Electron Acceptor Complex of Photosystem I in Spinach Chloroplasts." *Nature* 256: 668–670.

Fajer, Jack, Daniel C. Brune, M. S. Davis, A. Forman, and L. D. Spaulding. 1975. "Primary Charge Separation in Bacterial Photosynthesis: Oxidized Chlorophylls and Reduced Pheophytin." *Proceedings of the National Academy of Sciences USA* 72: 4956–4960.

Feher, George. 1971. "Some Chemical and Physical Properties of a Bacterial Reaction Center Particle and Its Primary Photochemical Reactants." *Photochemistry and Photobiology* 14: 373–387.

Feher, George, James P. Allen, Melvin Y. Okamura, and Douglas C. Rees. 1989. "Structure and Function of Bacterial Photosynthetic Reaction Centres." *Nature* 339: 111–116.

Fernández-Morán, Humberto. 1962. "Cell-Membrane Ultrastructure. Low-Temperature Electron Microscopy and X-Ray Diffraction Studies of Lipoprotein Components in Lamellar Systems." *Circulation* 26: 1039–1065.

Fernández-Morán, Humberto, T. Oda, P. V. Blair, and David E. Green. 1964. "A Macromolecular Repeating Unit of Mitochondrial Structure and Function." *Journal of Cell Biology* 22: 63–100.

Fish, Leonard E., Ulrich Kück, and Lawrence Bogorad. 1985. "Two Partially Homologous Adjacent Light-Inducible Maize Chloroplast Genes Encoding Polypeptides of the P700 Chlorophyll *a*-Protein Complex of Photosystem I." *Journal of Biological Chemistry* 260: 1413–1421.

Fiske, Cyrus H. and Yellapragada Subbarow. 1927. "The Nature of the 'Inorganic Phosphate' in Voluntary Muscle." *Science* 65: 401–403.

Fletcher, Walter M. and Frederic G. Hopkins. 1907. "Lactic Acid in Amphibian Muscle." *Journal of Physiology* 35: 247–309.

Florkin, M. 1972. *History of Biochemistry. Part I Proto-Biochemistry. Part II Proto-Biochemistry to Biochemistry. Comprehensive Biochemistry Vol. 30.* Amsterdam: Elsevier.

―――― 1975a. *History of Biochemistry. Part III History of the Identification of the Sources of Free Energy in Organisms. Comprehensive Biochemistry Vol. 31.* Amsterdam: Elsevier.

―――― 1975b. "Glycolysis as a Source of Free Energy." In *Proceedings of the Conference of the Historical Development of Bioenergetics*, John T. Edsall (ed.), pp. 115–161. Boston: American Academy of Arts and Sciences.

Floyd, Robert A., Britton Chance, and Don Devault. 1971. "Low Temperature Photo-Induced Reactions in Green Leaves and Chloroplasts." *Biochimica et Biophysica Acta* 226: 103–112.

Frank, Robert G. Jr. 1980. *Harvey and the Oxford Physiologists.* Berkeley: University of California Press.

French, C. Stacy and M. L. Anson. 1941. "Oxygen Production by Isolated Chloroplasts." *American Journal of Botany* 28: 12s.

French, C. Stacy and Violet K. Young. 1952. "The Fluorescence Spectra of Red Algae and the Transfer of Energy From Phycoerythrin to Phycocyanin and Chlorophyll." *Journal of General Physiology* 35: 873–890.

Frenkel, Albert. 1954. "Light Induced Phosphorylation by Cell-Free Preparations of Photosynthetic Bacteria." *Journal of the American Chemical Society* 76: 5568–5569.

——— 1956. "Photophosphorylation of Adenine Nucleotides by Cell-Free Preparations of Purple Bacteria." *Journal of Biological Chemistry* 222: 823–834.

Friedmann, Herbert C. 1997. "From Friedrich Wöhler's Urine to Eduard Buchner's Alcohol." In *New Beer in an Old Bottle. Eduard Buchner and the Growth of Biochemical Knowledge*, Athel Cornish-Bowden (ed.), pp. 67–122. València: Universitat de València.

Fruton, Joseph S. 1972. *Molecules and Life*. New York: Wiley-Interscience.

——— 1985. "Contrasts in Scientific Style. Emil Fischer and Franz Hofmeister: Their Research Groups and Their Theory of Proteins Structure." *Proceedings of the American Philosophical Society* 129: 313–370.

——— 1999. *Proteins, Enzymes, Genes. The Interplay of Chemistry and Biology*. New Haven, CT: Yale University Press.

Futai, Masamitsu and Hiroshi Omote. 1999. "Mutational Analysis of ATP Synthase. An Approach to Catalysis and Energy Coupling." In *Frontiers of Cellular Bioenergetics*, S. Papa, F. Guerrieri, and J. M. Tager (eds.), pp. 399–421. New York: Kluwer Academic/Plenum.

Gabellini, Nadia, John R. Bowyer, Eduard Hurt, B. Andreaelamdri, and Günter Hauska. 1982. "A Cytochrome b/c_1 Complex With Ubiquinol-Cytochrome c_2 Oxidoreductase Activity From *Rhodopseudomonas sphaeroides* GA." *European Journal of Biochemistry* 126: 105–111.

Gaffron, Hans. 1946. "Photosynthesis and the Production of Organic Matter on Earth." In *Currents in Biochemical Research*, David Green (ed.), pp. 25–48. New York: Interscience.

Gaffron, Hans and K. Wohl. 1936. "Zur Theorie die Assimilation." *Naturwissenschaften* 24: 81–90, 103–107.

Gamble, James L. 1957. "Potassium Binding and Oxidative Phosphorylation in Mitochondria and Mitochondrial Fragments." *Journal of Biological Chemistry* 228: 955–971.

Garlid, Keith D., David E. Orosz, Martin Modriansky, Stefano Vassanelli, and Petr Ježek. 1996. "On the Mechanism of Fatty Acid-Induced Proton Transport by Mitochondrial Uncoupling Protein." *Journal of Biological Chemistry* 271: 2615–2620.

Gay, Nicholas and John E. Walker. 1981a. "The ATP Operon: Nucleotide Sequence of the Promoter and the Genes for the Membrane Proteins and the δ Subunit of *Escherichia coli* ATP-Synthase." *Nucleic Acids Research* 9: 3919–3926.

——— 1981b. "The atp Operon: Nucleotide Sequence of the Region Encoding the α-Subunit of *Escherichia coli* ATP-Synthase." *Nucleic Acids Research* 9: 2187–2194.

Gest, Howard. 1991. "The Legacy of Hans Molisch (1856–1937), Photosynthesis Savant." *Photosynthesis Research* 30: 49–59.

——— 1997. "A 'Misplaced Chapter' in the History of Photosynthesis Research; the Second Publication (1796) on Plant Processes by Dr Jan Ingen-Housz, MD, Discoverer of Photosynthesis." *Photosynthesis Research* 53: 65–72.

——— 2000. "Bicentenary Homage to Dr Jan Ingen-Housz, MD (1730–1799), Pioneer of Photosynthesis Research." *Photosynthesis Research* 63: 183–190.

——— 2002. "History of the Word *Photosynthesis* and Evolution of Its Definition." *Photosynthesis Research* 73: 7–10.

——— 2004. "Samuel Ruben's Contributions to Research on Photosynthesis and Bacterial Metabolism With Radioactive Carbon." *Photosynthesis Research* 80: 77–83.

Gilbert, G. Nigel and Michael Mulkay. 1984. *Opening Pandora's Box. A Sociological Analysis of Scientists' Discourse*. Cambridge: Cambridge University Press.

Glenman, Stuart S. 1996. "Mechanisms and the Nature of Causation." *Erkenntnis* 44: 49–71.

González-Halphen, Diego, Margaret A. Lindorfer, and Roderick A. Capaldi. 1988. "Subunit Arrangement in Beef Heart Complex III." *Biochemistry* 27: 7021–7031.

Govindjee, and David Krogmann. 2004. "Discoveries in Oxygenic Photosynthesis (1727–2003): A Perspective." *Photosynthesis Research* 80: 15–57.

Govindjee, Rajni, Jan B. Thomas, and Eugene Rabinowitch. 1961. "'Second Emerson Effect' in the Hill Reaction of *Chlorella* Cells With Quinone as Oxidant." *Science* 132: 421.

Green, David E. 1939. "Reconstruction of the Chemical Events in Living Cells." In *Perspectives in Biochemistry*, Joseph Needham and David E. Green (eds.), pp. 175–186. Cambridge: Cambridge University Press.

——— 1959. "Electron Transport and Oxidative Phosphorylation." *Advances in Enzymology* 21: 73–129.

——— 1974. "The Electromechanical Model for Energy Coupling in Mitochondria." *Biochimica et Biophysica Acta* 346: 27–78.

Green, David E. and Harold Baum. 1970. *Energy and the Mitochondrion*. New York: Academic.

Green, David E., P. V. Blair, and T. Oda. 1963. "Isolation and Characterization of the Unit of Electron Transfer in Heart Mitochondria." *Science* 140: 382.

Green, David E. and David H. MacLennan. 1967. "The Mitochondrial System of Enzymes." In *Metabolic Pathways*, David M. Greenberg (ed.), 3rd ed., pp. 48–111. New York: Academic.

Greenwood, Colin, Michael T. Wilson, and Maurizio Brunori. 1974. "Studies on Partially Reduced Mammalian Cytochrome Oxidase. Reactions With Carbon Monoxide and Oxygen." *Biochemical Journal* 137: 205–215.

Gresser, Michael J., Jill A. Myers, and Paul D. Boyer. 1982. "Catalytic Site Cooperativity of Beef Heart Mitochondrial F_1 Adenosine Triphosphatase. Correlations of Initial Velocity, Bound Intermediate, and Oxygen Exchange Measurements With an Alternating Three-Site Model. *Journal of Biological Chemistry* 257: 12030–12038.

Greville, Guy D. 1969. "A Scrutiny of Mitchell's Chemiosmotic Hypothesis of Respiratory Chain and Photosynthetic Phosphorylation." *Current Topics in Bioenergetics* 3: 1–78.

Griffiths, David E. 1976. "Studies of Energy-Linked Reactions. Net Synthesis of Adenosine Triphosphate by Isolated Adenosine Triphosphate Synthase Preparations: A Role for Lipoic Acid and Unsaturated Fatty Acids." *Biochemical Journal* 160: 809–812.

Grote, Mathias. 2013. "Purple Matter, Membranes and 'Molecular Pumps' in Rhodopsin Research (1960s–1980s)." *Journal of the History of Biology* 46: 331–368.

Guerlac, Henry. 1957. "Joseph Black and Fixed Air. A Bicentenary Retrospective, With Some New or Little Known Material." "Joseph Black and Fixed Air: Part II." *Isis* 48: 124–151, 433–456.

Hackenbrock, Charles R. 1968a. "Ultrastructural Bases for Metabolically Linked Mechanical Activity in Mitochondria: II. Electron Transport-Linked Ultrastructural Transformations in Mitochondria." *Journal of Cell Biology* 37: 345–369.

——— 1968b. "Chemical and Physical Fixation of Isolated Mitochondria in Low-Energy and High-Energy States." *Proceedings of the National Academy of Sciences* 61: 598–605.

Hackenberg, Heinz and Martin Klingenberg. 1980. "Molecular Weight and Hydrodynamic Parameters of the Adenosine 5'-Triphosphate Carrier in Triton X-100." *Biochemistry* 19: 548–555.

Haldane, John Burden Sanderson. 1927. "CXLII Carbon Monoxide as a Tissue Poison." *Biochemical Journal* 21: 1068–1075.

Haldane, John and J. Lorrain Smith. 1896. "The Oxygen Tension of Arterial Blood." *Journal of Physiology* 20: 497–520.

Hales, Stephen. 1727. *Vegetable Staticks*. London: W. and J. Innys.

Hangarter, Roger P. and Howard Gest. 2004. "Pictorial Demonstrations of Photosynthesis." *Photosynthesis Research* 80: 421–425.

Hankamer, Ben, James Barber, and Egbert J. Boekema. 1997. "Structure and Membrane Organization of Photosystem II in Green Plants." *Annual Review of Plant Physiology and Plant Molecular Biology* 48: 641–671.

Harden, Arthur. 1932. *Alcoholic Fermentation*, 4th ed. London: Longmans, Green.

Harden, Arthur and William J. Young 1906. "The Alcoholic Ferment of Yeast Juice." *Proceedings of the Royal Society B* 77: 405–420.

——— 1908. "The Alcoholic Ferment of Yeast Juice. Part III. The Function of Phosphates in the Fermentation of Glucose by Yeast Juice." *Proceedings of the Royal Society B* 80: 299–311.

Harold, Franklin M. 2001. *The Way of the Cell. Molecules, Organisms and the Order of Life.* Oxford: Oxford University Press.

Harris, E. J., R. Cockrell, and Bert C. Pressman. 1966. "Induced and Spontaneous Movements of Potassium Ions Into Mitochondria." *Biochemical Journal* 99: 200–213.

Harrop Jr., George A. and Eleazar S. G. Barron. 1928. "Studies on Blood Cell Metabolism I. The Effect of Methylene Blue and Other Dyes Upon the Oxygen Consumption of Mammalian and Avian Erythrocytes." *Journal of Experimental Medicine* 48: 207–223.

Hart, Helen. 1930. "Nicholas Théodore de Saussure." *Plant Physiology* 5: 424–429.

Hatch, Marshall D. and Charles R. Slack. 1966. "Photosynthesis by Sugar-Cane Leaves. A New Carboxylation Reaction and the Pathway of Sugar Formation." *Biochemical Journal* 101: 103–111.

Hatefi, Youssef. 1959. "Studies on the Electron Transport System XVII. Effects of Adenosine Diphosphate and Inorganic Phosphate on the Steady-State Oxidoreduction Level of Coenzyme Q." *Biochimica et Biophysica Acta* 31: 502–512.

——— 1963. "Coenzyme Q, (Ubiquinone)." *Advances in Enzymology* 25: 275–328.

——— 1966. "The Functional Complexes of the Mitochondrial Electron-Transfer System." In *Comprehensive Biochemistry*, M. Florkin and E. H. Stotz (eds.), Vol. 14, *Biological Oxidations*, pp. 199–231. Amsterdam: Elsevier.

——— 1985. "The Mitochondrial Electron Transport and Oxidative Phosphorylation System." *Annual Review of Biochemistry* 54: 1015–1069.

——— 1999. "The Mitochondrial Enzymes of Oxidative Phosphorylation." In *Frontiers of Cellular Bioenergetics*, Sergio Papa, Ferrucio Guerrieri, and Joseph M. Tager (eds.), pp. 23–47. New York: Kluwer Academic/Plenum.

Hatefi, Youssef, A. G. Haavik, L. R. Fowler, and David E. Griffiths. 1962. "Studies on the Electron Transfer System: XLII. Reconstitution of the Electron Transfer Chain." *Journal of Biological Chemistry* 237: 2661–2669.

Hatefi, Youssef, A. G. Haavik and David E. Griffiths. 1962. Studies on The Electron Transfer System. XL Preparation and Properties of Mitochondrial DPNH-Coenzyme Q Reductase. *Journal of Biological Chemistry* 237: 1676–1680.

Hatefi, Youssef, Robert L. Lester, Frederic L. Crane, and Carl Widmer. 1959. "Studies on the Electron Transport System XVI. Enzymic Oxidoreduction Reactions of Coenzyme Q." *Biochimica et Biophysica Acta* 31: 490–501.

Haupts, Ulrich, Jörg Tittor, and Dieter Oesterhelt. 1999. "Closing in on Bacteriorhodopsin: Progress in Understanding the Molecule." *Annual Review of Biophysics and Biomolecular Structure* 28: 367–399.

Hauska, Günter, Eduard Hurt, Nadia Gabellini, and Wolfgang Lockau. 1983. "Comparative Aspects of Quinol-Cytochrome *c*/Plastocyanin Oxidoreductases." *Biochimica et Biophysica Acta* 726: 97–133.

Haxo, Francis T. and Lawrence R. Blinks. 1950. "Photosynthetic Action Spectra of Marine Algae." *Journal of General Physiology* 33: 389–422.

Helmholtz, Hermann von. 1847. "The Conservation of Force: A Physical Memoir." In *Selected Writings of Hermann von Helmholtz*, Russell Kahl (ed.), 1971, pp. 3–55. Middletown, CT: Wesleyan University Press.

——— 1861. "The Application of the Law of the Conservation of Force." In *Selected Writings of Hermann von Helmholtz*, Russell Kahl (ed.), 1971, pp. 109–121. Middletown, CT: Wesleyan University Press.

Henderson, Peter J. F. and Henry A Lardy. 1970. "Bongkrekic Acid: An Inhibitor of the Adenine Nucleotide Translocase of Mitochondria." *Journal of Biological Chemistry* 245: 1319–1326.

Henderson, Richard. 1977. "The Purple Membrane From *Halobacterium halobium*." *Annual Review of Biophysics and Bioengineering* 6: 87–109.

Henderson, Richard, J. M. Baldwin, T. A. Ceska, F. Zemlin, E. Beckmann, and Kenneth H. Downing. 1990. "Model for the Structure of Bacteriorhodopsin Based on High-Resolution Electron Cryo-Microscopy." *Journal of Molecular Biology* 213: 899–929.

Henderson, Richard and P. N. T. Unwin. 1975. "Three-Dimensional Model of Purple Membrane Obtained by Electron Microscopy." *Nature* 257: 28–32.

Hill, Archibald V. 1914. "The Oxidative Removal of Lactic Acid." *Journal of Physiology* 48: x–xi.

——— 1932. "The Revolution in Muscle Physiology." *Physiological Reviews* 12: 56–67.

Hill, Robert. 1939. "Oxygen Produced by Isolated Chloroplasts." *Proceedings of the Royal Society B* 127: 192–210.

——— 1940. "The Reduction of Ferric Oxalate by Isolated Chloroplasts." *Proceedings of the Royal Society B* 129: 238–255.

——— 1951. "Oxidoreduction in Chloroplasts." *Advances in Enzymology* 12: 1–39.

——— 1954. "The Cytochrome *b* Component of Chloroplasts." *Nature* 174: 501–503.

——— 1965. "The Biochemists' Green Mansions: The Photosynthetic Electron-Transport Chain in Plants." *Essays in Biochemistry* 1: 121–151.

Hill, Robert and Fay Bendall. 1960. "Function of the Two Cytochrome Components in Chloroplasts: A Working Hypothesis." *Nature* 186: 136–137.

Hill, Robert and R. Scarisbrick. 1951. "The Haematin Compounds of Leaves." *New Phytologist* 50: 98–111.

Hind, Geoffrey. 1968. "The Site of Action of Plastocyanin in Chloroplasts Treated With Detergent." *Biochimica et Biophysica Acta* 153: 235–240.

Hinkle, Peter and Lawrence L. Horstman. 1971. "Respiration-Driven Proton Transport in Submitochondrial Particles." *Journal of Biological Chemistry* 246: 6024–6028.

Hoerr, Normand. 1957. "Robert Russell Bensley. 1867–1956." *Anatomical Record* 128: 1–18.

Hofmeister, Franz 1901. "Die chemische Organisation der Zelle." *Naturwissenschaftliche Rundschau* 16: 581–583, 600–602, 612–614.

Hogeboom, George H., Albert Claude, and Rollin D. Hotchkiss. 1946. "The Distribution of Cytochrome Oxidase and Succinoxidase in the Cytoplasm of the Mammalian Liver Cell." *Journal of Biological Chemistry* 165: 615–629.

Hogeboom, George H., Walter C. Schneider, and George E. Palade. 1948. "Cytochemical Studies of Mammalian Tissues. I. Isolation of Intact Mitochondria From Rat Liver; Some Biochemical Properties of Mitochondria and Submicroscopic Particulate Material." *Journal of Biological Chemistry* 172: 619–635.

Holmes, Frederic L. 1974. *Claude Bernard and Animal Chemistry.* Cambridge, MA: Harvard University Press.

——— 1985. *Lavoisier and the Chemistry of Life.* Madison: University of Wisconsin Press.

——— 1991, 1993. *Hans Krebs.* 2 vols. New York: Oxford University Press.

Hooke, Robert. 1666. "An Account of an Experiment Made by Mr. Hook, of Preserving Animals Alive by Blowing Through Their Lungs With Bellows." *Philosophical Transactions of the Royal Society* 2: 539–540.

Hopkins, Frederick Gowland. 1913. "The Dynamic Side of Biochemistry." In *Hopkins and Biochemistry. 1861–1947*, Joseph Needham and Ernest Baldwin (eds.), 1949, pp. 136–159. Cambridge: W. Heffer & Sons.

Hunter, F. Edmund Jr. 1951. "Oxidative Phosphorylation During Electron Transport." In *Phosphorus Metabolism*, W. D. McElroy and H. B. Glass (eds.), Vol. 1, pp. 297–330. Baltimore: Johns Hopkins University Press.

Hurt, Eduard and Günter Hauska. 1982. "Identification of the Polypeptides in the Cytochrome b_6/f Complex From Spinach Chloroplasts With Redox-Center-Carrying Subunits." *Journal of Bioenergetics and Biomembranes* 14: 405–423.

Ingen-Housz, Jan. 1779. *Experiments Upon Vegetables. Discovering Their Great Power of Purifying Common Air in the Sun-Shine, and of Injuring It in the Shade and at Night.* London: Elmsly and Payne.

Iwata, So, Christian Ostermeier, Bernd Ludwig, and Hartmut Michel. 1995. "Structure at 2.8 Å Resolution of Cytochrome c Oxidase From *Paracoccus denitrificans.*" *Nature* 376: 660–669.

Izawa, S. and Norman E. Good. 1968. "The Stoichiometric Relation of Phosphorylation to Electron Transport in Isolated Chloroplasts." *Biochimica et Biophysica Acta* 162: 380–391.

Jagendorf, André T. 1998. "Chance, Luck and Photosynthesis Research: An Inside Story." *Photosynthesis Research* 57: 215–229.

——— 2002. "Photophosphorylation and the Chemiosmotic Perspective." *Photosynthesis Research* 73: 233–241.

Jagendorf, André T. and Ernest Uribe. 1966. "ATP Formation Caused by Acid-Base Transition of Spinach Chloroplasts." *Proceedings of the National Academy of Sciences* 55: 170–177.

Joliot, Pierre. 2004. "Period-Four Oscillations of the Flash-Induced Oxygen Formation in Photosynthesis." *Photosynthesis Research* 76: 65–72.

Joliot, Pierre and Bessel Kok. 1975. "Oxygen Evolution and Photosynthesis." In *Bioenergetics of Photosynthesis*, Rajni Govindjee (ed.), pp. 388–412. New York: Academic.

Jordan, Patrick, Petra Fromme, Horst T. Witt, Olaf Klukas, Wolfram Saenger, and Norbert Krauss. 2001. "Three-Dimensional Structure of Cyanobacterial Photosystem I at 2.5 Å Resolution." *Nature* 411: 909–917.

Judah, J. D. 1951. "The Action of 2:4-Dinitrophenol on Oxidative Phosphorylation." *Biochemical Journal* 49: 271–285.

Judson, Horace Freeland. 1996. *The Eighth Day of Creation*, 25th anniversary edition. Cold Spring Harbor, NY: Cold Spring Harbor Press.

Junge, Wolfgang and Winfried Ausländer. 1973. "The Electric Generator in Photosynthesis of Green Plants. I. Vectorial and Protolytic Properties of the Electron Transport Chain." *Biochimica et Biophysica Acta* 333: 59–70.

Junge, Wolfgang and A. William Rutherford. 2007. "Horst Tobias Witt (1922–2007)." *Nature* 448: 425.

Junge, Wolfgang, D. Sabbert, and Siegfried Engelbrecht. 1996. "Rotary Catalysis by F-ATPase: Real-Time Recording of Intersubunit Rotation." *Berichte der Bunsengesellschaft für physikalische Chemie* 100: 2014–2019.

Kagawa, Yasuo and Efraim Racker. 1966. "Partial Resolution of the Enzymes Catalysing Oxidative Phosphorylation. X. Correlation of Morphology and Function in Submitochondrial Particles." *Journal of Biological Chemistry* 241: 2475–2482.

——— 1971. "Partial Resolution of the Enzymes Catalyzing Oxidative Phosphorylation. XXV. Reconstitution of Vesicles Catalyzing ^{32}Pi-Adenosine Triphosphate Exchange." *Journal of Biological Chemistry* 246: 5477–5487.

Kagawa Yasuo, Nobuhito Sone, Masasuke Yoshida, Hajime Hirata, and Harumasa Okamoto. 1976. "Proton Translocating ATPase of a Thermophilic Bacterium. Morphology, Subunits and Chemical Composition." *Journal of Biochemistry* 80: 141–151.

Kahl, Russell. 1971. *Selected Writings of Hermann von Helmholtz*. Middletown, CT: Wesleyan University Press.

Kalckar, Herman M. 1937. "Phosphorylation in Kidney Tissue." *Enzymologia* 2: 47–52.

——— 1939a. "The Nature of Phosphoric Esters Formed in Kidney Extracts." *Biochemical Journal* 33: 631–641.

——— 1939b. "Coupling Between Phosphorylations and Oxidations in Kidney Extracts." *Enzymologia* 6: 209–212.

——— 1941. "The Nature of Energetic Coupling in Biological Synthesis." *Chemical Reviews* 28: 71–178.

——— 1966. "Lipmann and the 'Squiggle.'" In *Current Aspects in Biochemical Energetics*, N. O. Kaplan and E. P. Kennedy (eds.), pp. 1–8. New York: Academic.

——— 1969. *Biological Phosphorylations: Development of Concepts*. New Haven, CT: Prentice-Hall.

——— 1991. "50 Years of Biological Research—From Oxidative Phosphorylation to Energy Requiring Transport Regulation." *Annual Review of Biochemistry* 60: 1–37.

Kamen, Martin D. 1986. "On Creativity of Eye and Ear: A Commentary on the Career of T. W. Engelmann." *Proceedings of the American Philosophical Society* 130: 232–246.

Katoh, Sakae. 1960. "A New Copper Protein From *Chlorella ellipsoidea*." *Nature* 186: 533–534.

Kaufman, Seymour. 1951. "Soluble α-Ketoglutaric Dehydrogenase From Heart Muscle and Coupled Phosphorylation." In *Phosphorus Metabolism*, W. D. McElroy and H. B. Glass (eds.), Vol. 1, pp. 370–373. Baltimore: Johns Hopkins University Press.

Kayalar, Celik, Jan Rosing, and Paul D. Boyer. 1977. "An Alternating Site Sequence for Oxidative Phosphorylation Suggested by Measurement of Substrate Binding Patterns and Exchange Reaction Inhibitions." *Journal of Biological Chemistry* 252: 2486–2491.

Keilin, David. 1925. "On Cytochrome, a Respiratory Pigment, Common to Animals, Yeast and Higher Plants." *Proceedings of the Royal Society B* 98: 312–339.

——— 1927. "Influence of Carbon Monoxide and Light on Indophenol Oxidase of Yeast Cells." *Nature* 119: 670–671.

——— 1930. "Cytochrome and Intracellular Oxidase." *Proceedings of the Royal Society B* 106: 418–444.

——— 1966. *The History of Cell Respiration and Cytochrome*. Cambridge: Cambridge University Press.

Keilin, David and Edward F. Hartree. 1937. "Preparation of Pure Cytochrome *c* From Heart Muscle and Some of Its Properties." *Proceedings of the Royal Society B* 122: 298–308.

——— 1938. "Cytochrome Oxidase." *Proceedings of the Royal Society B* 125: 171–186.

——— 1939. "Cytochrome and Cytochrome Oxidase." *Proceedings of the Royal Society B* 127: 167–191.

———— 1940. Succinic Dehydrogenase-Cytochrome System of Cells. Intracellular Respiratory System Catalysing Aerobic Oxidation of Succinic Acid." *Proceedings of the Royal Society B* 129: 277–306.

———— 1949. "Effect of Low Temperature on the Absorption Spectra of Haemoproteins; With Observations on the Absorption Spectrum of Oxygen." *Nature* 164: 254–258.

———— 1955. "Relationship Between Certain Components of the Cytochrome System." *Nature* 176: 200–206.

Keilin, David and Edward C. Slater. 1953. "Cytochrome." *British Medical Bulletin* 9: 89–97.

Kennedy, Eugene P. and Albert L. Lehninger. 1949. "Oxidation of Fatty Acids and Tricarboxylic Acid Cycle Intermediates by Isolated Rat Liver Mitochondria." *Journal of Biological Chemistry* 179: 957–972.

Khorana, H. Gobind, Gerhard E. Gerber, Walter C. Herlihy, Christopher P. Gray, Robert J. Anderegg, Kayoko Nihei, and Klaus Biemann. 1979. "Amino Acid Sequence of Bacteriorhodopsin." *Proceedings of the National Academy of Sciences USA* 76: 5046–5050.

Kielley, W. Wayne and Ruth K. Kielley. 1951. Myokinase and Adenosine triphosphatase in Oxidative Phosphorylation." *Journal of Biological Chemistry* 191: 485–500.

Klingenberg, Martin. 1970. "Mitochondria Metabolite Transport." *FEBS Letters* 6: 145–154.

———— 1986. "A Biochemist's View of His Struggle for Knowledge: Review of Forty Years' Service to Science." In G. Semenza (ed.), *Selected Topics in Biochemistry: Personal Recollections*, G. Semenza (ed.), Vol. 36, *Comprehensive Biochemistry*, pp. 327–398. Amsterdam: Elsevier.

Klingenberg, Martin, Paolo Riccio, and Heinrich Aquila. 1978. "Isolation of the ADP, ATP Carrier as the Carboxyatractylate-Protein Complex From Mitochondria." *Biochimica et Biophysica Acta* 503: 193–210.

Klingenberg, Martin and Edith Winkler. 1985. "The Reconstituted Isolated Uncoupling Protein Is a Membrane Potential Driven H⁺ Translocator." *The EMBO Journal* 4: 3087–3092.

Kluyver, Albert Jan and Hendrick J. L. Donker. 1926. "Die Einheit in der Biochemie." *Chemie der Zelle und Gewebe* 13: 134–190.

Knaff, David B., Richard Malkin, J. Clark Myron, and Marshall Stoller. 1977. "The Role of Plastoquinone and β-Carotene in the Primary Reaction of Plant Photosystem II." *Biochimica et Biophysica Acta* 459: 402–411.

Knowles, Aileen F. and Harvey S. Penefsky. 1972. "The Subunit Structure of Beef Heart Mitochondrial Adenosine Triphosphatase: Isolation Procedures." *Journal of Biological Chemistry* 247: 6617–6623.

Kohler, Robert E. 1971. "The Background to Eduard Buchner's Discovery of Cell-Free Fermentation." *Journal of the History of Biology* 4: 35–61.

———— 1972. "The Reception of Eduard Buchner's Discovery of Cell-Free Fermentation." *Journal of the History of Biology* 5: 327–353.

———— 1973. "The Enzyme Theory and the Origin of Biochemistry." *Isis* 64: 181–196.

———— 1982. *From Medical Chemistry to Biochemistry. The Making of a Biomedical Discipline.* Cambridge: Cambridge University Press.

Kok, Bessel. 1957. "Absorption Changes Induced by the Photochemical Reaction of Photosynthesis." *Nature* 179: 583–584.

———— 1959. "Light Induced Absorption Changes in Photosynthetic Organisms. II. A Split-Beam Difference Spectrophotometer." *Plant Physiology* 34: 184–192.

———— 1961. "Partial Purification and Determination of Oxidation Reduction Potential of the Photosynthetic Chlorophyll Complex Absorbing at 700 mμ." *Biochimica et Biophysica Acta* 48: 527–533.

Kok, Bessel, Bliss Forbush, and Marion McGloin. 1970. "Cooperation of Charges in Photosynthetic Oxygen Evolution—I. A Linear Four Step Mechanism." *Photochemistry and Photobiology* 11: 457–475.

Krab, Klaas and Márten Wikström. 1978. "Proton-Translocating Cytochrome *c* Oxidase in Artificial Phospholipid Vesicles." *Biochimica et Biophysica Acta* 504: 200–214.

Krämer, Reinhard and Martin Klingenberg. 1977. "Reconstitution of Adenine Nucleotide Transport With Purified ADP, ATP-Carrier Protein." *FEBS Letters* 82: 363–367.

Krebs, Hans A. 1953. "Some Aspects of the Energy Transformation in Living Matter." *British Medical Bulletin* 9: 97–104.

——— 1972. "Otto Heinrich Warburg. 1883–1970." *Biographical Memoirs of Fellows of the Royal Society* 18: 629–699.

——— 1981. *Reminiscences and Reflections*. Oxford: Oxford University Press.

Krebs, Hans A. and William A. Johnson. 1937. "The Role of Citric Acid in Intermediate Metabolism in Animal Tissues." *Enzymologia* 4: 148–156. (See also *FEBS Letters* 117, Supplement, pp. K2–K10.)

Krebs, Mark P. and H. Gobind Khorana. 1993. "Mechanism of Light-Dependent Proton Translocation by Bacteriorhodopsin." *Journal of Bacteriology* 175: 1555–1560.

Kresge, Nicole, Robert D. Simoni, and Robert L. Hill. 2009. "Bringing Electron Paramagnetic Resonance (EPR) to Biochemistry: The Work of Helmut Beinert." *Journal of Biological Chemistry* 284: p. e4.

——— 2011. "Bacterial Chimeras and Reversible Phosphorylation: The Work of Walther Stoeckenius." *Journal of Biological Chemistry* 284: pp. e7–8.

Kretovich, W. L. 1983. "A. N. Bach, Founder of Soviet School of Biochemistry." In *Selected Topics in the History of Biochemistry: Personal Recollections*, G. Semenza, (ed.), Vol. 35, 353–364. *Comprehensive Biochemistry*. Amsterdam: Elsevier.

Krogmann, David W., André T. Jagendorf, and Mordhay Avron. 1959. "Uncouplers of Spinach Chloroplast Photosynthetic Phosphorylation." *Plant Physiology* 34: 272–277.

Kruhøffer, Poul and Christian Crone. 1972. "Einar Lundsgaard, 1899–1968." *Ergebnisse der Physiologie* 65: 1–14.

Kühlbrandt, Werner, Da Neng Wang, and Yoshinori Fujiyoshi. 1994. "Atomic Model of Plant Light-Harvesting Complex by Electron Crystallography." *Nature* 367: 614–621.

Kuhn, Thomas S. 1970. *The Structure of Scientific Revolutions*, 2nd ed. Chicago: University of Chicago Press.

——— 1977. *The Essential Tension. Selected Studies in Scientific Tradition and Change*. Chicago: University of Chicago Press.

LaNoue, Kathryn, S. M. Mizani, and Martin Klingenberg. 1978. "Electrical Imbalance of Adenine Nucleotide Transport Across the Mitochondrial Membrane." *Journal of Biological Chemistry* 253: 191–198.

LaNoue, Kathryn F. and Anton C. Schoolwerth. 1979. "Metabolite Transport in Mitochondria." *Annual Review of Biochemistry* 48: 871–922.

Lanyi, Janos K. 1978. "Light Energy Conversion in *Halobacterium halobium*." *Microbiological Reviews* 42: 682–706.

——— 1998. "The Local-Access Mechanism of Proton Transport by Bacteriorhodopsin." *Biochimica et Biophysica Acta* 1365: 17–22.

Lardy, Henry A., Diane Johnson, and W. C. McMurray. 1958. "Antibiotics as Tools for Metabolic Studies. I. A Survey of Toxic Antibiotics in Respiratory. Phosphorylative and Glycolytic Systems." *Archives of Biochemistry and Biophysics* 78: 587–597.

Lardy, Henry A. and Harlene Wellman. 1952. "Oxidative Phosphorylations: Rôle of Inorganic Phosphate and Acceptor Systems in Control of Metabolic Rates." *Journal of Biological Chemistry* 195: 215–224.

——— 1953. "The Catalytic Effect of 2,4-Dinitrophenol on Adenosine triphosphate Hydrolysis by Cell Particles and Soluble Enzymes." *Journal of Biological Chemistry* 201: 357–370.

Lavoisier, Antoine and Pierre S. Laplace. 1783. *Mémoire sur la chaleur*. Reproduced as the original and in English translation in Henry Guerlac, 1982, *Memoir on Heat*. New York: Neale Watson Academic.

Lehninger, Albert L. 1951a. "Oxidative Phosphorylation in Diphosphopyridine Nucleotide-Linked Systems." In *Phosphorus Metabolism*, W. D. McElroy and H. B. Glass (eds.), Vol. 1, pp. 344–366. Baltimore: Johns Hopkins University Press.

——— 1951b. "Phosphorylation Coupled to Oxidation of Dihydrodiphosphopyridine Nucleotide." *Journal of Biological Chemistry* 190: 345–359.

——— 1955. "Oxidative Phosphorylation." *Harvey Lectures* 49: 176–215.

——— 1959. "Reversal of Various Types of Mitochondrial Swelling by Adenosine Triphosphate." *Journal of Biological Chemistry* 234: 2465–2471.

——— 1960. "The Enzymic and Morphologic Organization of the Mitochondria." *Pediatrics* 26: 466–475.

——— 1961. "Components of the Energy-Coupling Mechanism and Mitochondrial Structure." In *Biological Structure and Function II*, T. W. Goodwin and O. Lindberg (eds.), pp. 31–51. London: Academic.

——— 1964. *The Mitochondrion*. New York: Benjamin.

——— 1965. *Bioenergetics*. New York: Benjamin.

——— 1971 *Bioenergetics*, 2nd ed. New York: Benjamin.

Lehninger, Albert L., Ernesto Carafoli, and Carlo S. Rossi. 1967. "Energy-Linked Ion Movements in Mitochondrial Systems." *Advances in Enzymology* 29: 259–320.

Lehninger, Albert L. and Charles L. Wadkins. 1962. "Oxidative Phosphorylation." *Annual Review of Biochemistry* 31: 47–78.

Liebig, Justus. 1842. *Animal Chemistry, or Organic Chemistry in Its Application to Physiology and Pathology*, William Gregory (ed.). Facsimile of 1842 Cambridge edition, New York: Johnson Reprint Corporation, 1964.

——— 1843. *Chemistry in Its Applications to Agriculture and Physiology*, Lyon Playfair (ed.), 3rd ed. London: Taylor & Walton.

——— 1852. *Animal Chemistry*, William Gregory (ed.), 3rd London ed., Part 1, *The Chemical Process of Respiration and Nutrition*. Ann Arbor: University of Michigan Library facsimile edition.

Lipmann, Fritz. 1939. "Coupling Between Pyruvic Acid Dehydrogenation and Adenylic Acid Phosphorylation." *Nature* 143: 281.

——— 1940. "A Phosphorylated Oxidation Product of Pyruvic Acid." *Journal of Biological Chemistry* 134: 463–464.

——— 1941. "Metabolic Generation and Utilization of Phosphate Bond Energy." *Advances in Enzymology* 1: 99–162.

——— 1946. "Metabolic Process Patterns." In *Currents in Biochemical Research*, D. E. Green (ed.), pp. 137–148. New York: Interscience.

——— 1971. *Wanderings of a Biochemist*. New York: Wiley.

Lohmann, Karl. 1931. "Untersuchungen über die chemische Natur des Koferments der Milchsäurebildung." *Biochemische Zeitschrift* 237: 445–482.

Loomis, William F. and Fritz Lipmann. 1948. "Reversible Inhibition of the Coupling Between Phosphorylation and Oxidation." *Journal of Biological Chemistry* 173: 807–808.

Losada, Manuel M., Achim V. Trebst, S. Ogata, and Daniel I. Arnon. 1960. "Equivalence of Light and Adenosine Triphosphate in Bacterial Photosynthesis." *Nature* 186: 753–760.

Losada, Manuel M., F. Robert Whatley, and Daniel I. Arnon. 1961. "Separation of Two Light Reactions in Noncyclic Photophosphorylation of Green Plants." *Nature* 190: 601–610.

Lozier, Richard H., Roberto A. Bogomolni, and Walther Stoeckenius. 1975. "Bacteriorhodopsin: A Light-Driven Proton Pump in *Halobacterium halobium.*" *Biophysical Journal* 15: 955–962.

Ludwig, Bernd and Gottfried Schatz. 1980. "A Two-Subunit Cytochrome *c* Oxidase (Cytochrome aa_3) From *Paracoccus denitrificans.*" *Proceedings of the National Academy of Sciences USA* 77: 196–200.

Luecke, Hartmut, Brigitte Schobert, Hans-Thomas Richter, Jean-Philippe Cartailler, and Janos K. Lanyi. 1999. "Structural Changes in Bacteriorhodopsin During Ion Transport at 2 Angstrom Resolution." *Science* 286: 255–260.

Lundsgaard, Einar. 1930. "Weitere Untersuchungen über Muskelkontraktionen ohne Milchsäurebildung." *Biochemische Zeitschrift* 227: 51–83.

——— 1932. "The Significance of the Phenomenon 'Alactacid Muscle Contraction' for an Interpretation of the Chemistry of Muscle Contraction." *Danske Hospitalstidende* 75: 84–95. Reproduced in English in Kalckar 1969.

Machamer, Peter, Lindley Darden, and Carl F. Craver 2000. "Thinking About Mechanisms." *Philosophy of Science* 67: 1–25.

MacMunn, Charles A. 1886. "Researches on Myohaematin and the Histohaematins." *Philosophical Transactions of the Royal Society* 177: 267–298.

——— 1914. *Spectrum Analysis Applied to Biology and Medicine.* London: Longman, Green.

Maley, Gladys F. and Henry A. Lardy. 1954. "Phosphorylation Coupled With the Oxidation of Reduced Cytochrome *c.*" *Journal of Biological Chemistry* 210: 903–909.

Malkin, Richard. 1982. "Photosystem I." *Annual Review of Plant Physiology* 33: 455–479.

Malkin, Richard and Alan J. Bearden. 1971. "Primary Reactions of Photosynthesis: Photoreduction of a Bound Chloroplast Ferredoxin at Low Temperature as Detected by EPR Spectroscopy." *Proceedings of the National Academy of Sciences USA* 68: 16–19.

Malmström, Bo G. 1979. "Cytochrome Oxidase." *Biochimica et Biophysica Acta* 549: 281–303.

——— 2000. "Mitchell Saw the New Vista, If Not the Details." *Nature* 403: 356.

Marres, Carla A. M. and Edward C. Slater. 1977. "Polypeptide Composition of Purified QH_2:Cytochrome *c* Oxidoreductase From Beef Heart Mitochondria." *Biochimica et Biophysica Acta* 462: 531–548.

Martius, Carl. 1937. "Über den Abbau der Citronensäure." *Zeitschrift für physiologische Chemie* 247: 104–110.

Maruyama, Koscak. 1991. "The Discovery of Adenosine Triphosphate and the Establishment of Its Structure." *Journal of the History of Biology* 24: 145–154.

Mason, Thomas L. and Gottfried Schatz. 1973. "Cytochrome *c* Oxidase From Bakers Yeast: II. Site of Translation of the Protein Components." *Journal of Biological Chemistry* 248: 1355–1360.

Matlin, Karl S. 2016. "The Heuristic of Form: Mitochondrial Morphology and the Explanation of Oxidative Phosphorylation." *Journal of the History of Biology* 49: 37–94.

Mayer, Julius Robert. 1845. *Die Organische Bewegung in ihren Zusammenhang mit dem Stoffwechserl* [*The Organic Motion in Its Relation to Metabolism*]. Heilbronn: n.p.

——— 1851. *Bemerkungen über das mechanische Aequivalent der Wäme*. Heilbronn: Johann Ulrich Landherr.

Mayow, John. 1674. *Tractatus quinque medico-physici*. Translated A. Crum Brown and Leonard Dobbin, 1907. Edinburgh: The Alembic Club.

McAlister, Edward D. and Jack Myers. 1940. "Time Course of Photosynthesis and Fluorescence." *Science* 92: 241–143.

Meijer, A. J. and Karel van Dam. 1974. "The Metabolic Significance of Anion Transport in Mitochondria." *Biochimica et Biophysica Acta* 346: 213–244.

Mendelsohn, Everett. 1964. *Heat and Life*. Cambridge, MA: Harvard University Press.

Menke, Wilhelm. 1962. "Structure and Chemistry of Plastids." *Annual Review of Plant Physiology* 13: 27–44.

Meyer, Arthur. 1883. *Das Chlorophyllkorn in chemischer, morphologischer und biologischer Beziehung*. Leipzig, Germany: Arthur Felix.

Meyerhof, Otto. 1924. *Chemical Dynamics of Life Phenomena*. London: J. B. Lippincott.

——— 1926. "Ueber die enzymatische Milchsäurebildung im Muskelextrakt." *Biochemische Zeitschrift* 178: 395–418, 462–490.

——— 1927. "Recent Investigation on the Aerobic and Anaerobic Metabolism of Carbohydrates." *Journal of General Physiology* 8: 531–542.

——— 1937. "Über die Synthese der Kreatinphosphorsäure im Muskel und die 'Reaktionsform' des Zuckers." *Naturwissenschaften* 25: 443–446.

Meyerhof, Otto and Karl Lohmann. 1932. "Über energetische Wechselbeziehungen zwischen dem Umsatz der Phosphorsäureester im Muskelextrakt." *Biochemische Zeitschrift* 253: 431–461.

Michel, Hartmut and Johann Deisenhofer. 1988. "Relevance of the Photosynthetic Reaction Center From Purple Bacteria to the Structure of Photosystem II." *Biochemistry* 27: 1–7.

Mitchell, Peter. 1959. "Structure and Function in Microorganisms." *Biochemical Society Symposium* 16: 73–93.

——— 1961. "Coupling of Phosphorylation to Electron and Hydrogen Transfer by a Chemi-Osmotic Type of Mechanism." *Nature* 191: 144–148.

——— 1966a. "Chemiosmotic Coupling in Oxidative and Photosynthetic Phosphorylation." *Biological Reviews* 41: 445–502.

——— 1966b. *Chemiosmotic Coupling in Oxidative and Photosynthetic Phosphorylation*. Bodmin, Cornwall, UK: Glynn Research.

——— 1975a. "Protonmotive Redox Mechanism of the Cytochrome b-c_1 Complex in the Respiratory Chain: Protonmotive-Ubiquinone Cycle." *FEBS Letters* 56: 1–6.

——— 1975b. "The Protonmotive Q Cycle: A General Formulation." *FEBS Letters* 59: 137–139.

——— 1976. "Possible Molecular Mechanisms of the Protonmotive Function of Cytochrome Systems." *Journal of Theoretical Biology* 62: 327–367.

——— 1985. "Molecular Mechanics of Protonmotive $F_o F_1$ ATPases. Rolling Well and Turnstile Hypothesis." *FEBS Letters* 182: 1–7.

——— 1991. "Foundations of Vectorial Metabolism and Osmochemistry." *Bioscience Reports* 11: 297–346.

Mitchell, Peter, Roy Mitchell, A. John Moody, Ian C. West, Harold Baum, and John Wrigglesworth. 1985. "Chemiosmotic Coupling in Cytochrome Oxidase. Possible Protonmotive O Loop and O Cycle Mechanisms." *FEBS Letters* 188: 1–7.

Mitchell, Peter and Jennifer Moyle. 1959. "Coupling of Metabolism and Transport by Enzymic Translocation of Substrates Through Membranes." *Proceedings of the Royal Physical Society of Edinburgh* 28: 19–27.

———— 1965. "Stoichiometry of Proton Translocation Through the Respiratory Chain and Adenosine Triphosphatase Systems of Rat Liver Mitochondria." *Nature* 208: 147–151.

———— 1967. "Respiration-Driven Proton Translocation in Rat Liver Mitochondria." *Biochemical Journal* 105: 1147–1162.

Mitchell, Roy, Ian C. West, A. John Moody, and Peter Mitchell. 1986. "Measurement of the Proton-Motive Stoichiometry of the Respiratory Chain of Rat Liver Mitochondria: The Effect of N-Ethylmaleimide." *Biochimica et Biophysica Acta* 849: 229–235.

Mohl, Hugo von. 1837. *Untersunchungen über die anatomischen Verhältnisse des Chlorophylls.* PhD Dissertation, University of Tübingen, Germany.

Molisch, Hans. 1925. "Über Kohlensäure-Assimilation toter Blätter." *Zeitschrift für Botanik* 17: 577–593.

Morange, Michael. 1998. *A History of Molecular Biology.* Cambridge, MA: Harvard University Press.

Morton, Richard A. 1961. "Isolation and Characterization of Ubiquinone (Coenzyme Q) and Ubichromenol." In *CIBA Foundation Symposium on Quinones in Electron Transport*, G. E. W. Wolstenholme and C. M. O'Connor (eds.), pp. 5–25. London: C and A Churchill.

Mowery, Patrick C. and Walther Stoeckenius. 1981. "Photoisomerization of the Chromophore in Bacteriorhodopsin During the Proton Pumping Photocycle." *Biochemistry* 20: 2302–2306.

Musser, Siegfried M., Michael H. B. Stowell, and Sunney I. Chan. 1995. "Cytochrome *c* Oxidase: Chemistry of a Molecular Machine." *Advances in Enzymology and Related Areas of Molecular Biology* 71: 79–208.

Myers, Jack and C. Stacy French. 1960. "Evidences From Action Spectra for a Specific Participation of Chlorophyll *b* in Photosynthesis." *Journal of General Physiology* 43: 723–736.

Nanba, Osamu and Kimiyuki Satoh. 1987. "Isolation of a Photosystem II Reaction Center Consisting of D-1 and D-2 Polypeptides and Cytochrome *b-559*." *Proceedings of the National Academy of Sciences* 84: 109–112.

Nass, Margit M. K., Sylvan Nass, and B. A. Afzelius. 1965. "The General Occurrence of Mitochondrial DNA." *Experimental Cell Research* 37: 516–539.

Needham, Dorothy M, 1971. *Machina Carnis.* Cambridge: Cambridge University Press.

Needham, Dorothy M. and Raman K. Pillai. 1937a. "The Coupling of Oxido-Reductions and Dismutations With Esterification of Phosphate in Muscle." *Biochemical Journal* 31: 1837–1851.

———— 1937b. "Coupling of Dismutations With Esterification of Phosphate in Muscle." *Nature* 140: 64–65.

Negelein, Erwin and Heinz Brömel. 1939. "R-Diphosphoglycerinsäure, ihre Isolierung und Eigenschaften." *Biochemische Zeitschrift* 303: 132–144.

Nicholls, David G. 1979. "Brown Adipose Tissue Mitochondria." *Biochimica et Biophysica Acta* 549: 1–29.

Nicholls, David G. and Eduardo Rial. 1999. "A History of the First Uncoupling Protein, UCP1." *Journal of Bioenergetics and Biomembranes* 31: 399–406.

Nickelsen, Kärin. 2009. "The Construction of a Scientific Model: Otto Warburg and the Building Block Strategy." *Studies in History and Philosophy of Biological and Biomedical Sciences* 40: 73–86.

Nielsen, Sigurd O. and Albert L. Lehninger. 1955. "Phosphorylation Coupled to the Oxidation of Ferrocytochrome *c*." *Journal of Biological Chemistry* 215: 555–570.

Noji, Hiroyuki, Ryohei Yasuda, Masasuke Yoshida, and Kazuhiko Kinosita. 1997. "Direct Observation of the Rotation of F_1-ATPase." *Nature* 386: 299–302.

Norris, James R., Hugo Scheer, M. E. Druyan, and Joseph J. Katz. 1974. "An Electron-Nuclear Double Resonance (Endor) Study of the Special Pair Model for Photo-Reactive Chlorophyll in Photosynthesis." *Proceedings of the National Academy of Sciences USA* 71: 4897–4900.

Ochoa, Severo. 1940. "Nature of Oxidative Phosphorylation in Brain Tissue." *Nature* 145: 267.

——— 1943. "Efficiency of Aerobic Phosphorylation in Cell-Free Heart Extracts." *Journal of Biological Chemistry* 151: 493–505.

Oesterhelt, Dieter and Walther Stoeckenius. 1971. "Rhodopsin-Like Protein From the Purple Membrane of *Halobacterium halobium*." *Nature New Biology* 233: 149–152.

——— 1973. "Functions of a New Photoreceptor Membrane." *Proceedings of the National Academy of Sciences USA* 70: 2853–2857.

Okamoto, Harumasa, Nobuhito Sone, Hajime Hirata, Masasuke Yoshida, and Yasuo Kagawa. 1977. "Purified Proton Conductor in Proton Translocating Adenosine Triphosphatase of a Thermophilic Bacterium." *Journal of Biological Chemistry* 252: 6125–6131.

Okamura, Melvin Y., R. A. Isaacson, and George Feher. 1975. "Primary Acceptor in Bacterial Photosynthesis: Obligatory Role of Ubiquinone in Photoactive Reaction Centers of *Rhodopseudomonas spheroides*." *Proceedings of the National Academy of Sciences USA* 72: 3491–3495.

Okamura, Melvin Y., L. A. Steiner, and George Feher. 1974. "Characterization of Reaction Centers From Photosynthetic Bacteria. I. Subunit Structure of the Protein Mediating the Primary Photochemistry in *Rhodopseudomonas spheroides* R-26." *Biochemistry* 13: 1394–1402.

Oleszko, Susan and Evangelos N. Moudrianakis. 1974. "The Visualization of the Photosynthetic Coupling Factor in Embedded Spinach Chloroplasts." *Journal of Cell Biology* 63: 936–948.

Olson, John M. and Britton Chance. 1960. "Oxidation-Reduction Reactions in the Photosynthetic Bacterium *Chromatium*. I. Absorption Spectrum Changes in Whole Cells." *Archives of Biochemistry and Biophysics* 88: 26–39.

Packer, Lester. 1961. "Metabolic Control of Structural States of Mitochondria." In *Biological Structure and Function II*, T. W. Goodwin and O. Lindberg (eds.), pp. 85–92. London: Academic.

——— 1973. "Membrane Particles of Mitochondria." In *Mechanisms in Bioenergetics*, G. F. Azzone, L. Ernster, S. Papa, E. Quagliariello, and N. Siliprandi (eds.), pp. 33–52. New York: Academic.

Palade, George E. 1952. "The Fine Structure of the Mitochondrion." *Anatomical Record* 114: 427–451.

——— 1956. "Electron Microscopy of Mitochondria and Other Cytoplasmic Structures." In *Enzymes: Units of Biological Structure and Function*, Oliver H. Gaebler (ed.), pp. 185–215. New York: Academic.

Palmieri, Ferdinando and Ben van Ommen. 1999. "The Mitochondrial Carrier Protein Family." In *Frontiers of Cellular Bioenergetics*, S. Papa, F. Guerrieri, and J. M. Tager (eds.), pp. 489–519. New York: Kluwer Academic/Plenum.

Paolillo, Dominick J. Jr. 1970. "The Three-Dimensional Arrangement of Intergranal Lamellae in Chloroplasts." *Journal of Cell Science* 6: 243–255.

Parascandola, John. 1975. "Dinitrophenol and Bioenergetics: An Historical Perspective." In *Proceedings of the Conference on the Historical Development of Bioenergetics*, J. T. Edsall (ed.), pp. 194–202. Boston: American Academy of Arts and Sciences.

Park, R. B. and A. O. Pfeifhofer. 1969. "Ultrastructural Observations on Deep-Etched Thylakoids." *Journal of Cell Science* 5: 299–311.

Parnas, Jacob. 1929. "Über die Ammoniabildung im Muskel und ihren Zusammenhangmit Function und Zustandsänderung. VI." *Biochemische Zeitschrift* 206: 16–38.

Parsons, Donald F., G. Ronald Williams, and Britton Chance. 1966. "Characteristics of Isolated and Purified Preparations of the Outer and Inner Membranes of Mitochondria." *Annals of the New York Academy of Sciences* 137: 643–666.

Pasteur, Louis. 1879. *Examen critique d'un écrit posthume de Claude Bernard sur la fermentation*. Paris: Gauthier-Villars.

Pebay-Peyroula, Eva, Cécile Dahout-Gonzalez, Richard Kahn, Véronique Trézéguet, Guy J.-M. Lauquin, and Gérard Brandolin. 2003. "Structure of Mitochondrial ADP/ATP Carrier in Complex With Carboxyatractyloside." *Nature* 426: 39–44.

Penefsky, Harvey S., Maynard E. Pullman, Anima Datta, and Efraim Racker. 1960. "Partial Resolution of the Enzymes Catalyzing Oxidative Phosphorylation. II. Participation of a Soluble Adenosine Triphosphatase in Oxidative Phosphorylation." *Journal of Biological Chemistry* 235: 3330–3336.

Perrin, Carleton E. 1990. "The Chemical Revolution." In *Companion to the History of Modern Science*, Robert Olby, G. N. Cantor, John R. R. Christie, and M. J. S. Hodge (eds.), pp. 264–277. London: Routledge.

Peters, Rudolph A. 1954. "Otto Meyerhof 1884–1951." *Obituary Notices of Fellows of the Royal Society* 9: 175–200.

Pfaff, Erich and Martin Klingenberg. 1968. "Adenine Nucleotide Translocation of Mitochondria. 1. Specificity and Control." *European Journal of Biochemistry* 6: 66–79.

Pfaff, Erich, Martin Klingenberg, and Hans-Walter Heldt. 1965. "Unspecific Permeation and Specific Exchange of Adenine Nucleotides in Liver Mitochondria." *Biochimica et Biophysica Acta* 104: 312–315.

Pflüger, Eduard W. 1875. "Beiträge zur Lehre von der Respiration, 1: Ueber die physiologische Verbrennung in dem lebendigen Organismen." *Pflügers Archiv* 10: 251–369.

Prebble, John N. 1981. *Mitochondria, Chloroplasts and Bacterial Membranes*. London: Longman.

——— 1988. "Mitochondria and Chloroplasts." In *Solute Transport in Plant Cells and Tissues*, D. A. Baker and J. L. Hall (eds.), pp. 28–82. Harlow, Essex, UK: Longman Scientific and Technical.

——— 1996. "Successful Theory Development in Biology: A Consideration of the Theories of Oxidative Phosphorylation Proposed by Davies and Krebs, Williams and Mitchell." *Bioscience Reports* 16: 207–215.

——— 2001. "The Philosophical Origins of Mitchell's Chemiosmotic Concepts. The Personal Factor in Scientific Theory Formulation." *Journal of the History of Biology* 34: 433–460.

——— 2002. "Peter Mitchell and the Ox Phos Wars." *Trends in Biochemical Sciences* 27: 209–212.

——— 2010. "The Discovery of Oxidative Phosphorylation: A Conceptual Off-Shoot From the Study of Glycolysis." *Studies in History and Philosophy of Biological and Biomedical Sciences* 41: 253–262.

——— 2013. "Contrasting Approaches to a Biological Problem: Paul Boyer, Peter Mitchell and the Mechanism of the ATP Synthase, 1961–1985." *Journal of the History of Biology* 46: 699–737.

Prebble, John N. and Bruce Weber. 2003. *Wandering in the Gardens of the Mind. Peter Mitchell and the Making of Glynn*. New York: Oxford University Press.

Priestley, Joseph. 1772. "Observations on Different Kinds of Air." *Philosophical Transactions of the Royal Society* 62:147–264.

——— 1775. *Experiments and Observations on Different Kinds of Air,* 2nd ed. London: J. Johnson.

Prince, Roger C. and Graham N. George. 1995. "Cytochrome *f* Revealed." *Trends in Biochemical Sciences* 20: 217–218.

Pullman, Maynard E., Harvey S. Penefsky, and Efraim Racker. 1958. "A Soluble Protein Fraction Required for Coupling Phosphorylation to Oxidation in Submitochondrial Fragments of Beef Heart Mitochondria." *Archives of Biochemistry and Biophysics* 76: 227–230.

Pullman, Maynard. E., Harvey S. Penefsky, Anima Datta, and Efraim Racker. 1960. "Partial Resolution of the Enzymes Catalyzing Oxidative Phosphorylation. I. Purification and Properties of Soluble Dinitrophenol-Stimulated Adenosine Triphosphatase." *Journal of Biological Chemistry* 235: 3322–3329.

Quayle, J. Rodney, R.C. Fuller, Andrew A. Benson, and Melvin Calvin. 1954. "Enzymatic Carboxylation of Ribulose Diphosphate." *Journal of the American Chemical Society* 76: 3610–3611.

Rabinowitch, Eugene I. 1945. *Photosynthesis and Related Processes I.* New York: Interscience.

——— 1952. "Photosynthesis." *Annual Review of Plant Physiology* 3: 229–264.

——— 1961. "Robert Emerson 1903–1959." In *Biographical Memoirs*, pp. 111–131. Washington, DC: National Academy of Sciences.

Racker, Efraim. 1961. "Mechanisms of Synthesis of Adenosine Triphosphate." *Advances in Enzymology* 23: 323–399.

——— 1965. *Mechanisms in Bioenergetics.* New York: Academic.

——— 1970a, "Function and Structure of the Inner Membrane of Mitochondria and Chloroplasts." In *Membranes of Mitochondria and Chloroplasts*, E. Racker (ed.), pp. 127–171. New York: Van Nostrand Reinhold.

——— 1970b. "The Two Faces of the Inner Mitochondrial Membrane." *Essays in Biochemistry* 6: 1–22.

——— 1976. "A Pathway From Psychiatry to Cancer." In *Reflections on Biochemistry*, A. Kornberg, B. L. Horecker, L. Cornudella, and J. Oro (eds.), pp. 39–44. Oxford: Pergamon.

Racker, Efraim and Anne Kandrach. 1973. "Partial Resolution of the Enzymes Catalysing Oxidative Phosphorylation. XXXIX. Reconstitution of the Third Segment of Oxidative Phosphorylation." *Journal of Biological Chemistry* 248: 5841–5847.

Racker, Efraim and I. Krimsky. 1952. "The Mechanism of Oxidation of Aldehydes by Glyceraldehyde-3-Phosphate Dehydrogenase." *Journal of Biological Chemistry* 248: 731–743.

Racker, Efraim and Franziska W. Racker. 1981. "Resolution and Reconstitution: A Dual Autobiographical Sketch." In *Of Oxygen, Fuels and Living Matter Part 1*, G. Semenza (ed.), pp. 265–287. Chichester, UK: Wiley.

Racker, Efraim and Walther Stoeckenius. 1974. "Reconstitution of Purple Membrane Vesicles Catalyzing Light-Driven Proton Uptake and Adenosine Triphosphate Formation." *Journal of Biological Chemistry* 249: 662–663.

Ragan, C. Ian. 1976. "NADH-Ubiquinone Oxidoreductase." *Biochimica et Biophysica Acta* 456: 249–290.

Ragan, C. Ian and Efraim Racker. 1973. "Partial Resolution of the Enzymes Catalyzing Oxidative Phosphorylation. XXVIII. The Reconstitution of the First Site of Energy Conservation." *Journal of Biological Chemistry* 248: 2563–2569.

Reed, Dan W. 1969. "Isolation and Composition of a Photosynthetic Reaction Center Complex From *Rhodopseudomonas spheroides.*" *Journal of Biological Chemistry* 244: 4936–4941.

Reed, Dan W. and Roderick K. Clayton. 1968. "Isolation of a Reaction Center Fraction From *Rhodopseudomonas spheroides.*" *Biochemical and Biophysical Research Communications* 30: 471–475.

Reed, Dan W. and Gerald A. Peters. 1972. "Characterization of the Pigments in Reaction Center Preparations From *Rhodopseudomonas spheroides.*" *Journal of Biological Chemistry* 247: 7148–7152.

Reeves, S. G. and David O. Hall. 1978. "Photophosphorylation in Chloroplasts." *Biochimica et Biophysica Acta* 463: 275–297.

Rich, Peter. 1999. "Mechanism of Proton-Motive Activity of Heme-Copper Oxidases." In *Frontiers of Cellular Bioenergetics*, S. Papa, F. Guerrieri, and J. M. Tager (eds.), pp. 179–192. New York: Kluwer Academic/Plenum.

Rieske, John S. 1976. "Composition, Structure and Function of Complexes III of the Respiratory Chain." *Biochimica et Biophysica Acta* 456: 195–247.

Rieske, John, Waldo S. Zaugg, and Raymond E. Hansen. 1964. "Studies on the Electron Transfer System: LIX. Distribution of Iron and of the Component Giving a Paramagnetic Resonance Signal at g=1.90 in Subfractions of Complex III." *Journal of Biological Chemistry* 239: 3023–3030.

Ringler, Robert A., Shigeki Minakami, and Thomas P. Singer. 1963. "Studies on the Respiratory Chain-Linked Reduced Nicotinamide Adenine Nucleotide Dehydrogenase." *Journal of Biological Chemistry* 238: 801–810.

Robertson, David. 1981. "Membrane Structure." *The Journal of Cell Biology* 91: 189s–204s.

Robertson, Rutherford N. 1960. "Ion Transport and Respiration." *Biological Reviews* 35: 231–264.

——— 1992. "A Dilettante Australian Plant Physiologist." *Annual Review of Plant Physiology and Plant Molecular Biology* 43: 1–24.

Rosenberg, Steven A. and Guido Guidotti. 1968. "The Protein of Human Erythrocyte Membranes: I. Preparation, Solubilisation, and Partial Characterization." *Journal of Biological Chemistry* 243: 1985–1992.

Ruben, Samuel. 1943. "Photosynthesis and Photophosphorylation." *Journal of the American Chemical Society* 65: 279–282.

Ruben, Samuel, Martin D. Kamen, W. Z. Hassid, and D. DeVault. 1939. "Photosynthesis With Radiocarbon." *Science* 90: 570–571.

Ruben, Samuel, Merle Randall, Martin Kamen, and James L. Hyde. 1941. "Heavy Oxygen (O^{18}) as a Tracer in the Study of Photosynthesis." *Journal of the American Chemical Society* 63: 877–879.

Rubner, Max. 1894. "Die quelle der thierischen Wärme," *Zeitschrift für Biologie* 30: 73–142.

Rumberg, Bernd, P. Schmidt-Mende, J. Weikard, and Horst T. Witt. 1963. "Correlation Between Absorption Changes and Electron Transport in Photosynthesis." In *Photosynthetic Mechanisms of Green Plants*, B. Kok and A. T. Jagendorf (eds.), pp. 18–34. Washington, DC: National Academy of Sciences National Research Council Publication 1145.

Sachs, Julius. 1862. "Ueber den Einfluss des Lichtes auf die Bildung des Amylum in den Chlorphyllkörnern." *Botanische Zeitung* 20: 365–373.

Sacks, Jacob. 1943–4. "The Absence of Phosphate Transfer in Oxidative Muscular Contraction." *American Journal of Physiology* 140: 316–320.

Sager, Ruth and George E. Palade. 1957. "Structure and Development of the Chloroplast in *Chlamydomonas.*" *Journal of Biophysical and Biochemical Cytology* 3: 463–488.

Saraste, Matti, Nicholas J. Gay, Alex Eberle, Michael J. Runswick, and John E. Walker. 1981. "The atp Operon: Nucleotide Sequence of the Genes for the γ, β and ε Subunits of *Escherichia coli* ATP Synthase." *Nucleic Acids Research* 9: 5287–5296.

Saraste, Matti, Timo Pentillä, John R. Coggins, and Mårten Wikström. 1980. "Structure of Bovine Cytochrome Oxidase." *FEBS Letters* 114: 35–38.

Schachman, Howard K., Arthur B. Pardee, and Roger Y. Stanier. 1952. "Studies on the Macromolecular Organization of Microbial Cells." *Archives of Biochemistry and Biophysics* 38: 245- 260.

Schatz, Gottfried. 1997. "Efraim Racker: 28 June 1913 to 9 September 1991." In *Selected Topics in the History of Biochemistry: Personal Recollections V*, G. Semenza and R. Jaeniche (eds.), Vol. 40, *Comprehensive Biochemistry*, pp. 253–276. Amsterdam: Elsevier.

Schrödinger, Erwin. 1944. *What Is Life*. Cambridge: Cambridge University Press.

Schwann, Theodor. 1847. *Microscopical Researches Into the Accordance in the Structure and Growth of Animals and Plants*. Trans. Henry Smith. London: Sydenham Society.

Senior, Alan E. 1973. "The Structure of Mitochondrial ATPase." *Biochimica et Biophysica Acta* 301: 249–277.

——— 1988. "ATP Synthesis by Oxidative Phosphorylation." *Physiological Reviews* 68: 177–231.

Singer, Seymour Jonathan and Garth L. Nicolson. 1972. "The Fluid Mosaic Model of the Structure of Cell Membranes." *Science* 175: 720–731.

Singer, Thomas P. 1954. "Solubilization, Assay and Purification of Succinic Dehydrogenase." *Biochimica et Biophysica Acta* 15: 151–153.

Sjöstrand, Fritiof S. 1953. "Electron Microscopy of Mitochondrial and Cytoplasmic Double Membranes." *Nature* 171: 30–32.

Skulachev, Vladimir P. 1999. "Uncoupling of Respiration and Phosphorylation." In *Frontiers of Cellular Bioenergetics*, Sergio Papa, Ferrucio Guerrieri, and Joseph M. Tager (eds.), pp. 89–118. New York: Kluwer Academic/Plenum.

Skulachev, Vladimir P., A. A. Sharaf, and E. A. Liberman. 1967. "Proton Conductors in the Respiratory Chain and Artificial Membranes." *Nature* 216: 718–719.

Slater, Edward C. 1949. "Cytochrome c_1 of Yakushiji and Okuniki," *Nature* 163: 532.

——— 1953. "Mechanism of Phosphorylation in the Respiratory Chain." *Nature* 172: 975–978.

——— 1958. "Mechanism of Oxidative Phosphorylation." *Reviews of Pure and Applied Chemistry* 8: 221–264.

——— 1966. "Oxidative Phosphorylation." In *Comprehensive Biochemistry 14*, M. Florkin and E. Stotz (eds.), pp. 327–396. Amsterdam: Elsevier.

——— 1967. "The Respiratory Chain and Oxidative Phosphorylation: Some of the Unsolved Problems." In *Biochemistry of Mitochondria*, E. C. Slater, Z. Kaniuga, and L. Wojtczak (eds.), pp. 1–10. London: Academic.

——— 1974. "From Cytochrome to Adenosine Triphosphate and Back." *Biochemical Society Transactions* 2: 1149–1163.

——— 1981. "The Discovery of Oxidative Phosphorylation." *Trends in Biochemical Sciences* 6: 226–227.

——— 1984. "Vladimir Alexandrovich Engelhardt." *Trends in Biochemical Sciences* 9: 504–505.

——— 1985. "The BAL-Labile Factor in the Respiratory Chain." In *Comprehensive Biochemistry*, G. Semenza (ed.), Vol. 36, *Selected Topics in the History of Biochemistry: Personal Recollections*, pp. 197–253. Amsterdam: Elsevier.

——— 1997. "An Australian Biochemist in Four Countries." In *Comprehensive Biochemistry*, G. Semenza and R. Jaenicke (eds.), Vol. 40, *Selected Topics in the History of Biochemistry: Personal recollections V*, pp. 69–203. Amsterdam: Elsevier.

Slater, Edward C. and F. A. Holton. 1954. "Oxidative Phosphorylation Coupled With Oxidation of α-Ketoglutarate by Heart-Muscle Sarcosomes." *Biochemical Journal* 56: 28–40.

Slater, Edward C., C. P. Lee, J. A. Berden, and H. J. Wegdam. 1970. "High-Energy Forms of Cytochrome *b*." *Nature* 226: 1248–1249.

Slater, Edward C., Ernest Quagliariello, Sergio Papa, and Joseph M. Tager. 1970. "Introduction." In *Energy Transport and Energy Conservation*, J. Tager, S. Papa, E. Quagliariello, and E. Slater (eds.), pp. 1–4. Bari, Italy: Adriatica Editrice.

Smith, Lucile and Elmer Stotz. 1954. "Purification of Cytochrome *c* Oxidase." *Journal of Biological Chemistry* 209: 819–828.

Sottocasa, Gian Luigi, Bo Kuylenstierna, Lars Ernster, and Anders Bergstrand. 1967. "An Electron-Transport System Associated With the Outer Membrane of Liver Mitochondria." *Journal of Cell Biology* 32: 415–438.

Spallanzani, Lazaro. 1807. *Rapports de l'air avec les être orgsanisées.* Ed. J. Senebier. Geneva: Paschoud.

Spitzer, W. 1897. "Die Bedeutung gewisser Nucleoproteride für die oxidative Leistung der Zelle." *Pflüger's Archiv für die gesamte Physiologie* 67: 615–656.

Staehelin, L. Andrew. 2003. "Chloroplast Structure: From Chlorophyll Granules to Supra-Molecular Architecture of Thylakoid Membranes." *Photosynthesis Research* 76: 185–196.

Steinmann, E. and Fritiof S. Sjöstrand, 1955. "The Ultrastructure of Chloroplasts." *Experimental Cell Research* 8: 15–23.

Stiehl, H. H. and Horst T. Witt. 1969. "Quantitative Treatment of the Function of Plastoquinone in Photosynthesis." *Zeitschrift für Naturforschung* 24b: 1588–1598.

Stokes, George G. 1852. "On the Change of Refrangibility of Light." *Philosophical Transactions of the Royal Society of London* 142: 463–562.

——— 1863. "On the Supposed Identity of Biliverdin With Chlorophyll, With Remarks on the Constitution of Chlorophyll." *Proceedings of the Royal Society London* 13: 144–145.

——— 1864. "On the Reduction and Oxidation of the Colouring Matter of the Blood." *Proceedings of the Royal Society London* 13: 355–364.

Strain, Harold H. and Joseph Sherma. 1967. "Translation of Tswett 1906." *Journal of Chemical Education* 44: 238–242.

Straley, Susan C., William A. Parson, David C. Mauzerall, and Roderick K. Clayton. 1973. "Pigment Content and Molar Extinction Coefficients of Photochemical Reaction Centers From *Rhodopseudomonas spheroides*." *Biochimica et Biophysica Acta* 305: 597–609.

Straub, Ferenc B. 1939. "Isolation and Properties of a Flavoprotein From Heart Muscle." *Biochemical Journal* 33: 787–792.

Sybesma, Christiaan and John M. Olson. 1963. "Transfer of Chlorophyll Excitation Energy in Green Photosynthetic Bacteria." *Proceedings of the National Academy of Sciences USA* 49: 248–253.

Szent-Györgyi, Albert. 1939. "Oxidation and Fermentation." In *Perspectives in Biochemistry,* J. Needham and D. E. Green (eds.), pp. 165–174. Cambridge: Cambridge University Press.

Tagawa, Kunio and Daniel I. Arnon. 1962. "Ferredoxins as Electron Carriers in Photosynthesis and in the Biological Production and Consumption of Hydrogen Gas." *Nature* 195: 537–543.

Teich, Mikuláš. 1992. *A Documentary History of Biochemistry 1770–1940.* Leicester, UK: Leicester University Press.

Theorell, A. Hugo. 1936. "Reines Cytochom c." *Biochemische Zeitschrift* 285: 207–218.

——— 1947. "Heme-Linked Groups and Mode of Action of Some Hemoproteins." *Advances in Enzymology* 7: 265–303.

Theorell, A. Hugo and Å. Åkesson. 1941. "Studies on Cytochrome *c.* I. Electrophoretic Purification of Cytochrome *c* and Its Amino Acid Composition. II. The Optical Properties of Pure Cytochrome *c* and Some of Its Derivatives. III. Titration Curves." *Journal of the American Chemical Society* 63: 1804–1820.

Thunberg, Thorsten L. 1918. "Zur Kenntnis der Einwirkung tierischer Gewebe auf Methylenblau." *Skandinavisches Archiv für Physiologie* 35: 163–195.

——— 1923. "Über einen neuen Weg von der Kohlensäure zum Formaldehyd. Ein Beitrag zur Theorie der Kohlensäureassimilation." *Zeitschrift für physikalische Chemie* 106: 305–312.

——— 1930. "The Hydrogen Activating Enzymes of the Cells." *Quarterly Review of Biology* 5: 318–347.

Tolmach, L. J. 1951. "Effects of Triphosphopyridine Nucleotide Upon Oxygen Evolution and Carbon Dioxide Fixation by Illuminated Chloroplasts." *Nature* 167: 946–948.

Traube, Moritz, 1861. "Ueber die Beziehung der Respiration zur Muskeltätigkeit und die Bedeutung der Respiration überhaupt." *Virchows Archiv* 21: 386–414.

Trebst, Achim. 1974. "Energy Conservation in Photosynthetic Electron Transport of Chloroplasts." *Annual Review of Plant Physiology* 25: 423–458.

Trebst, Achim V., Harry Y. Tsujimoto, and Daniel I. Arnon. 1958. "Separation of Light and Dark Phases in the Photosynthesis of Isolated Chloroplasts." *Nature* 182: 351–355.

Trumpower, Bernard L. 1981. "Function of the Iron-Sulfur Protein of the Cytochrome b-c_1 Segment in Electron-Transfer and Energy-Conserving Reactions of the Mitochondrial Respiratory Chain." *Biochimica et Biophysica Acta* 639: 129–155.

Tsukihara, Tomitake, Hiroshi Aoyama, Eiki Yamashita, Takashi Tomizaki, Hiroshi Yamaguchi, Kyoko Shinzawa-Itoh, Ryosuke Nakashima, Rieko Yaono, and Shinya Yoshikawa. 1996. "The Whole Structure of the 13 Subunit Oxidized Cytochrome c Oxidase at 2.8 Å." *Science* 272: 1136–1144.

Tswett, Mikhail S. 1906. "Adsorptionsanalyse und chromatographische Methode. Anwendung auf die Chemie des Chlorophylls." *Berichte der Deutschen botanischen Gesellschaft* 24: 384–393.

University of Wisconsin. 1942. *Symposium on Respiratory Enzymes*. Madison: University of Wisconsin Press.

Vambutas, Vida K. and Efraim Racker. 1965. "Partial Resolution of the Enzymes Catalyzing Photophosphorylation. I Stimulation of Photophosphorylation by a Preparation of a Latent Ca⁺⁺-Dependent Adenosine Triphosphatase From Chloroplasts." *Journal of Biological Chemistry* 240: 2660–2667.

van Gelder, Bob F. and Helmut Beinert. 1969. "Studies of the Heme Components of Cytochrome *c* Oxidase by EPR Spectroscopy." *Biochimica et Biophysica Acta* 189: 1–24.

van Niel, Cornelius B. 1935. "Photosynthesis of Bacteria." *Cold Spring Harbor Symposia* 3: 138–150.

——— 1941. "The Bacterial Photosyntheses and Their Importance for the General Problem of Photosynthesis." *Advances in Enzymology* 1: 263–328.

——— 1944, "The Culture, General Physiology, Morphology, and Classification of the Non-Sulphur Purple and Brown Bacteria." *Bacteriological Reviews* 8: 1–118.

Vasington, Frank D. and Jerome V. Murphy. 1962. "Ca++ Uptake by Rat Kidney Mitochondria and Its Dependence on Respiration and Phosphorylation." *Journal of Biological Chemistry* 237: 2670–2677.

Vishniac, Wolf and Severo Ochoa. 1951. "Photochemical Reduction of Pyridine Nucleotides by Spinach Grana and Coupled Carbon Dioxide Fixation." *Nature* 167: 768–769.

——— 1952. "Phosphorylation Coupled to Photochemical Reduction of Pyridine Nucleotides by Chloroplast Preparations." *Journal of Biological Chemistry* 198: 501–505.

Von Jagow, Gebhard and Walter Sebald. 1980. "*b*-Type Cytochromes." *Annual Review of Biochemistry* 49: 281–314.

Vredenberg, Wim J. and Louis N. M. Duysens. 1963. "Transfer of Energy From Bacteriochlorophyll to a Reaction Centre During Bacterial Photosynthesis." *Nature* 197: 355–357.

Walker, David A. 2002. "'And Whose Bright Presence'—An Appreciation of Robert Hill and His Reaction." *Photosynthesis Research* 73: 51–54.

Walker, John E., I. M. Fearnley, Nicholas J. Gay, B. W. Gibson, F. D. Northrop, S. J. Powell, Michael J. Runswick, Matti Saraste, and V. L. J. Tybulewicz. 1985. "Primary Structure and Subunit Stoichiometry of F_1-ATPase From Bovine Mitochondria." *Journal of Molecular Biology* 184: 677–701.

Walker, John E. and Michael J. Runswick. 1993. "The Mitochondrial Transport Protein Superfamily." *Journal of Bioenergetics and Biomembranes* 25: 435–446.

Walker, John E., Michael J. Runswick, and Linda Poulter. 1987. "ATP Synthase From Bovine Mitochondria. The Characterization and Sequence Analysis of Two Membrane Associated Sub-Units and of the Corresponding cDNAs." *Journal of Molecular Biology* 197: 89–100.

Wang, Jui H., Vijay Joshi, and Joe C. Wu. 1986. "Geometric Isomers of Covalently Labelled Mitochondrial F_1-Adenosinetriphosphatase With Different Properties." *Biochemistry* 25: 7996–8001.

Warburg, Otto H. 1913. "Über sauerstoffatmende Körnchen aus Leberzellen und über Sauerstoffatmung in Berkefeld-Filtraten wässriger Leberextrakte." *Pflügers Archive für Physiologie* 154: 599–617.

——— 1914. "Über die Rolle des Eisens in der Atmung des Seeigeleis nebst Bemerkungen über einige durch Eisen beschleeunigten Oxydationen." *Zeitschrift für physiologische Chemie* 92: 231–256.

——— 1930. "The Enzyme Problem and Biological Oxidations." *Bulletin of the Johns Hopkins Hospital* 46: 341–358.

——— 1932. "The Oxygen-Transferring Ferment of Respiration." In *Nobel Lectures: Including Presentation Speeches and Laureates' Biographies. Physiology or Medicine 1922–1941*, pp. 254–270. Amsterdam: Elsevier, 1965.

——— 1949. *Heavy Metal Prosthetic Groups and Enzyme Action.* Trans. A. Lawson. Oxford: Clarendon.

Warburg, Otto and Walter Christian. 1936. "Pyridin, der Wasserstoffübertragende Bestandteil von Gärungsfermenten (Pyridin-Nucleotide)." *Biochemische Zeitschrift* 287: 291–328.

——— 1939. "Isolierung und Kristallisation des Proteins des oxydierenden Garungsferments." *Biochemische Zeitschrift* 303: 40–68.

Warburg, Otto and Günter Krippahl. 1960. "Notwendigkeit der Kohlensäure für die Chinon- und ferricyanid-Reactionen in grünen Grana." *Zeitschrift für Naturforschung* 15B: 367–369.

Warburg, Otto and Erwin Negelein. 1922. "Über den Energieumsatz bei der Kohlensäureassimilation." *Zeitschrift für physikalische Chemie* 102: 235–266.

Wassink, Evert Christiaan. 1951. "Chlorophyll Fluorescence and Photosynthesis." *Advances in Enzymology* 11: 91–199.

Weber, Bruce and John N. Prebble. 2006 "An Issue of Originality and Priority: The Correspondence and Theories of Oxidative Phosphorylation of Peter Mitchell and Robert J. P. Williams." *Journal of the History of Biology* 39: 125–163.

Weier, T. E., C. R. Stocking, W. W. Thomson, and H. Drever. 1963. "The Grana as Structural Units of Chloroplasts of Mesophyll of *Nicotiana rustica* and *Phaseolus vulgaris*." *Journal of Ultrastructure Research* 8: 122–143.

Werkman, Chester H. and Harland G. Wood. 1942. "Heterotrophic Assimilation of Carbon Dioxide." *Advances in Enzymology* 2: 135–182.

Widger, William R., William A. Cramer, Reinhold G. Herrmann, and Achim Trebst. 1984. "Sequence Homology and Structural Similarity Between Cytochrome *b* of Mitochondrial Complex III and the Chloroplast b_6-*f* Complex: Position of the Cytochrome *b* Hemes in the Membrane." *Proceedings of the National Academy of Sciences USA* 81: 674–678.

Wieland, Heinrich. 1932. *On the Mechanism of Oxidation.* New Haven, CT: Yale University Press.

Wikström, Mårten, K. F. 1971. "Properties of Three Cytochrome *b*-Like Species in Mitochondria and Submitochondrial Particles." *Biochimica et Biophysica Acta* 253: 332–345.

——— 1973. "The Different Cytochrome *b* Components in the Respiratory Chain of Animal Mitochondria and Their Role in Electron Transport and Energy Conservation." *Biochimica et Biophysica Acta* 301: 155–193.

——— 1977. "Proton Pump Coupled to Cytochrome *c* Oxidase in Mitochondria." *Nature* 266: 271–273.

——— 1989. "Identification of the Electron Transfers in Cytochrome Oxidase That Are Coupled to Proton-Pumping." *Nature* 338: 776–778.

——— 2008. "How I Became a Biochemist." *IUBMB Life* 60: 414–417.

Wikström, Mårten, K. F. and Jan A. Berden. 1972. "Oxidoreduction of Cytochrome *b* in the Presence of Antimycin." *Biochimica et Biophysica Acta* 283: 403–420.

Wikström, Mårten, K. F. and Robert Casey. 1985. "The Oxidation of Exogenous Cytochrome *c* by Mitochondria. Resolution of a Long-Standing Controversy." *FEBS Letters* 183: 293–298.

Wikström, Mårten, K. F. and Klaas Krab. 1979. "Proton-Pumping Cytochrome *c* Oxidase." *Biochimica et Biophysica Acta* 549: 177–222.

Wildman, Sam G. 2002. "Along the Trail From Fraction I Protein to Rubisco (Ribulose Bisphosphate Carboxylase-Oxygenase)." *Photosynthesis Research* 73: 243–250.

Williams, Robert J. P. 1959. "Coordination, Chelation and Catalysis." In *The Enzymes*, 2nd ed., vol. 1, P. Boyer, H. Lardy, and H. Myrbäck (eds.), Vol. 1, pp. 391–441. New York: Academic.

——— 1961. "Possible Functions of Chains of Catalysts." *Journal of Theoretical Biology* 1: 1–17.

——— 1962. "Possible Functions of Chains of Catalysts II." *Journal of Theoretical Biology* 3: 209–229.

Willstätter, Richard and Arthur Stoll. 1913. *Untersuchungen über Chlorophyll.* Berlin: Springer.

——— 1918. *Untersuchungen über die Assimilation der Kohlensäure.* Berlin: Springer.

Witt, Horst T., A. Müller, and B. Rumberg. 1961. "Experimental Evidence for the Mechanism of Photosynthesis." *Nature* 191: 194–195.

Woodward, Robert B., W. A. Ayer, J. M. Beaton, F. Bickelhaupt, R. Bonnett, P. Buchschacher, G. L. Closs, H. Dutler, J. Hannah, F. P. Hauck, S. Itô, A. Langemann, E. Le Goff, W. Leimgruber, W. Lwowski, J. Sauer, Z. Valenta, and H. Volz. 1960. "The Total Synthesis of Chlorophyll." *Journal of the American Chemical Society* 82: 3800–3802.

Wydrzynski, Thomas, Nick Zumbulyadis, Paul G. Schmidt, Herbert S. Gutowsky, and Rajni Govindjee. 1976. "Proton Relaxation and Charge Accumulation During Oxygen Evolution in Photosynthesis." *Proceedings of the National Academy of Sciences USA* 73: 1196–1198,

Xia, Di, Chang-An Yu, Hoeon Kim, Jia-Zhi Xia, Anatoly M. Kachurin, Li Zhang, Linda Yu, and Johann Deisenhofer. 1997. "Crystal Structure of the Cytochrome bc_1 Complex From Bovine Heart Mitochondria." *Science* 277: 60–66.

Yakushiji, Eijiro and Kazuo Okunuki. 1940. "Über eine neue Cytochromkomponente und ihre Funktion." *Proceedings of the Imperial Academy Japan* 16: 299–302.

Young, Frank G. 1953. "Prof. T. Thunberg." *Nature* 172: 1079.

Yu, Chang-An, Di Xia, Hoeon Kim, Johann Deisenhofer, Li Zhang, Anatoly M. Kachurin, and Linda Yu. 1998. "Structural Basis of Functions of the Mitochondrial Cytochrome bc_1 Complex." *Biochimica et Biophysica Acta* 1365: 151–158.

Zallen, Doris. 1993a. "Redrawing the Boundaries of Molecular Biology: The Case of Photosynthesis." *Journal of the History of Biology* 26: 65–87.

——— 1993b. "The 'Light' Organism for the Job: Green Algae and Photosynthesis Research." *Journal of the History of Biology* 26: 269–279.

Ziegler, Daniel, David E. Green, and K. A. Doeg. 1959. "Studies on the Electron Transfer System. XXV. The Isolation and Properties of a Lipoprotein With Diaphorase Activity From Beef Heart Mitochondria." *Journal of Biological Chemistry* 234: 1916–1921.

Ziegler, Daniel, R. L. Lester, and David E. Green. 1956. "Oxidative Phosphorylation by an Electron Transport Particle From Beef Heart." *Biochimica et Biophysica Acta* 21: 80–85.

Zouni, Athina, Horst-Tobias Witt, Jan Kern, Petra Fromme, Norbert Krauss, Wolfram Saenger, and Peter Orth. 2001. "Crystal Structure of Photosystem II From *Synechococcus elongates* at 3.8Å Resolution." *Nature* 409: 739–743.

Index

Berzelius, Jöns Jacob, 34
binding change mechanism, 210–211
biochemistry, foundation of, 40
bioenergetics, creation of, 217
Black, Joseph, 23
Blackman, Frederick, 141
Blackman reaction, 142, 149, 153
blood, 17–19, 21, 30, 35
Boardman, Keith, 166
Boekema, Egbert, 203
Boltzmann, Ludwig, 132–133, 150
bongkrekate, 206
Bonnet, Charles, 129
bot fly, 55
Boyer, Paul, 14, 116–117, 122–126, 208, 210–212, 230, 234
Boyle, Robert, 18, 20
Briggs, George, 142
brown fat mitochondria, 189–190
Buchner, Eduard, 38–41, 43–44, 222
Buchner, Hans, 38
Bunsen, Robert, 35

C3 pathway, 147
C4 pathway, 148
calcium, uncoupling activity, 189
calcium, uptake by mitochondria, 119–120
Calvin cycle, 147–149
Calvin, Melvin, 147, 149
Capaldi, Roderick, 196
carbohydrate oxidation, 43–44, 51
carbon dioxide assimilation, 137–138, 141, 146–147
carbon dioxide (fixed air), 23, 25
carbon monoxide, 54–57
carbon of plants, source of, 131–132
carbon reduction cycle, 147–149, 155, 158
carbon-11, 146
carbon-14, 144, 146–148
carbonic acid, 131, 135
carboxydismutase, 147n
carotene, 137
carotenoid, 136, 159, 160, 170, 203–204
catalase, 59, 79
cation transport, 119–120
Caventou, Joseph, 134
cell structure, 33
cell theory, 27
centrifugation, differential, 74–75
CF_1, 109, 178

CF_o, 178
Chance, Britton, 57, 76, 78, 80, 82, 84–87, 90, 98, 107–108, 112, 123–127, 222, 234
charcoal, iron, 52–53, 58
chemical revolution, 6
chemical theory, 96–98, 111, 116, 122, 125–126, 186, 218, 224
    and crisis, 229–232
    description of, 89–91
    in photosynthesis, 152, 175–177
chemiosmotic theory (hypothesis), 12–13, 95, 111–113, 116, 191, 208, 224
    acceptance of, 229, 231–235
    description of, 100–104, 182–183
    discussion of, 122–127
    evidence for, 104
    in membrane transport, 119–121, 205–207
    in photosynthesis, 177–179
    postulates for, 123
Cheniae, George, 175
Chlorella, 138, 144, 147, 158, 161, 174
chloride ion transport, 189
Chlorobium, 140, 148, 167, 171
    reaction center, 175
chlorophyll, 14, 133–139, 144, 149, 151–154, 159–161, 165, 203, 204
    chemical synthesis of, 137
    energised state of, 158
    role of, 152
chlorophyll a, 137, 160, 162, 166
chlorophyll b, 137, 160–162, 166
chlorophyll-CO_2 complex, 137–139, 141, 153, 215, 218
chlorophyll-proteins, 203–204
chloroplast, 109, 133–135, 154, 165, 174, 200
    ATP synthase, 212
    complete photosynthesis in, 150–151
    electron transport chain, 146, 150, 153, 155, 162, 171–175, 184
    envelope, 157
    genome, 205
    and the Hill reaction, 143–146
    isolation of, 215
    preparation of, 143
    structure, 155–158
Chodat, Robert, 34
cholate, 93
Chromatium, 140, 153, 167
chromatography, column, 136
chromatography, paper, 147–148